Fungal Biology

Series Editors

Vijai Kumar Gupta
Department of Chemistry and Biotechnology
Tallinn University of Technology
Akadeemia tee, Tallinn, Estonia

Maria G. Tuohy
School of Natural Sciences
National University of Ireland Galway
Galway, Ireland

About the Series

Fungal biology has an integral role to play in the development of the biotechnology and biomedical sectors. It has become a subject of increasing importance as new fungi and their associated biomolecules are identified. The interaction between fungi and their environment is central to many natural processes that occur in the biosphere. The hosts and habitats of these eukaryotic microorganisms are very diverse; fungi are present in every ecosystem on Earth. The fungal kingdom is equally diverse, consisting of seven different known phyla. Yet detailed knowledge is limited to relatively few species. The relationship between fungi and humans has been characterized by the juxtaposed viewpoints of fungi as infectious agents of much dread and their exploitation as highly versatile systems for a range of economically important biotechnological applications. Understanding the biology of different fungi in diverse ecosystems as well as their interactions with living and non-living is essential to underpin effective and innovative technological developments. This series will provide a detailed compendium of methods and information used to investigate different aspects of mycology, including fungal biology and biochemistry, genetics, phylogenetics, genomics, proteomics, molecular enzymology, and biotechnological applications in a manner that reflects the many recent developments of relevance to researchers and scientists investigating the Kingdom Fungi. Rapid screening techniques based on screening specific regions in the DNA of fungi have been used in species comparison and identification, and are now being extended across fungal phyla. The majorities of fungi are multicellular eukaryotic systems and therefore may be excellent model systems by which to answer fundamental biological questions. A greater understanding of the cell biology of these versatile eukaryotes will underpin efforts to engineer certain fungal species to provide novel cell factories for production of proteins for pharmaceutical applications. Renewed interest in all aspects of the biology and biotechnology of fungi may also enable the development of "one pot" microbial cell factories to meet consumer energy needs in the 21st century. To realize this potential and to truly understand the diversity and biology of these eukaryotes, continued development of scientific tools and techniques is essential. As a professional reference, this series will be very helpful to all people who work with fungi and should be useful both to academic institutions and research teams, as well as to teachers, and graduate and postgraduate students with its information on the continuous developments in fungal biology with the publication of each volume.

More information about this series at http://www.springer.com/series/11224

R. Naraian
Editor

Mycodegradation of Lignocelluloses

Springer

Editor
R. Naraian
Department of Biotechnology
Faculty of Science
Veer Bahadur Singh Purvanchal University
Jaunpur, Uttar Pradesh, India

ISSN 2198-7777 ISSN 2198-7785 (electronic)
Fungal Biology
ISBN 978-3-030-23833-9 ISBN 978-3-030-23834-6 (eBook)
https://doi.org/10.1007/978-3-030-23834-6

This Springer imprint is published by the registered company Springer Nature Switzerland AG
The registered company address is: Gewerbestrasse 11, 6330 Cham, Switzerland

Preface

Lignocellulose is the most abundantly available biomass of complex biopolymers originated from plants growing on earth. This is a biological material that comprises several agricultural residues, woods, grasses, and forest wastes chemically composed of different concentrations of components varying in different plant species from which it is originated. Lignocellulosic residues are basically composed of two major carbohydrate polymers, viz., cellulose and hemicellulose, and a non-carbohydrate closed ring-bearing phenolic polymer lignin. Cellulose is considered as the principal structural polysaccharide constituent of primary plant cell wall by accounting for 30–50% of lignocellulose on dry weight basis. The hemicellulose is considered as the second most significant polysaccharide constituent of lignocellulose, which generally accounts for 15–30% in all plant cell walls. Hemicelluloses are found to be imbedded in the complex structure of plant cell walls with a significant function of cellulose microfibrils binding to strengthen the plant cell wall. Distinct to the cellulose, hemicellulose structurally constitutes a random and amorphous structure made up of several heteropolymer units including xylan, glucomannan, glucuronoxylan, xyloglucan, and the arabinoxylan. However, the third constituent is a highly complex component of lignocellulose known as phenolic lignin accounting the amount of 15–30%. The lignin, as the part of all vascular plants, represents the second top most carbon source after the available cellulose on the earth. Lignin indeed provides strength and rigidity to cell wall and protects it from microbial attacks causing decomposition because of the presence of closed phenolic rings in its complex structure.

Nowadays, huge amount of lignocellulose in rural areas is either burnt or negligently thrown, which creates serious post-disposal problems of health and environment. Thus, it is necessary to implement a method of efficiently degrading lignocelluloses to a non-polluting level. The microbial bioconversion of lignocelluloses into several valuable products of human usages is an efficient technological solution. Among several microbial classes, fungi are the most efficient at this bioconversion because of the potential of their strongest lignocellulolytic enzyme-producing versatile machinery. The degradation of lignocelluloses accomplished by the fungi is termed as "mycodegradation," which can be executed by the huge diversity of fungal species.

The content of this book is comprised of the contributions of chapters from several eminent academicians and researchers of reputed institutions providing recent updates on mycodegradation of lignocelluloses. The chapters of the book cover almost all the aspects of mycodegradation such as the biochemistry of lignocellulose, involved fungal biodiversity and several enzymes playing a role in the process of mycodegradation integrating hemicellulolysis, cellulolysis, and ligninolysis. The book also covers the information in respect to mechanism of lignocellulose mycodegradation both under natural and in vitro conditions. The aspect of biofuel production using the process of lignocellulose degradation and enhancement of lignocelluloses employing nanotechnological approaches has also been incorporated to make the content advanced and informative. Consequently, the book has a comprehensive and up-to-date information on mycodegradation of lignocellulose with clear and easy illustration with simplified elucidations.

I am thankful to many people who helped me to bring this book to its genuine structure. I extremely wish to express my hearty thanks to Dr. Eric Stannard, senior editor, Botany, Springer; Dr. Vijai Kumar Gupta and Maria G. Tuohy, series editors, Fungal Biology, Springer; Ms. Sanjana M. Sundaram, project coordinator, Springer; and Mrs. Cathrine Selvaraj, project manager for their unified generous assistance, constant support, and patience in finalizing the book. I give my special thanks to Prof. (Dr.) Raja Ram Yadav, Hon'ble Vice-Chancellor, V. B. S. Purvanchal University, Jaunpur (UP), India, for constant motivations. My sincere thanks also go to all chapter contributors, without which the structure of the book would not have been possible. From the depth of my heart, I extend my thanks to my entire team of research scholars including Dr. Abhishek Kr. Bharadwaj, Mr. Roshan L. Gautam, Ms. Shweta Singh, Ms. Nahida Arif, Mr. Tripuresh P. Shahee, Ms. Madhumita Singh, and Mr. Pramneel Kumar for their substantial support during editing the book in a very prompt manner. I also give special thanks to my esteemed friends, well-wishers, colleagues, and senior faculty members of V. B. S. Purvanchal University, Jaunpur (UP), India.

Jaunpur, Uttar Pradesh, India R. Naraian

Contents

Chapter 1
Basic Mechanism of Lignocellulose Mycodegradation

Roshan Lal Gautam, Shweta Singh, Simpal Kumari, Archana Gupta, and R. Naraian

1 Introduction

Lignocellulose refers to a dry plant biomass abundantly available on earth in huge amount, which is annually synthesized and microbially degraded in nature. Tremendous amount of lignocellulosic wastes is also additionally produced by different industries including forestry, pulp and paper, agriculture, food from municipal solid waste, and animal wastes (Champagne 2007; Wen et al. 2004). The woody plant cell wall consists of major structural polymer components, namely, cellulose, hemicellulose, and lignin (Scheller and Ulvskov 2010; Gibson 2012). The lignocellulosic complex in plant cell wall contains approximately 40–60% cellulose, 20–40% hemicellulose, and 10–25% lignin (Kubicek 2012a, b), which can vary from one plant species to another. Sometimes huge biomass of lignocelluloses remains unused, which is either dumped or burned negligently creating big environmental issues. However, lignocellulosic biomass stores tremendous amount of energy that can be converted into another form and be employed for different valuable purposes. The microorganisms are the only predominantly responsible factors for lignocellulose degradation including fungi, and the most rapid degraders in fungi are the members of basidiomycetes (Sanchez 2009). White-rot fungi (WRF) have a versatile potential to secrete a wide range of wood-degrading enzymes involved in the breakdown of carbohydrate components (cellulose and hemicellulose) and aromatic constituent (lignin) (Kameshwar and Qin 2016). There are reports of approximately 10,000 species of WRF, having different aptitude for depolymerization of lignocellulosic materials and further mineralization to carbon and water (Kirk 1984; Halis et al. 2012). WRF such

R. L. Gautam · S. Singh · S. Kumari · R. Naraian (✉)
Department of Biotechnology, Faculty of Science, Veer Bahadur Singh Purvanchal University, Jaunpur, Uttar Pradesh, India

A. Gupta
Department of Botany, C.M.P. Degree College, University of Allahabad, Prayagraj, Uttar Pradesh, India

R. Naraian (ed.), *Mycodegradation of Lignocelluloses*, Fungal Biology,
https://doi.org/10.1007/978-3-030-23834-6_1

as *Phanerochaete chrysosporium*, is highly preferred, which simultaneously degrade all cell wall polymers, while several others selectively decay lignin ahead of cellulose and hemicellulose (Akhtar et al. 1992; Behrendt and Blanchette 1997; Srebotnik and Messner 1994; Hori and Cullen 2016). These abilities of white-rot fungi are because of that they can grow under less restrictive conditions to degrade lignocellulosic wastes (Salvachua et al. 2011).

The cellulose-degrading organisms very luxuriantly use lignocellulose as feedstock for their growth, reproduction, and consequently bioconversions (Chen 2014). The microbial conversion of lignocellulosic complexes to monomers involves the synergistic action of multiple enzymes, such as peroxidases, laccases, xylanases, endoglucanases, exoglucanases, β-xylosidase, and β-glucosidases (Hasunuma et al. 2013). The extracellular enzymes are much crucial for depolymerization of cellulose and degradation of lignin into simple metabolites (Akhtar et al. 1997). Moreover, enzymatic hydrolysis of lignocelluloses is one of the most efficient options for the fruitful conversion of polysaccharide into monosaccharide by avoiding toxic metabolite formation (Pramanik and Sahu 2017).

The efficient biodegradation of lignocellulosic biomass with the use of versatile microbial communities has been regarded as an excellent and efficient approach for multitude of biotechnological applications (Fang et al. 2012; Li et al. 2012). The potential applications of lignocellulolytic enzymes for a variety of uses such as industrial and environmental technologies need huge quantities of these enzymes at lowest costs (Elisashvili et al. 2008). Cellulases and xylanases are characterized as inducible enzymes, usually synthesized by microbes when grown on polysaccharide substrates, with lignocellulosic biomass (Howard et al. 2003; Iqbal et al. 2013; Fatokun et al. 2016).

2 Lignocellulose

Lignocellulosic materials including wood and grass along with forest and agricultural residues are chemically composed of cellulose, hemicellulose, and lignin, with varying ratios of these components according to the plant species from which they belong (Rennie and Scheller 2014; Bosetto et al. 2016). Lignocelluloses are the structural polysaccharides of plants composed of cellulose (~50%), hemicellulose (~30%), and lignin (~20%), which are widely distributed among the vascular plants (Coral et al. 2002; Sajith et al. 2016).

3 Cellulose

Cellulose originated from plants as the major constituent of lignocelluloses is the most abundant organic polymer on earth (Naika et al. 2007). This is the main building material of which plants are made as a very important polymer substance

(Lavanya et al. 2011). Cellulose is a high-molecular-weight-containing (Klemm et al. 2005), linear polymer of glucose linked through α-1,4-linkages and is usually arranged in microcrystalline structures (Kuhad et al. 1997; Leschine 1995). Its structure consists of extensive intramolecular hydrogen bonding networks, which tightly binds the glucose units. This is a water-insoluble polymer with a rigid linear structure (Festucci-Buselli et al. 2007). This polymer known to be an important renewable source of bioenergy is involved in several microbial treatments for its significant conversion into soluble and fermentable glucose (Bhalla et al. 2013).

4 Hemicellulose

Hemicelluloses are the second most plentiful constituent of lignocellulosic biomass (Okeke and Lu 2011) after celluloses. Hemicelluloses are generally found associated with cellulose in primary and secondary walls of all plants (Jeffries 1994). In the structure of cell wall, hemicelluloses function as an adhesive between the stiff reinforcing cellulose microfibrils and the matrix lignin (Salmen and Burgert 2009; Kulasinski et al. 2016). The most common form of hemicellulose is the xylan, composed from the backbone of β-(1-4)-linked xylose monomers (Scheller and Ulvskov 2010). Hemicellulose may primarily contain the mannose and galactose but can also contain pentose, xylose, and arabinose. The hemicellulose as a hetero-polysaccharide can be composed of different water-soluble sugars including hexoses, pentoses, and glucuronic acid monomers (Kuhad et al. 1997), interacting with cellulose and lignin through either hydrogen bonding or covalent linkages (Ebringerova et al. 2005). Basically, xylan-type of hemicelluloses are known to be the hemicellulosic constituent of the cell walls of hardwoods and of any herbaceous plants constituting approximately 20–35% weight of the total biomass (Chen et al. 2014). Hemicelluloses are composed of three groups of amorphous polymers such as xylans, mannans, and galactans (Fan et al. 1987).

5 Lignin

Lignin is a major structural component of plants with high-molecular-weight, three-dimensional aromatic hetero-polymer, associated to other cell wall components cellulose and hemicellulose (Jeffries 1990). This is the most abundant source of carbon in soil after cellulose (Kubicek 2012a, b). Lignin is highly irregular in structure and is an insoluble polymer composed of phenylpropanoid subunits, namely, p-hydroxyphenyl (H-type), guaiacyl (G-type), and syringyl (S-type) units (Malherbe and Cloete 2002). Lignin is an aromatic polymer constituted of three basic units such as p-coumaryl alcohol (4-hydroxycinnamyl alcohol), coniferyl alcohol, and sinapyl alcohol, which are collectively pronounced as monolignols derived from phenylalanine (amino acid) (Whetten and Sederoff 1995).

6 Biodegradation Mechanism

The biodegradation lignocellulosic complex substrates not only depend on environmental factors alone, but also on the degradative potential of the microbial class and population (Waldrop et al. 2000). The complex polymers of lignocelluloses are degraded by the variety of microorganisms and with the potential synergistic action of various enzymes (Parani and Eyini 2010) (Table 1.1). The cellulose and hemicelluloses are selectively degraded by "soft-rot" fungi (Dix and Webster 1995). The commercially available enzymes can effectively degrade lignocelluloses into their monomers (Okeke and Lu 2011). Most of the fungi have two specific types of extracellular enzymatic systems, viz., the hydrolytic system, which produces hydrolases responsible for polysaccharide degradation, and a unique oxidative and extracellular ligninolytic system that degrades lignin and opens phenyl rings (Sanchez 2009). Several microorganisms including fungi are strongly capable to produce lignocellulolytic enzymes involved in the degradation of lignocellulosic biomass. Mycodegradation of distinct lignocellulosic components (cellulose, hemicellulose, and lignin) occurs variably with diverse group of fungi and with the attack of specific enzymes independently (Fig. 1.1). The breakdown of hemicellulose, cellulose, and lignins can be termed as hemicellulolysis, cellulolysis, and ligninolysis simultaneously.

6.1 *Cellulolysis*

Cellulolysis is the term which explains the process of cellulose hydrolysis employing efficient cellulosic enzymes. As the cellulose is a polymer of β (1-4)-linked glucose, individually associating with hydrogen bonds (Sjostrom 1993), which are hydrolyzed by enzymes of microbial origin. Fungi are potentially recognized as the strongest group of cellulose-degradable microorganisms producing high level of cellulases (Chen 2014). The efficient cellulolytic fungus genus can be sequentially represented as the species of *Aspergillus, Penicillium, Chaetomium, Trichoderma, Fusarium, Stachybotrys, Cladosporium, Alternaria, Acremonium, Ceratocystis, Myrothecium, Humicola*, etc. (Wood 1985; Mehrotra and Aneja 1990; Sajith et al. 2016). All the above-mentioned genus of fungi have excellent capability of enzymes production involved in cellulolysis (Fig. 1.2).

6.1.1 Enzymes Involved in Cellulolysis

Cellulase as a complex mixture of enzymes is produced by most fungi and other microorganism naturally, which act synergistically to decompose celluloses. Cellulases are able to break β-1, 4-glycosidic bond in the structure of cellulose by attacking as multicomponent enzyme system comprised of three major classes of

Table 1.1 Different recent studies performed with several fungal species for the degradation of variable lignocellulosic substrates

S.N.	Fungal genus	Lignocellulosic substrate	References
1.	*Aspergillus, Gliocladium*	Rice straw	Shruti et al. (2015)
2.	*Ganoderma lucidum* C7016	Birch wood	Xiao et al. (2013)
3.	*Pleurotus sajor-caju*	Corncob	Menezes et al. (2017) Naraian et al. (2010)
4.	*Hypholoma fasciculare, Stropharia rugosoannulata, Stachybotrys*	Wheat straw	Singh et al. (2014)
5.	*Trichoderma viride, Ganoderma lucidum*	Corn stover	Shahzadi et al. (2014)
6.	*Pleurotus pulmonarius*	Corncob	Correa et al. (2016)
7.	*Agaricus bisporus*	Wheat straw	Kabel et al. (2017)
8.	*Pleurotus ostreatus*	Sorghum and millet chaff	Ryden et al. (2017)
9.	*Flammulina velutipes*	Ramie stalk	Xie et al. (2017)
10.	*Auricularia thailandica*	Sawdust, wheat husk, sugarcane bagasse	Bandara et al. (2017)
11.	*Inonotus obliquus*	Rice straw, wheat straw, bagasse	Xu et al. (2014)
12.	*Aspergillus niger*	Rice straw	Yang et al. (2018)
13.	*Penicellium oxalicum* GZ-2	Corn stover	Liao et al. (2014)
14.	*Chaetomium globosum* BCC5776	Rice straw	Wanmolee et al. (2016)
15.	*Ganoderma lucidum*	Bean stalks, cotton stalk, maize straw, rice straw, sugarcane bagasse, wheat straw	Rashad et al. (2019)
16.	*Lentinula edodes*	Birch-wood	Xiong et al. (2019)
17.	*Pleurotus tuber-regium*	Sugarcane, corncob	Lam et al. (2018)
18.	*Pleurotus florida*	Mixed leafy biomass	Malayil et al. (2016)
19.	*Auricularia auricular*	Spent mushroom, rice husks	Meng et al. (2018)
20.	*Pleurotus ostreatus*	Sawdust, cottonseed hull, corncob	Xiao et al. (2019)
21.	*Gymnopilus pampeanus*	Populus, eucalyptus sawdust	Colavolpe and Alberto (2014)
22.	*Trichoderma reesei*	Wheat straw, rice straw, rice husk	Taherzadeh-Ghahfarokhi and Mokhtarani (2019)
23.	*Fusarium* sp.	Sugarcane bagasse	Bertonha et al. (2018)
24.	*Fusarium verticillioides*	Gamba grass	de Almeida et al. (2019)
25.	*Humicola insolens, Trichoderma reesei*	Rice straw	Kogo et al. (2017)
26.	*Myrothecium verrucaria*	Birch sawdust	Wang et al. (2017)
27.	*Acremonium cellulolyticus, Trichoderma reesei*	Eucalyptus, Douglas fir, rice straw	Fujii et al. (2009)

(continued)

Table 1.1 (continued)

S.N.	Fungal genus	Lignocellulosic substrate	References
28.	*Acremonium strictum*	Sugarcane bagasse	Goldbeck et al. (2013)
29.	*Ceratocystis fimbriata*	Coffee husk	Mussatto and Teixeira (2010)
30.	*Cladosporium oxysporum* GQ-3	Wheat bran	Guan et al. (2016)
31.	*Alternaria alternate*	Wheat bran	Faten and Abeer (2012)

enzyme: endoglucanases (endo-β-1,4-glucanase, CMCase, EC 3.2.1.4), exoglucanase (cellobiohydrolases, EC 3.2.1.91), and β-glucosidase (cellobiase, EC 3.2.1.21) (Dincer and Telefoncu 2006) (Fig. 1.2). The detailed information of these three enzymes have been accounted as below.

Endoglucanase

Endoglucanase (endo-β1-4-d-glucan 4-glucanohydrolase, EC 3.2.1.4) acts randomly to cleave beta-linked bonds within the cellulose chain (Saha 2004). Endoglucanase also attacks on cellodextrins, which is an intermediate product after cellulose hydrolysis, and consequently converts them to cellobiose and glucose units (Sajith et al. 2016). These enzymes are generally found active against the acid-swollen amorphous cellulose, soluble derivatives of cellulose such as carboxymethylcellulose and cellooligosaccharides (Wood 1989). The endoglucanases are the most economical enzymes employed at large scale for the hydrolysis of celluloses (Henriksson et al. 1999) (Fig. 1.2).

Exoglucanase

Structurally, exoglucanases (exo-β1-4-d-glucan 4-cellobiohydrolase, EC 3.2.1.91) have long loops and affinity to act with crystalline sites of cellulose chains and primarily result in cellodextrin (Segato et al. 2014). Exoglucanases become usually active on crystalline part of cellulose and cleave disaccharide units one from the both ends either at the non-reducing or reducing end (Mrudula and Murugammal 2011). These are found in two forms; first attack upon the reducing end or working on the non-reducing end; second the cellobiohydrolases acting unidirectionally on the long chain oligomers (Juturu and Wu 2014) (Fig. 1.2).

β-Glucosidase

Beta-glucosidase (EC 3.2.1.21) is the key enzyme, which plays a significant role in the complete conversion of cellobiose to simpler glucose (Berlin et al. 2007). Such enzymes possess a rigid structure with highly active sites present in a large cavity,

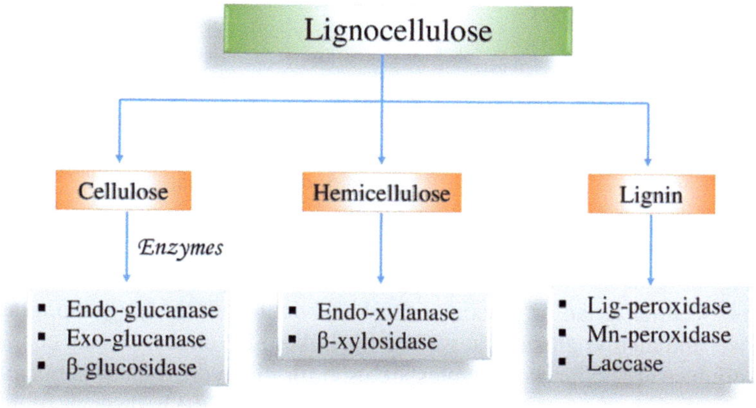

Fig. 1.1 The three integral components of lignocellulose (cellulose, hemicellulose, and lignin) and cluster of enzymes involved in hydrolysis of respective component

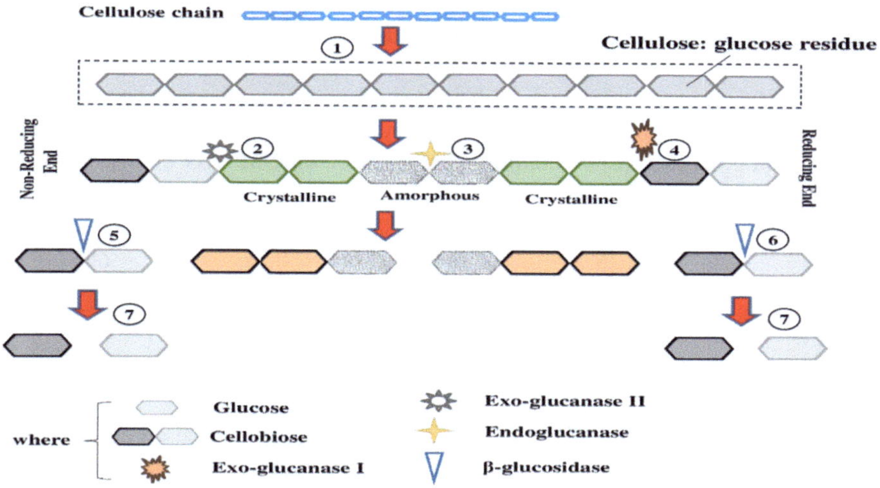

Fig. 1.2 Major events involved in the process of enzymatic cellulolysis through the catalytic action of several cellulolytic fungal enzymes generating monomers (glucose) by breaking linear polymer of cellulose complex: (1) linear chain of cellulose, (2) attack of exoglucanase I at the non-reducing end of crystalline region, (3) attack of endoglucanase within ends at amorphous region, (4) attack of exoglucanase II at the reducing end of crystalline region, (5 and 6) attack of β-glucosidase on the cellobiose, (7) formation of glucose after breaking of cellobiose

basically termed as the active site pocket, that indeed favours the entry of disaccharides (Nam et al. 2010). It is absolutely inactive against the crystalline or amorphous region of cellulose (Sadhu and Maiti 2013). The saccharification of cellulose seems to be inefficient without the β-glucosidases from any source (Sternberg et al. 1977) (Fig. 1.2). Thus, this enzyme is considered to be a very efficient enzyme without which saccharification of celluloses cannot be accomplished successfully.

6.2 Hemicellulolysis

The term hemicellulolysis can be explained as the breakdown of hemicelluloses into their oligomer units. Hemicellulose is a component of lignocelluloses connected to lignin and cellulose through covalent bonds, but because of the involvement of few hydrogen bonds, it is much promptly broken down in comparison to crystalline cellulose (Jacobsen and Wyman 2000). Hemicellulose comprised of the branching chains of sugars, which consists of (1) aldopentoses (xylose and arabinose) and (2) aldohexoses (glucose, mannose, and galactose). Most biomass of hemicelluloses consists of either xylan or glucomannan backbones with acetyl group, arabinose, galactose, and methyl glucuronic acid on the linked side chains (Zhang et al. 2011). Hemicellulase and xylanase enzymes are used for the hydrolysis of hemicelluloses (El-Zawawy et al. 2011; Hu et al. 2011). In consequence, the hydrolysis of hemicelluloses yields sugars and organic acids as their products (Steinbach et al. 2017) (Fig. 1.3).

6.2.1 Enzymes Involved in Hemicellulolysis

Many bacteria, fungi, and yeast can degrade hemicellulose through the production of a cluster of hydrolytic enzymes, including xylanases, xylosidases, arabinofuranosidases, and glucuronidases (Shallom and Shoham 2003; Maayer et al. 2014). The

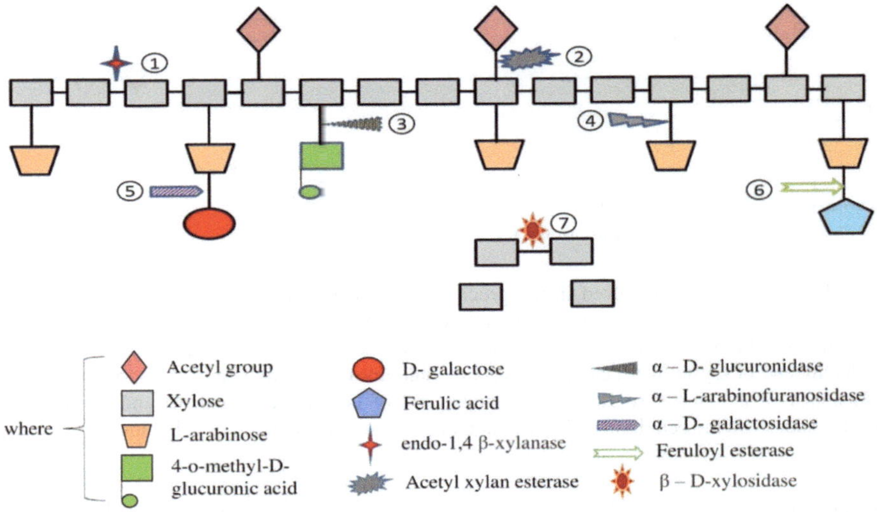

Fig. 1.3 Major events involved in enzymatic hemicellulolysis through the catalytic action of several hemicellulolytic fungal enzymes generating monomers (xylose, ferulic, and 4-*o* methyl-D-glucuronic acid) by breaking polymer of hemicellulose structure. Attack of (1) endo 1,4 β-xylanase, (2) acetyl xylan esterase, (3) α-D-glucuronidase, (4) α-L-arabinofuranosidase, (5) α-D-galactosidase, (6) feruloyl esterase, and (7) β-D-xylosidase

quick and complete degradation of hemicellulose needs the synergistic action of a broader group of hydrolytic enzymes (Polizeli et al. 2005). The enzymes selectively remove undesired branches in hemicellulose structure and keep the desired polymer backbone intact (Sedlmeyer 2011). During most of the bioconversion processes, xylan is firstly converted to xylose or into xylooligosaccharides, by the catalytic action of xylanolytic enzymes (Chivero et al. 2001) (Fig. 1.3).

Xylan

Xylan is the group of hemicellulose that represents the most abundant biopolymers present on earth. Xylan is primarily formed in the secondary cell wall together with cellulose (1,4-β-glucan) and lignin as the major constituents of plant cell walls (Kulkarni et al. 1999). This is a complex polymer consisting of 1,4-linked xylosyl residues that can be acetylated or it can also have covalently attached arabinosyl and glucuronic acid as side groups (De Groot et al. 1998). The complete enzymatic hydrolysis of xylans is performed by the action of multiple enzyme such as endo-β-1,4 xylanase and β-xylosidase and additionally through the help of several accessory enzyme, viz., α-L-arabinofuranosidase, acetyl xylan esterase, α-glucuronidase, ferulic acid esterase, and p-coumaric acid esterase (Guais et al. 2010). The detailed information of these enzymes is accounted separately as below.

Xylanase

Xylanases (EC 3.2.1.8) are the enzyme complexes capable for the hydrolysis of xylan (Shallom and Shoham 2003). These enzymes are produced mainly by a variety of microorganisms that take part in breakdown of xylans (Polizeli et al. 2005), fungi being the majority of them. The exploitation of xylanases has been extensively proved in paper industries (Sohpal et al. 2010). Xylanases are classified into glycosyl hydrolase (GH) families 5, 7, 8, 10, 11, and 43 based on their amino acid sequences, structural foldings, and mechanisms of action on substrate (Cantarel et al. 2009).

Endo-1,4-β-Xylanase

The endo-1,4-β-xylanase (EC 3.2.1.8) is a crucial enzyme that is capable of randomly cleaving the β-1,4-glycosidic linkages in the backbone of xylan (Conejo-Saucedo et al. 2017). The attack of endo-1,4-β-xylanase produces oligosaccharides, after the cleavage of xylan (Jeffries 1994), releasing smaller xylooligosaccharides (Collins et al. 2005). Therefore, endo-1, 4-β-xylanases are very promising through direct involvement by cleaving glycosidic bonds and liberating structurally smaller xylooligosaccharides (Verma and Satyanarayana 2012; Anand et al. 2018). On the basis of end product obtained, the working behaviours of endoxylanase are categorized into two major types: (a) non-debranching xylanase and (b) debranching xylanase. The non-debranching xylanase cannot split the 1,3-α-L-arabinofuranosyl branch of arabinoxylans, while debranching xylanase do not hydrolyze side branches to liberate arabinose (Polizeli et al. 2005)

β-Xylosidase

β-Xylosidases (EC 3.2.1.37) are promising hydrolytic enzymes, which play an important role in degradation (Knob et al. 2010), and they hydrolyze small xylooligosaccharides and xylobiose, releasing β-D-xylopyranosyl residues from the non-reducing terminus (Biely 1985, 1993; Polizeli et al. 2005), and 1,4-β-xylosidase produces xylose from oligosaccharides (Jeffries 1994).

On the basis of variation in amino acid sequences, these enzymes are classified into five GH families: 3, 39, 43, 52, and 54. The β-xylosidases from GH43 family hydrolyze the non-reducing ends of xylooligomers (Graciano et al. 2012).

Arabinofuranosidases

Alpha-L-arabinofuranosidases (α-L-arabinofuranoside arabinofuranohydrolases, EC 3.2.1.55) are exo-enzymes that remove terminal non-reducing end from the arabinose containing poly- and oligosaccharides (Kozlova et al. 2015). Arabinoxylans basically composed of a linear backbone with β-1,4-linked D-xylopyranosyl units, which are partially substituted with arabinofuranosyls (Sorensen et al. 2006). During catalysis, arabinofuranosidases remove L-arabinose residues substituted at positions 2 and 3 of the β-D-xylopyranosyl (Anand et al. 2018). Therefore, α-L-arabinofuranosidases have attracted attention due to their significant role in degradation of lignocelluloses as well as effect on the activity of several other enzymes acting on lignocelluloses (Numan and Bhosle 2006).

α-Glucuronidase

Alpha-1, 2-glucuronidases (EC 3.2.1.131) are enzymes belonging to GH family 67 (Lombard et al. 2014), which act only on short xylan oligosaccharides that carry the (Me) GlcA substitution at their reducing ends (Golan et al. 2004; Nurizzo et al. 2002).

Acetyl Xylan Esterase

Acetyl xylan esterases (AXE, EC 3.1.1.72) are also very important enzymes that cleave attached acetyl side groups in the structure of xylan (Till et al. 2013). These enzymes have the potential to remove the *O*-acetyl groups from both positions 2 and 3 on the β-D-xylopyranosyl residues of acetyl xylan (Bajpai 2014).

Ferulic Acid Esterase and p-Coumaric Acid Esterase

Ferulic acid esterase (EC 3.1.1.73) is an enzymes that catalytically cleaves the feruloyl groups substituted at the 5′-OH group of arabinosyl residues present in arabinoxylans, which is known to modulate their functional properties (Latha et al. 2007). However, the ester bonds are cleaved by ferulic acid and p-coumaric acid esterases (Qi et al. 2011).

6.3 Ligninolysis

The term ligninolysis is defined as the degradation of lignin, that has been extensively researched with wood-rotting organisms, especially white-rot basidiomycetes fungi (Martinez et al. 2004; Wan and Li 2012; Falade et al. 2017). Under the process of ligninolysis, the recalcitrant lignin with the help of lignolytic enzymes is catabolized into simpler metabolites. However, multiple strategies exist in nature for the conversion of lignin, executed by a limited range of microorganisms than cellulose degradation (Cragg et al. 2015). Though lignin degradation is performed by group of microorganisms, WRF have stronger potency than that of other organisms and they are most responsible for the lignin decomposition (Tuomela et al. 2000).

The lignin is a highly insoluble complex branched organic polymer constituted by the phenylpropane units linked by carbon–carbon and ether linkages, which forms an extensive cross-linked network within plant cell wall (Rodrigo de Souza 2013). Lignin is also an amorphic three-dimensional structure composed of phenylpropanoid subunits (Datta et al. 2017). These subunits can be accounted as (1) coniferyl alcohol (guaiacyl propanol), (2) coumaryl alcohol (p-hydroxyphenyl propanol), and (3) sinapyl alcohol (syringyl alcohol) (Bajpai 2016) (Fig. 1.4).

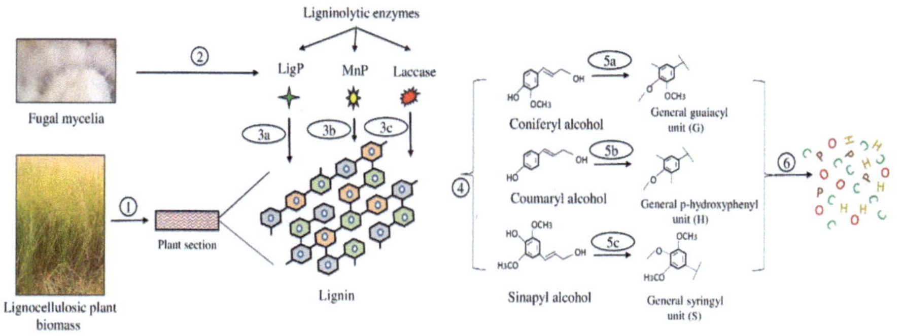

Fig. 1.4 Important events of ligninolysis during depolymerization of complex structure of lignin: (1) lignocellulosic biomass, (2) mycelial biomass, (3) secretion of lignolytic enzymes (LiP, MnP, and lac), (4a) attack of LiP, (4b) attack of MnP, (4c) attack of laccase on lignin polymer (coniferyl alcohol, coumaryl alcohol, and sinapyl alcohol), (5) formation of resultant metabolites after attack of LiP, MnP, and laccase

6.3.1 Enzymes Involved in Ligninolysis

Ligninolytic enzymes secreted from fungal mycelia capable of degrading lignin are collectively termed as ligninases. The extracellular lignin-degrading enzymes such as lignin peroxidase (LiP), manganese-dependent peroxidase (MnP), and laccase are responsible for decomposing a variety of recalcitrant organic compounds ranging from simple phenolic molecules to many xenobiotics (Hofrichter 2002; Steffen et al. 2002; Baldrian 2006). At the time of catalytic action, laccases use molecular oxygen as electron acceptors, while both peroxidases (LiP and MnP) use hydrogen peroxide as a co-substrate (Mai et al. 2004). Both the LiP and MnP were discovered during mid of 1980s in industrially preferred *Phanerochaete chrysosporium* and were described as true lignases because of their high potential redox value (Gold et al. 2000; Martinez 2002). The major fungal members *P. chrysosporium, Ceriporiala cerata, Cyathus stercoreus, C. subvermispora, Pycnoporus cinnabarinus,* and *Pleurotus ostreatus* produce enzymes potentially able to degrade lignin (Kumar et al. 2015; Kumar and Wyman 2009; Madadi and Abbas 2017).

Lignin Peroxidase

Lignin peroxidase (EC 1.11.1.14) is a biotechnologically very important and potentially delignifying enzyme degrading lignocellulosic materials (Harley et al. 1988; Valaskova et al. 2007). LiP is one of the key enzyme, which acts in the oxidation of non-phenol lignin structures to cation radicals (Gubernatorova and Dolgonosov 2010). Lignin peroxidase resembles horse radish peroxidase, and as highly studied this contains Fe (III) as a cofactor which is structurally pentacoordinated with four heme tetrapyrrole nitrogens and with a histidine (Hammel and Cullen 2008). It is LiP which is thought to directly attack at the Cα–Cβ bonds of propyl side groups attached with aromatic subunits of lignin (MacDonald et al. 2012). LiPs are known to catalyze the H_2O_2-dependent oxidative depolymerization of several non-phenolic lignin (diarylpropane), β-O-4 non-phenolic lignin model compounds, and many phenolic compounds such as guaiacol, vanillyl alcohol, catechol, syringic acid, and acetosyringone redox (Wong 2009). In consequence of LiP catalysis, formation of intermediate radicals like phenoxy radicals and veratryl alcohol radical cations through oxidizing substrates takes place in multistep electron transfers of lignin peroxidase (Fig. 1.4).

Manganese Peroxidase

The manganese peroxidase (MnP; EC 1.11.1.13) enzyme is catalytically most versatile lignin-modifying peroxidase produced by most wood-colonizing basidiomycetes causing modification in lignin (Hofrichter 2002). This was first discovered in *P. chrysosporium* white-rot fungus (Glenn and Gold 1985; Paszczynski et al. 1985). MnP attacks on phenolic and non-phenolic units of lignin through catalyzing lipid peroxidation reactions (Binod et al. 2011; Kuila et al. 2016). MnP contains glycosylated heme that can catalyze the oxidation of Mn^{2+} to Mn^{3+} in a H_2O_2-dependent reaction (Zhao et al. 2015). During the enzymatic activity of MnP, Mn works like a mediator (Datta et al. 2017). Manganese ions chelated by organic acids are lactate and malonate (Wariishi et al. 1989). This oxidizes the phenolic compounds of lignin to its phenoxy radicals by chelating Mn^{3+} organic acid complex (Aehle 2007). During lignin degradation process, Mn stimulates and enhances the MnP enzymatic activity at higher levels of Mn (Rothschild et al. 1999) (Fig. 1.4).

Laccase

Laccase (EC 1.10.3.2; benzenediol–oxygen oxidoreductases) are multicopper enzymes belonging to the group of blue oxidases that catalyzes oxidation of a wide range of organic and inorganic compounds, such as diphenols, polyphenols, diamines, and aromatic amines (Gianfreda et al. 1999; More et al. 2011). This enzyme was first isolated from the extract of Japanese lacquer tree *Rhus vernicifera* (Yoshida 1883), which is widely distributed in most plants (Mayer and Staples 2002), fungi (Baldrian 2006), bacteria (Claus 2003), and in insects (Hoegger et al. 2006). Laccases have the potential to cleave β-1 and β-O-4 dimer linkages between Cα–Cβ and cause Cα oxidation and aryl–alkyl cleavages (Dwivedi et al. 2011). The ability of laccases for oxidizing phenolic compounds and to reduce molecular oxygen into water has led to intensive studies (Jolivalt et al. 1999; Xu 1996; Thurston 1994). The fungal laccases catalyze the polymerization, depolymerization, methylation, and demethylation processes of phenols (Leonowicz et al. 1985). The oxidation reactions catalyzed by laccases result in the formation of free radicals, which act as intermediate substrates for the enzymes (Ferraroni et al. 2007). Laccase also catalyzes the generation of phenoxyl radicals and their unspecific reactions, consequently leading to Cα-hydroxyl oxidation to ketone, alkyl-aryl cleavage, demethoxylation, and Cα–Cβ cleavage in phenolic sub-compound (Isroi et al. 2011) (Fig. 1.4). The physiological function of laccase is lignolysis by laccase-mediator system (LMS).

Almost all genus of WRF typically produce this potent enzyme the laccase (Kaarik 1965; Bollag and Leonowicz 1984). In fungi, laccases execute various important functions such as participation in pigmentation (Clutterbuck 1990), fruiting body initiation in mushroom, sporulation, and efficient detoxification (Thurston 1994), and degradation of lignin and pathogenesis (Thurston 1994; Mayer and Staples 2002). It is believed that fungal laccases are involved not only in the degradation of lignin but also in the intoxication of toxic phenols produced during the degradation (Thurston 1994; Solomon et al. 1996; Youn et al. 1995).

7 Conclusion

Based on the findings of the several studies regarding mycodegradation, it can be concluded that the fungi are the very efficient degraders and bioconverters of lignocelluloses. Several groups of fungi are able to secrete variety of lignocellulolytic enzymes under both natural and artificial conditions. The enzymes secreted by fungi attack on different bonds and linkages present in the chemical structure of lignocelluloses and simplify them into oligomeric metabolites. Therefore, the fungi in low-cost biodegradation have established an attractive biological machinery for bioconversion of lignocellulosic substances. The incorporation of this kind of technology with the capitalization of natural lignocellulosic residues into low-cost and environment-friendly high-bearing enzymatic processing of lignocelluloses establishes a sustainable bioconversion into many value-added by-products.

References

Aehle W (2007) Enzymes in industry: production and applications, 3th edn. Wiley, Hoboken, NJ, pp 314–320

Akhtar M, Attridge MC, Myers GC, Kirk TK, Blanchette RA (1992) Biomechanical pulping of loblolly pine with different strains of the white rot fungus *Ceriporiopsis subvermispora*. Tech Asso Pulp Pap Ind 75:105–109

Akhtar M, Blanchette RA, Kirk TK (1997) Fungal delignification and biomechanical pulping of wood. In: Scheper T (ed) Advances in biochemical engineering/biotechnology, vol 57. Springer, Berlin, pp 160–193

Anand D, Yadav S, Yadav D (2018) Screening of potential xylanase producing fungal strains under solid state fermentation condition. Int J Biol Res 3:348–354

Bajpai P (2014) Microbial xylanolytic systems and their properties. In: Xylanolytic enzymes, pp 19–36

Bajpai P (2016) Structure of lignocellulosic biomass. In: Pretreatment of lignocellulosic biomass for biofuel production. Springer Briefs in Green Chem Sustain, pp 7–12

Baldrian P (2006) Fungal laccases occurence and properties. FEMS Microbiol Rev 30:215–242

Bandara AR, Karunarathna SC, Mortimer PE, Hyde KD, Khan S, Kakumyan P, Xu J (2017) First successful domestication and determination of nutritional and antioxidant properties of the red ear mushroom *Auricularia thailandica* (Auriculariales, Basidiomycota). Mycol Prog 16:1029–1039

Behrendt CJ, Blanchette RA (1997) Biological processing of pine logs for pulp and paper production with *Phlebiopsis gigantean*. Appl Environ Microbiol 63:1995–2000

Berlin A, Maximenko V, Gilkes N, Saddler J (2007) Optimization of enzyme complexes for lignocellulose hydrolysis. Biotechnol Bioeng 97:287–296

Bertonha LC, Neto ML, Garcia JAA, Vieira TF, Castoldi R, Bracht A, Peralta RM (2018) Screening of *Fusarium* sp. for xylan and cellulose hydrolyzing enzymes and perspectives for the saccharification of delignified sugarcane bagasse. Biocatal Agric Biotechnol 16:385–389

Bhalla A, Bansal N, Kumar S, Bischoff KM, Sani RK (2013) Improved lignocellulose conversion to biofuels with thermophilic bacteria and thermostable enzymes. Bioresour Technol 128:751–759

Biely P (1985) Microbial xylanolytic systems. Trends Biotechnol 3:286–290

Biely P (1993) Biochemical aspects of the production of microbial hemicellulases. In: Cougland MP, Hazlewood GP (eds) Hemicelluloses and hemicellulases. Portland Press, London, pp 29–52

Binod P, Janu KU, Sindhu R, Pandey A (2011) Hydrolysis of lignocellulosic biomass for bioethanol production. In: Ashok P, Ricke LC, Steven C (eds) Biofuels: alternative feedstock's and conversion processes. Elsevier Inc, Burlington, MA, pp 229–250

Bollag JM, Leonowicz A (1984) Comparative studies of extracellular fungal laccases. Appl Environ Microbiol 48:849–854

Bosetto A, Justo PI, Zanardi B, Venzon SS, Graciano L, Santos EL, Simao RCG (2016) Research progress concerning fungal and bacterial β-xylosidases. Appl Biochem Biotechnol 178:766–795

Cantarel BL, Coutinho PM, Rancurel C, Bernard T, Lombard V, Henrissat B (2009) The carbohydrate-active enzymes database (CAZy): an expert resource for glycogenomics. Nucleic Acids Res 37:233–238

Champagne P (2007) Feasibility of producing bio-ethanol from waste residues: a Canadian perspective feasibility of producing bio-ethanol from waste residues in Canada. Resour Conserv Recycl 44:225–234

Chen H (2014) Biological fundamentals for the biotechnology of lignocellulose. In: Biotechnology of lignocellulose: theory and practice, pp 73–141

Chen W, Zhong L, Peng X, Lin J, Sun R (2014) Xylan-type hemicelluloses supported terpyridine-palladium (II) complex as an efficient and recyclable catalyst for Suzuki–Miyaura reaction. Cellulose 21:125–137

Chivero ET, Mutukumira AN, Zvauya R (2001) Partial purification and characterisation of a xylanase enzyme produced by a microorganism isolated from selected indigenous fruits of Zimbabwe. Food Chem 72:179–185

Claus H (2003) Laccases and their occurrence in prokaryotes. Arch Microbiol 179:145–150

Clutterbuck AJ (1990) The genetics of conidiophore pigmentation in *Aspergillus nidulans*. J Gen Microbiol 136:1731–1738

Colavolpe MB, Alberto E (2014) Cultivation requirements and substrate degradation of the edible mushroom *Gymnopilus pampeanus*-A novel species for mushroom cultivation. Sci Hortic 180:161–166

Collins T, Gerday C, Feller G (2005) Xylanases, xylanase families an extremophilic xylanases. FEMS Microbiol Rev 29:3–23

Conejo-Saucedo U, Cano-Camacho H, Villa-Rivera MG, Lara-Marquez A, Lopez-Romero E, Zavala-Paramo MG (2017) Protein homology modeling, docking, and phylogenetic analyses of an endo-1,4-β-xylanase GH11 of *Colletotrichum lindemuthianum*. Mycol Prog 16:577–591

Coral G, Arikan B, Unaldi MN, Guvenmez H (2002) Some properties of crude carboxymethyl cellulase of *Aspergillus niger* Z10 wild-type strain. Turk J Biochem 26:209–213

Correa RCG, da Silva BP, Castoldi R, Kato CG, de Sa-Nakanishi AB, Peralta RA, de Souza CGM, Bracht A, Peralta RM (2016) Spent mushroom substrate of *Pleurotus pulmonarius*: a source of easily hydrolyzable lignocellulose. Folia Microbiol 61:439–448

Cragg SM, Beckham GT, Bruce NC, Bugg TDH, Distel DL, Dupree P, Etxabe AG, Goodell BS, Jellison J, McGeehan JE, McQueen-Mason SJ, Schnorr K, Walton PH, Watts JEM, Zimmer M (2015) Lignocellulose degradation mechanisms across the Tree of Life. Curr Opin Chem Biol 29:108–119

Datta R, Kelkar A, Baraniya D, Molaei A, Moulick A, Meena RS, Formanek P (2017) Enzymatic degradation of lignin in soil: a review. Sustainability 9:2–18

de Almeida MN, Falkoski DL, Guimaraes VM, de Rezendea ST (2019) Study of gamba grass as carbon source for cellulase production by *Fusarium verticillioides* and its application on sugarcane bagasse saccharification. Ind Crop Prod 133:33–43

De Groot PWJ, Visser J, Griensven LJLD, Schaap PJ (1998) Biochemical and molecular aspects of growth and fruiting of the edible mushroom *Agaricus bisporus*. Mycol Res 102:1297–1308

Dincer A, Telefoncu A (2006) Improving the stability of cellulase by immobilization on modified polyvinyl alcohol coated chitosan beads. J Mol Catal B Enzym 45:10–14

Dix NJ, Webster J (1995) Fungal ecology. Chapman and Hall, London, p 549

Dwivedi UN, Singh P, Pandey VP, Kumar A (2011) Structure–function relationship among bacterial, fungal and plant laccases. J Mol Catal B Enzym 68:117–128

Ebringerova A, Hromadkova Z, Heinze T (2005) Hemicellulose. Adv Polym Sci 186:1–67

Elisashvili V, Kachlishvili E, Penninckx M (2008) Effect of growth substrate, method of fermentation, and nitrogen source on lignocellulose-degrading enzymes production by white-rot basidiomycetes. J Ind Microbiol Biotechnol 35:1531–1538

El-Zawawy WK, Ibrahim MM, Abdel-Fattah YR, Soliman NA, Mahmoud MM (2011) Acid and enzyme hydrolysis to convert pretreated lignocellulosic materials into glucose for ethanol production. Carbohydr Polym 84:865–871

Falade AO, Nwodo UU, Iweriebor BC, Green E, Mabinya LV, Okoh AI (2017) Lignin peroxidase functionalities and prospective applications. Microbiologyopen 6:1–14

Fan L, Gharpuray MM, Lee YH (1987) Cellulose hydrolysis. Springer, Berlin

Fang GR, Li JJ, Cheng X, Cui ZJ (2012) Performance and spatial succession of a full-scale anaerobic plant treating high-concentration cassava bioethanol wastewater. J Microbiol Biotechnol 22:1148–1154

Faten MA, Abeer AE (2012) Enzyme activities of the marine-derived fungus *Alternaria alternata* cultivated on selected agricultural wastes. J Appl Biol Sci 7:39–46

Fatokun EN, Nwodo UU, Okoh AI (2016) Classical optimization of cellulase and xylanase production by a marine *Streptomyces* species. Appl Sci 6:286

Ferraroni M, Myasoedova NM, Schmatchenko V, Leontievsky AA, Golovleva LA, Scozzafava A, Briganti F (2007) Crystal structure of a blue laccase from *Lentinus tigrinus*: evidences for intermediates in the molecular oxygen reductive splitting by multicopper oxidases. BMC Struct Biol 7:60

Festucci-Buselli RA, Otoni WC, Joshi CP (2007) Structure, organization, and functions of cellulose synthase complexes in higher plants. Braz J Plant Physiol 19:1–13

Fujii T, Fang X, Inoue H, Murakami K (2009) Enzymatic hydrolyzing performance of *Acremonium cellulolyticus* and *Trichoderma reesei* against three lignocellulosic materials. Biotechnol Biofuels 2:24

Gianfreda L, Xu F, Bollag JM (1999) Laccases: a useful group of oxidoreductive enzymes. Biorem J 3:1–25

Gibson LJ (2012) The hierarchical structure and mechanics of plant materials. J R Soc Interface 9:2749–2766

Glenn JK, Gold MH (1985) Purification and characterization of an extracellular Mn (II)-dependent peroxidase from the lignin-degrading basidiomycete, *Phanerochaete chrysosporium*. Arch Biochem Biophys 242:329–341

Golan G, Shallom D, Teplitsky A, Zaide G, Shulami S, Baasov T (2004) Crystal structures of *Geobacillus* stearothermophilus α-glucuronidase complexed with its substrate and products: mechanistic implications. J Biol Chem 279:3014–3024

Gold MH, Youngs HL, Gelpke MD (2000) Manganese peroxidase. Met Ions Biol Syst 37:559–586

Goldbeck R, Ramos MM, Pereira GAG, Maugeri-Filho F (2013) Cellulase production from a new strain *Acremonium strictum* isolated from the Brazilian biome using different substrates. Bioresour Technol 128:797–803

Graciano L, Correa JM, Gandra RF, Seixas FAV, Kadowaki MK, Sampaio SC, Silva JLC, Osaku CA, Simao RCG (2012) The cloning, expression, purification, characterization and modeled structure of *Caulobacter crescentus* b-Xylosidase I. World J Microbiol Biotechnol 28:2879–2888

Guais O, Tourrasse O, Dourdoigne M, Parrou JL, Francois JM (2010) Characterization of the family GH54 α-L-arabinofuranosidases in *Penicillium funiculosum*, including a novel protein bearing a cellulose-binding domain. Appl Microbiol Biotechnol 87:1007–1021

Guan GQ, Zhao PX, Zhao J, Wang MJ, Huo SH, Cui FJ, Jiang JX (2016) Production and partial characterization of an alkaline xylanase from a novel fungus *Cladosporium oxysporum*. Biomed Res Int 2016:1–7

Gubernatorova TN, Dolgonosov BM (2010) Modeling the biodegradation of multicomponent organic matterin an aquatic environment: 3. Analysis of lignin degradation mechanisms. Water Resour 37:332–346

Halis R, Tan HR, Ashaari Z, Mohamad R (2012) Biomodification of kenaf chip using white rot fungi. Bioresources 7:984–987

Hammel KE, Cullen D (2008) Role of fungal peroxidases in biological ligninolysis. Curr Opin Plant Biol 11:349–355

Harley BS, Brodo PMA, Senior PJ (1988) Proceeding of royal society discussion meeting on utilisation of lignocellulosic wastes. Cambridge University Press, Cambridge

Hasunuma T, Okazaki F, Okai N, Hara KY, Ishii J, Kondo A (2013) A review of enzymes and microbes for lignocellulosic biorefinery and the possibility of their application to consolidated bioprocessing technology. Bioresour Technol 135:513–522

Henriksson G, Nutt A, Henriksson H, Pettersson B, Stahlberg J, Johansson G, Pettersson G (1999) Endoglucanase 28 (*Cel12A*), a new *Phanerochaete chrysosporium* cellulase. Eur J Biochem 259:88–95

Hoegger PJ, Kilaru S, James TY, Thacker JR, Kues U (2006) Phylogenetic comparison and classification of laccase and related multicopper oxidase protein sequences. FEBS J 273:2308–2326

Hofrichter M (2002) Review: lignin conversion by manganese peroxidase (MnP). Enzym Microb Technol 30:454–466

Hori C, Cullen D (2016) Prospects for bioprocess development basedon recent genome advances in lignocellulose degrading basidiomycetes. In: Schmoll M, Dattenbock C (eds) Gene expression systems in fungi: advancements and applications. Fungal Biol USA. Springer, Berlin, pp 161–181

Howard RR, Abotsi E, Renshurg JEL, Howard S (2003) Lignocellulose biotechnology: issues of bioconversion and enzyme production. Afr J Biotechnol 2:602–619

Hu J, Arantes V, Saddler JN (2011) The enhancement of enzymatic hydrolysis of lignocellulosic substrates by the addition of accessory enzymes such as xylanase: is it an additive or synergistic effect. Biotechnol Biofuels 4:36

Iqbal HMN, Asgher M, Ahmed I, Hussain S (2013) Media optimization for hyper-production of carboxymethyl cellulase using proximally analyzed agro-industrial residue with *Trichoderma harzianum* under SSF. Int J Agric Vet Med Sci 4:47–55

Isroi, Millati R, Syamsiah S, Niklasson C, Cahyanto MN, Lundquist K (2011) Biological pretreatment of lignocelluloses with white-rot fungi and its applications: a review. Bioresources 6:5224–5259

Jacobsen SE, Wyman CE (2000) Cellulose and hemicellulose hydrolysis models for application to current and novel pretreatment processes. Appl Biochem Biotechnol 84:81–96

Jeffries TW (1990) Biodegradation of lignin–carbohydrate complexes. Biodegradation 1:163–176

Jeffries TW (1994) Biodegradation of lignin and hemicelluloses. In: Ratledge C (ed) Biochemistry of microbial degradation. Kluwer, Dordrecht, pp 233–277

Jolivalt C, Raynal A, Caminade E, Kokel B, Le Goffic F, Mougin C (1999) Transformation of N′, N′-dimethyl-N-(hydroxyphenyl) ureas by laccase from the white rot fungus *Trametes versicolor*. Appl Microbiol Biotechnol 51:676–681

Juturu V, Wu JC (2014) Microbial cellulases: engineering, production and applications. Renew Sust Energ Rev 33:188–203

Kaarik A (1965) The identification of the mycelia of wood decay fungi by their oxidation reactions with phenolic compounds. Stud For Suec 31:1–79

Kabel MA, Jurak E, Makela MR, de Vries RP (2017) Occurrence and function of enzymes for lignocelluloses degradation in commercial *Agaricus bisporus* cultivation. Appl Microbiol Biotechnol 101:4363–4369

Kameshwar AKS, Qin W (2016) Lignin degrading fungal enzymes. In: Fang Z, Smith RL (eds) Production of biofuels and chemicals from lignin. Biofuels and biorefineries, vol 6. Springer, Singapore, pp 81–130

Kirk TK (1984) Lignin-degrading enzyme from *Phanerochaete chrysosporium*: purification, characterization, and catalytic properties of a unique H_2O_2-requiring oxygenase. Proc Natl Acad Sci U S A 81:2280–2284

Klemm D, Heublein B, Fink HP, Bohn A (2005) Cellulose: fascinating biopolymer and sustainable raw material. Angew Chemie Int Eds 44:3358–3393

Knob A, Terrasan CRF, Carmona EC (2010) β-Xylosidases from filamentous fungi: an overview. World J Microbiol Biotechnol 26:389–407

Kogo T, Yoshida Y, Kogane K, Matsumoto H, Watanabe T, Ogihara J, Kasum T (2017) Production of rice straw hydrolysis enzymes by the fungi *Trichoderma reesei* and *Humicola insolens* using rice straw as a carbon source. Bioresour Technol 233:67–73

Kozlova LV, Gorshkov OV, Mokshina NE, Gorshkova TA (2015) Differential expression of a-L-arabinofuranosidases during maize (*Zea mays* L.) root elongation. Planta 241:1159–1172

Kubicek CP (2012a) Fungi and lignocellulosic biomass. Wiley, Ames, IA

Kubicek CP (2012b) The plant biomass. In: Kubicek CP (ed) Fungi and lignocellulosic biomass. Wiley-Blackwell, Oxford, pp 1–28

Kuhad RC, Singh A, Eriksson KEL (1997) Microorganisms and enzymes involved in the degradation of plant fiber cell walls. Adv Biochem Eng Biotechnol 57:45–125

Kuila A, Sharma V, Garlapati VK, Singh A, Roy L (2016) Present status on enzymatic hydrolysis of lignocellulosic biomass for bioethanol production. Adv Biofeedstocks Biofuels 1:85

Kulasinski K, Salmen L, Derome D, Carmeliet J (2016) Moisture adsorption of glucomannan and xylan hemicelluloses. Cellulose 23:1629–1637

Kulkarni N, Shendye A, Rao M (1999) Molecular and biotechnological aspects of xylanases. FEMS Microbiol Rev 23:411–456

Kumar R, Wyman CE (2009) Effects of cellulase and xylanase enzymes on the deconstruction of solids from pretreatment of poplar by leading technologies. Biotechnol Prog 25:302–314

Kumar G, Bakonyi P, Periyasamy S, Kim SH, Nemestothy N (2015) Lignocellulose biohydrogen: practical challenges and recent progress. Renew Sust Energ Rev 44:728–737

Lam KL, Kaiwei S, Xiyang W, Shuze T, Xiaohui S, Kwan HS, Cheunga PCK (2018) The diploid genome of the only sclerotia-forming wild-type species in the genus *Pleurotus-Pleurotus tuber-regium*—provides insights into the mechanism of its biomass conversion from lignocellulose substrates. J Biotechnol 283:22–27

Latha GM, Srinivas P, Muralikrishna G (2007) Purification and characterization of ferulic acid esterase from malted finger millet (*Eleusine coracana*, Indaf-15). J Agric Food Chem 55:9704–9712

Lavanya D, Kulkarni PK, Dixit M, Raavi PK, Krishna LNV (2011) Sources of cellulose and their applications-a review. Int J Drug Formul Res 2:19–38

Leonowicz A, Szklarz G, Wojtas WM (1985) The effect of fungal laccase on fractionated lignosylphonates (*Peritan* Na). Phytochemistry 24:393–396

Leschine SB (1995) Cellulose degradation in anaerobic environments. Annu Rev Microbiol 49:399–426

Li PP, Wang XJ, Cui ZJ (2012) Survival and performance of two cellulose-degrading microbial systems inoculated into wheat straw-amended soil. J Microbiol Biotechnol 22:126–132

Liao H, Li S, Wei Z, Shen O, Xu Y (2014) Insights into high-efficiency lignocellulolytic enzyme production by *Penicillium oxalicum* GZ-2 induced by a complex substrate. Biotechnol Biofuels 7:162

Lombard V, Ramulu GH, Drula E, Coutinho PM, Henrissat B (2014) The carbohydrate-active enzymes database (CAZy) in 2013. Nucleic Acids Res 42:490–495

Maayer PD, Brumm PJ, Mead DA, Cowan DA (2014) Comparative analysis of the *Geobacillus* hemicellulose utilization locus reveals a highly variable target for improved hemicellulolysis. BMC Genomics 15:836

MacDonald J, Suzuki H, Master ER (2012) Expression and regulation of genes encoding lignocellulose-degrading activity in the genus *Phanerochaete*. Appl Microbiol Biotechnol 94:339–351

Madadi M, Abbas A (2017) Lignin degradation by fungal pretreatment: a review. J Plant Pathol Microbiol 8:2

Mai C, Kues U, Militz H (2004) Biotechnology in the wood industry. Appl Microbiol Biotechnol 63:477–494

Malayil S, Chanakya HN, Ashwath R (2016) Biogas digester liquid—a nutrient supplement for mushroom cultivation. Environ Nanotechnol Monit Manage 6:24–31

Malherbe S, Cloete TE (2002) Lignocellulose biodegradation: fundamentals and applications. Rev Environ Sci Biotechnol 1:105–114

Martinez AT (2002) Molecular biology and structure-function of lignin-degrading heme peroxidases. Enzym Microb Technol 30:425–432

Martinez D, Larrondo LF, Putnam N, Gelpke MD, Huang K, Chapman J (2004) Genome sequence of the lignocellulose degradingfungus *Phanerochaete chrysosporium* strain RP78. Nat Biotechnol 22:695–700

Mayer AM, Staples RC (2002) Laccase: new function for an old enzyme. Phytochemistry 60:551–565

Mehrotra R, Aneja K (1990) An introduction to mycology. New Age International, New Delhi

Menezes DB, Brazil OAV, Romanholo-Ferreira LF, Polizeli MLTM, Ruzene DS, Silva DP, Costa LP, Hernandez-Macedo ML (2017) Prospecting fungal ligninases using corncob lignocellulosic fractions. Cellulose 24:4355–4365

Meng X, Liu B, Xi C, Luo X, Yuan X, Wang X, Zhu W, Wang H, Cui Z (2018) Effect of pig manure on the chemical composition and microbial diversity during co-composting with spent mushroom substrate and rice husks. Bioresour Technol 251:22–30

More SS, Renuka PS, Pruthvi K, Swetha M, Malini S, Veena SM (2011) Isolation, purification, and characterization of fungal laccase from *Pleurotus* sp. Enzyme Res 2011:1–7

Mrudula S, Murugammal R (2011) Production of cellulase by *Aspergillus niger* under submerged and solid state fermentation using coir waste as a substrate. Braz J Microbiol 42:1119–1127

Mussatto SI, Teixeira JA (2010) Lignocellulose as raw material in fermentation processes. In: Mendez-Vilas A (ed) Current research, technology and education topics in applied microbiology and microbial biotechnology. Formatex Research Center, Spain, pp 897–903

Naika GS, Kaul P, Prakash VJ (2007) Purification and characterization of a new endoglucanase from *Aspergillus aculeatus*. Agric Food Chem 55:7566–7572

Nam KH, Sung MW, Hwang KY (2010) Structural insights into the substrate recognition properties of b-glucosidase. Biochem Biophys Res Commun 391:1131–1135

Naraian R, Singh D, Verma A, Garg SK (2010) Studies on in vitro degradability of mixed crude enzyme extracts produced from *Pleurotus* sp. J Environ Biol 31:945–951

Numan MT, Bhosle NB (2006) a-L-Arabinofuranosidases: the potential applications in biotechnology. J Ind Microbiol Biotechnol 33:247–260

Nurizzo D, Nagy T, Gilbert HJ, Davies GJ (2002) The structural basis for catalysis and specificity of the *Pseudomonas cellulosa* α-glucuronidase, GlcA67A. Structure 10:547–556

Okeke BC, Lu J (2011) Characterization of a defined cellulolytic and xylanolytic bacterial consortium for bioprocessing of cellulose and hemicelluloses. Appl Biochem Biotechnol 163:869–881

Parani K, Eyini M (2010) Effect of co-fungal treatment on biodegradation of coffee pulp waste in solid state fermentation. Asian J Exp Biol Sci 1:352–359

Paszczynski A, Huynh VB, Crawford R (1985) Enzymatic activities of an extracellular, manganese-dependent peroxidase from *Phanerochaete chrysosporium*. FEMS Microbiol Lett 29:37–41

Polizeli MLTM, Rizzatti ACS, Monti R, Terenzi HF, Jorge JA, Amorim DS (2005) Xylanases from fungi: properties and industrial applications. Appl Microbiol Biotechnol 67:577–591

Pramanik K, Sahu S (2017) Biological treatment of lignocellulosic biomass to bioethanol. Adv Biotechnol Microbiol 5:1–3

Qi M, Wang P, Selinger LB, Yanke LJ, Forster RJ, McAllister TA (2011) Isolation and characterization of a ferulic acid esterase (*Fae1A*) from the rumen fungus *Anaeromyces mucronatus*. J Appl Microbiol 110:1341–1350

Rashad FM, El-Kattan MH, Fathy HM, El-Fattah ABD, El-Tohamy DA, Farahat MAA (2019) Recycling of agro-wastes for *Ganoderma lucidum* mushroom production and *Ganoderma* post mushroom substrate as soil amendment. Waste Manag 88:147–159

Rennie EA, Scheller HV (2014) Xylan biosynthesis. Curr Opin Biotechnol 26:100–107

Rodrigo de Souza W (2013) Chapter 9: microbial degradation of lignocellulosic biomass. In: Sustainable degradation of lignocellulosic biomass-techniques, applications and commercialization. pp 207–247

Rothschild N, Levkowitz A, Hadar Y, Dosoretz CG (1999) Manganese deficiency can replace high oxygen levels needed for lignin peroxidase formation by *Phanerochaete chrysosporium*. Appl Environ Microbiol 65:483–488

Ryden P, Efthymiou MN, Tindyebwa TAM, Elliston A, Wilson DR, Waldron KW, Malakar PK (2017) Bioethanol production from spent mushroom compost derived from chaff of millet and sorghum. Biotechnol Biofuels 10:195

Sadhu S, Maiti TK (2013) Cellulase production by bacteria: a review. Br Microbiol Res J 3:235–258

Saha BC (2004) Production, purification and properties of endoglucanase from a newly isolated strain of *Mucor circinelloides*. Process Biochem 39:1871–1876

Sajith S, Priji P, Sreedevi S, Benjamin S (2016) An overview on fungal cellulases with an industrial perspective. J Nutr Food Sci 6:1–10

Salmen L, Burgert I (2009) Cell wall features with regard to mechanical performance. A review cost action E35 2004–2008: wood machining-micromechanics and fracture. Holzforschung 63:121–129

Salvachua D, Prieto A, Lopez-Abelairas M, Lu-Chau T, Martınez A, Martınez AJ (2011) Fungal pretreatment: an alternative in second-generation ethanol from wheat straw. Bioresour Technol 102:7500–7506

Sanchez C (2009) Lignocellulosic residues: biodegradation and bioconversion by fungi. Biotechnol Adv 27:185–194

Scheller HV, Ulvskov P (2010) Hemicelluloses. Annu Rev Plant Biol 61:263–289

Sedlmeyer FB (2011) Xylan as by-product of biorefineries: characteristics and potential use for food applications. Food Hydrocol 25:1891–1898

Segato F, Damasio ARL, de Lucas RC (2014) Genome analyses highlight the different biological roles of cellulases. Microbiol Mol Biol Rev 78:588–613

Shahzadi T, Anwar Z, Iqbal Z, Anjum A, Aqil T, Bakhtawar, Afzal A, Kamran M, Mehmood S, Irshad M (2014) Induced production of exoglucanase, and β-glucosidase from fungal co-culture of *T. viride* and *G. lucidum*. Adv Biosci Biotechnol 5:426–433

Shallom D, Shoham Y (2003) Microbial hemicellulases. Curr Opin Microbiol 6:219–228

Shruti, Sharma A, Malik DK (2015) Lignocellulose biomass degradation by microbial consortium isolated from harvested rice field. Int J Curr Microbiol App Sci 4:274–280

Singh S, Harms H, Schlosser D (2014) Screening of ecologically diverse fungi for their potential to pretreat lignocellulosic bioenergy feedstock. Appl Microbiol Biotechnol 98:3355–3370

Sjostrom E (1993) The structure of wood. In: Wood chem, pp 1–20

Sohpal VK, Dey A, Singh A (2010) Investigate of process parameters on xylanase enzyme activity in Melanocarpus albomyces batch culture. In: Proc World Congr Eng London, UK

Solomon EI, Sundaram UM, Machonkin TE (1996) Multicopper oxidases and oxygenases. Chem Rev 96:2563–2605

Sorensen HR, Jorgensen CT, Hansen CH, Jorgensen CI, Pedersen S, Meyer AS (2006) A novel GH43 α-L-arabinofuranosidase from *Humicola insolens*: mode of action and synergy with GH51 α-L-arabinofuranosidases on wheat arabinoxylan. Appl Microbiol Biotechnol 73:850–861

Srebotnik E, Messner KA (1994) Simple method that uses differential staining and light microscopy to assess the selectivity of wood delignification by white rot fungi. Appl Environ Microbiol 60:1383–1386

Steffen KT, Hatakka A, Hofrichter M (2002) Degradation of humic acids by the litter decomposing basidiomycete *Collybia dryophila*. Appl Environ Microbiol 68:3442–3448

Steinbach D, Kruse A, Sauer J (2017) Pretreatment technologies of lignocellulosic biomass in water in view of furfural and 5-hydroxymethylfurfural production—a review. Biomass Convers Biorefin 7:247–274

Sternberg D, Vijayakumar P, Reese ET (1977) β-Glucosidase-microbial production and effect on enzymatic hydrolysis of cellulose. Can J Microbiol 23:139–147

Taherzadeh-Ghahfarokhi MPR, Mokhtarani B (2019) Optimizing the combination of conventional carbonaceous additives of culture media to produce lignocellulose-degrading enzymes by *Trichoderma reesei* in solid state fermentation of agricultural residues. Renew Energy 131:946–955

Thurston CF (1994) The structure and function of fungal laccases. Microbiology 140:19–26

Till M, Goldstone DC, Attwood GT, Moon CD, Kelly WJ, Arcus VL (2013) Structure and function of an acetyl xylan esterase (*Est2A*) from the rumen bacterium *Butyrivibrio proteoclasticus*. Proteins 81:911–917

Tuomela M, Vikman M, Hatakka A, Itavaara M (2000) Biodegradation of lignin in a compost environment: a review. Bioresour Technol 72:169–183

Valaskova V, Snajdr J, Bittner B, Hofrichter M (2007) Production of lignocellulose-degrading enzymes and degradation of leaf litter by saprotrophic basidiomycetes isolated from a *Quercus petraea* forest. Soil Biol Biochem 39:2651–2660

Verma D, Satyanarayana T (2012) Cloning, expression and applicability of thermo-alkali-stable xylanase of *Geobacillus thermoleovorans* in generating xylooligosaccharides from agro-residues. Bioresour Technol 107:333–338

Waldrop MP, Balser TC, Firestone MK (2000) Linking microbial community composition to function in a tropical soil. Soil Biol Biochem 32:1837–1846

Wan C, Li Y (2012) Fungal pretreatment of lignocellulosic biomass. Biotechnol Adv 30:1447–1457

Wang QF, Niua LL, Jiao J, Guo N, Zang YP, Gai QY, Fuab YJ (2017) Degradation of lignin in birch sawdust treated by a novel *Myrothecium verrucaria* coupled with ultrasound assistance. Bioresour Technol 244:969–974

Wanmolee W, Sornlake W, Rattanaphan N, Suwannarangsee S, Laosiripojana N, Champreda V (2016) Biochemical characterization and synergism of cellulolytic enzyme system from *Chaetomium globosum* on rice straw saccharification. BMC Biotechnol 16:82

Wariishi H, Valli K, Gold MH (1989) Oxidative cleavage of a phenolic diarylpropane lignin model dimer by manganese peroxidase from *Phanerochaete chrysosporium*. Biochemist 28:6017–6023

Wen Z, Liao W, Chen S (2004) Hydrolysis of animal manure lignocellulosics for reducing sugar production. Bioresour Technol 91:31–39

Whetten R, Sederoff R (1995) Lignin biosynthesis. Plant Cell 7:1001–1013

Wong DW (2009) Structure and action mechanism of ligninolytic enzymes. Appl Biochem Biotechnol 157:174–209

Wood TM (1985) Properties of cellulolytic enzyme systems. Biochem Soc Trans 13:407–410

Wood TM (1989) Mechanisms of cellulose degradation by enzymes from aerobic and anaerobic fungi. In: Coughlan MP (ed) Enzyme systems for lignocellulose degradation. Elsevier Appl Sci, London, pp 17–35

Xiao LP, Shi ZJ, Bai YY, Wang W, Zhang XM, Sun RC (2013) Biodegradation of lignocellulose by white-rot fungi: structural characterization of water-soluble hemicelluloses. Bioenergy Res 6:1154–1164

Xiao Q, Yu H, Zhang J, Lia F, Li C, Zhang X, Ma F (2019) The potential of cottonseed hull as biorefinery substrate after biopretreatment by *Pleurotus ostreatus* and the mechanism analysis based on comparative proteomics. Ind Crop Prod 130:151–161

Xie C, Gong W, Yan L, Zhu Z, Hu Z, Peng Y (2017) Biodegradation of ramie stalk by *Flammulina velutipes*: mushroom production and substrate utilization. AMB Express 7:171

Xiong S, Martin C, Eilertsen L, Wei M, Myronycheva O, Larsson SH, Lestander TA, Atterhem L, Jonsson LJ (2019) Energy-efficient substrate pasteurisation for combined production of shiitake mushroom (*Lentinula edodes*) and bioethanol. Bioresour Technol 274:65–72

Xu F (1996) Oxidation of phenols, anilines, and benzenethiols by fungal laccases: correlation between activity and redox potentials as well as halide inhibition. Biochemist 35:7608–7614

Xu X, Hu Y, Quan L (2014) Production of bioactive polysaccharides by *Inonotus obliquus* under submerged fermentation supplemented with lignocellulosic biomass and their antioxidant activity. Bioprocess Biosyst Eng 37:2483–2492

Yang P, Zhang H, Cao L, Zheng Z, Mu D, Jiang S, Cheng J (2018) Combining sestc engineered A. niger with sestc engineered S. cerevisiae to produce rice straw ethanol via step-by-step and in situ saccharification and fermentation. 3 Biotech 8:12

Yoshida H (1883) Chemistry of lacquer (Urishi) part 1. J Chem Soc 43:472–486

Youn HD, Hah YC, Kang SO (1995) Role of laccase in lignin degradation by white-rot fungi. FEMS Microbiol Lett 132:183–188

Zhang X, Tu M, Paice MG (2011) Routes to potential bioproducts from lignocellulosic biomass lignin and hemicelluloses. Bioenergy Res 4:246–257

Zhao X, Huang X, Yao J, Zhou Y, Jia R (2015) Fungal growth and manganese peroxidase production in a deep tray solid-state bioreactor, and in vitro decolorization of poly r-478 by MnP. J Microbiol Biotechnol 25:803–813

Chapter 2
Fungal Biodiversity Producing Cellulase Involved in Efficient Cellulolysis

Praveen Kumar Gupta, Shreya Choudhary, C. Chandrananthi, J. Sharon Eveline, and S. P. Sushmitha

1 Introduction

Cellulases are a class of lignocellulolytic enzymes involved in converting cellulose into simple sugars. Lignocellulose constitutes 60% of the plant cell wall and is one of its major components. Lignocellulose comprises a family of lignin, hemicelluloses, and cellulose which are sugar biopolymer and its derivative (Galbe and Zacchi 2012). Cellulase is a polymer chain of D-glucose as monomeric units. It is the most widely available organic polymer in the world having 4×10^{10} tonnes produced every year (Coughlan 1985). Cellulases from microbial sources have become the main focus due to varied biotechnological applications. These are synthesized on cellulosic material by various fungal and bacterial sources (aerobic, anaerobic, mesophilic, or thermophilic) during their growth. Fungal cellulases are known to be simpler than their bacterial counterpart having two separate domains. The two separate domains are (1) catalytic domain and (2) cellulose-binding module (CBM), having approximately 35 amino acids (Carvalho et al. 2003).

Cellulase is a 3-enzyme membered family each having a particular function. The members in this family are exo-(1,4)-β-D-glucanase which is responsible for attacking cellulose chain ends, endo-(1,4)-β-D-glucanase which acts on O-glycosidic bonds in a random order which gives different lengths of glucan chain. The third is β-glucosidases which produces cellulose by acting on β-cellobiose (Kuhad et al. 2011). The most widely used and studied microorganisms for cellulase production are *Trichoderma*, *Aspergillus*, *Cellulomonas*, *Penicillium*, *Clostridium*, etc. (Kuhad et al. 2011).

P. K. Gupta (✉) · S. Choudhary · C. Chandrananthi · J. Sharon Eveline · S. P. Sushmitha
Department of Biotechnology, R.V College of Engineering, Bangalore, India
e-mail: praveenkgupta@rvce.edu.in

R. Naraian (ed.), *Mycodegradation of Lignocelluloses*, Fungal Biology,
https://doi.org/10.1007/978-3-030-23834-6_2

2 Classification of Cellulases

Cellulases are lignocellulolytic enzymes responsible for converting cellulose into simple sugars. They are complex enzyme system comprises endoglucanase (endo-1,4-β-D-glucanase), exoglucanase (exo-1,4-β-D-glucanase), and β-D-glucoside glucanohydrolase(β-D-glucosidase) (Sajith et al. 2016).

2.1 *Endoglucanase*

Endoglucanase, which can be endo-β-1,4-D-glucanase or endo-β-1,4-D-glucan-4glucano-hydrolase, is responsible for hydrolyzing carboxymethyl cellulose (CMC) in a random fashion. Hence, they are often referred as CMCase. Accordingly, the decrease in the length of polymer results in the rise of reducing sugar concentration. Endoglucanase converts the intermediate product of cellulose hydrolysis, cellodextrins into cellobiose and glucose. These enzymes are found to be inactive when acted against crystalline cellulose (e.g. Avicel, cotton) (Sajith et al. 2016).

2.2 *Exoglucanase*

Exoglucanase, which can be exo-β-1,4-D glucanase or cellobiohydrolase, degrades cellulose by splitting off the cellobiose units present in the non-reducing end of the chain. These enzymes are unable to hydrolyze hydroxyethyl cellulose and CMC derivates; however, they are found to be active against cellodextrins and amorphous solids which are present in partially degraded form. Exo-1,4-D-glucan-4-glucohydrolase or glucohydrolase are found as minor components in some cellulase derivatives (Sajith et al. 2016).

2.3 *β-Glucosidase*

This enzyme is responsible for hydrolyzing alkyl and aryl β-glucosides. Also, they are found to cleave cellobiose and shown to remove glucose group from the non-reducing tail of oligosaccharides thereby resulting in hydrolysis of cellulose (Sajith et al. 2016).

3 Fungal Cellulases

The synergetic action of different cellulases like exoglucanases, β-glucosidases, and endoglucanases is necessary for hydrolyzation of insoluble cellulose by microbial species, thereby converting it to simple sugars. Among the various microorganisms,

Table 2.1 Fungal species with cellulolytic abilities

Species of fungi	Class of fungi
Coniophora puteana	Brown-rot
Fomitopsis sp.	
Lanzites trabea	
Poria placenta	
Tyromyces palustris	
***Aspergillus* species:** *Aspergillus niger; A. nidulans; A. oryzae; A. terreus*	Soft-rot
***Chaetomium* species:** *Chaetomium cellulyticum; C. thermophilum*	
***Fusarium* species:** *Fusarium solani; F. oxysporum*	
Humicola insolens; H. grisea	
***Penicillium* species:** *Penicillium brasilianum; P. occitanis; P. decumbans*	
***Trichoderma* species:** *Trichoderma reesei; T. longibrachiatum; T. harzianum; Trichoderma atroviride*	
Agaricus arvensis	White-rot
Phanerochaete chrysosporium	
Phlebia gigantea	
Pleurotus ostreatus	
Sporotrichum thermophile	
Trametes versicolor	

fungi bring about 80% of the total cellulose breakdown and are considered as dynamic cellulose decomposers. The members of the Ascomycota, Basidiomycota, Deuteromycota, and Chytrids are some of the cellulose-decomposing fungi.

Industries prefer aerobic fungal cellulases as they are highly adaptive, extracellular, and secreted in bulk during their growth. Whereas cellulase extracted from other variable types of bacteria and fungi present in the form of enzyme complexes, these cellulases are mostly occurred in membrane-bound structure such as cellulosomes. Aerobic fungi such as *Trichoderma* sp., *Penicillium*, *Aspergillus* sp., and *Sclerotium* are highly cellulolytic in nature. *T. reesei* is a widely studied aerobic fungi as it is capable of hydrolyzing native cellulose. Many thermophilic fungi such *as Talaromyces emersonii, Chaetomium thermophilum, Sporotrichum thermophile,* and *Thermoascus aurantiacus* are being widely studied as the cellulose fibers bulge at high temperatures making them accessible for the action of hydrolytic enzymes (Sajith et al. 2016) (Table 2.1).

4 Characteristics of Cellulases

Cellulases are lignocellulolytic enzymes that hydrolyse the β-1,4 linkages in cellulose chains. Cellulases contain non-catalytic carbohydrate-associated module/carbohydrate-bound modules (CBMs) and/or other functionally known or unknown units which are situated at the N- or C-terminus in a catalytic segment (Zhang et al.

2018; Zhang and Zhang 2013). The major fungus for industrial cellulase production is *Trichoderma reesei*. Its genome encodes 10 cellulases and 16 hemicellulases (Meenu et al. 2014).

5 Occurrence

Microbes such as aerobic bacteria, fungi, yeast, and actinomycetes are involved in cellulase production. Fungi are the major producers of cellulase among the microbes and accounts for approximately 80% of the cellulose hydrolysis on Earth (Singh et al. 2007). Due to the constant contribution of cellulosic carbon in the form of clutter which acts as a substrate for disintegration and decomposition by the microbes, the residues of mangrove timberlands are found to be appropriate for discovering cellulose-disintegrating microorganisms (Behera et al. 2017).

6 Structure

The chemical assembly of cellulose molecule involves the β-anhydroglucose units with dominant hydroxyl groups. Each monomer component in the polymer involves primary and secondary hydroxyl assemblies. Each of the anhydroglucose units is linked together by β-1-4 glycosidic bonds (Fig. 2.1). Cellulose is a linear homopolysaccharide and has the capacity to form wide-ranging intra- and intermolecular hydrogen bonds due to its specific chemical arrangements (Sahin and Arslan 2008).

7 Cellulase Extraction from Fungal Sources

7.1 Aspergillus

Both the submerged (SmF) and solid-state fermentation (SSF) conditions are useful in the production of cellulase by using *Aspergillus niger*. The maximal production of cellulose was observed after incubating for 72 h in SSF and 96 h in SmF. The extracellular cellulose productivity is 14.6-fold high in SSF compared to that of SmF. The optimal conditions for maximum production of cellulase by *A. niger* was found to be with pH 6, temperature 30 °C, and additives like lactose, peptone, and coir waste as substrate (Kuhad et al. 2011). In this sense, *Aspergillus fumigatus* produces cellulase by using substrates like lignocellulosic materials under the controlled condition of SSF. In order to produce the maximum cellulase production from *A. fumigatus* Z5, the use of carboxymethyl cellulase (CMCase) in cultivation environments is acknowledged. Other encouraging conditions are identified as tem-

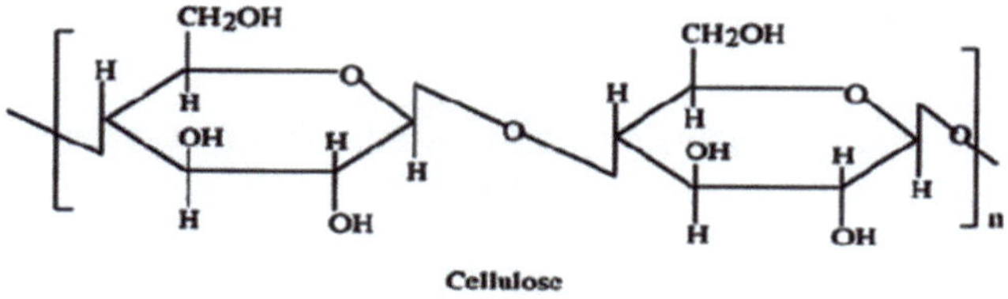

Fig. 2.1 Structure of cellulose (Chawla et al. 2014)

perature condition at 50 °C, initial moisture 80%, pH 4.0, and initial inoculum 7% (Liu et al. 2011).

7.2 *Trichoderma*

Like *Aspergillus niger*, *Trichoderma reesei* is extensively used for cellulase production. With the help of controlled overexpression of the gene *Trvib-1*, the outcome of cellulase production can be improved (Zhang et al. 2018). It was observed that the recombinant *T. reesei* Vib-1 is helpful in the increasing production of cellulase as compared to similar production from the parent strain. In addition, it has been also noted that the cellulase enzyme is produced by *T. harzianum* if it is supplemented with corn cob. Apart from these, *T. viride, T. asperellum, T. koningii, T. atroviride, T. longibrachiatum,* and *T. virens* are also capable of cellulose production (Pandey et al. 2015).

7.3 *Penicillium*

The submerged cultures of *Penicillium echinulatum* 9A02S1, with sorbitol and cellulose, can be used for cellulase production (Ritter et al. 2013). *Penicillium funiculosum* ATCC11797 is also capable of cellulase production. It was found that the optimum composition of culture media for increased production uses Avicel, urea, KH_2PO_4, $MgSO_4{\cdot}7H_2O$, and yeast extract (Lins et al. 2014). *Penicillium janthinellum* is also capable of cellulase production (Singh et al. 2017).

7.4 *Sclerotium*

Sclerotium rolfsii is an efficient producer of cellulolytic and hemicellulolytic enzymes (Sachslehner et al. 1997). On ultraviolet mutagenesis of crushed sclerotia, a mutant of *Sclerotium rolfsii* CPC 142 is formed, which is known to secrete a solution that has an effective filter paper-degrading activity in NM-2 growth medium cultures compared to the parent strain (Sadana et al. 1979).

7.5 Thermophilic Fungi

The thermophilic enzyme is gaining a lot of importance in recent times. This is because they produce thermostable cellulases. The best-known sources of cellulases such as that from species *Trichoderma*, *Penicillium*, or *Aspergillus* species are thermostable at temperature 50 °C. Chances of contamination of reactors are high by microorganisms using sugars liberated by the hydrolytic reactions, and at this temperature and the sugar recovery rate is very low. Hence to overcome this, the reaction temperature can be increased (Durand et al. 1984).

Search for new thermophilic having been taken up by many researchers to fill this need. Species like *Thermoascus aurantiacus* (Shepherd et al. 1981), *Sporotrichum thermophile* (Coutts and Smith 1976), and *Chaetomium thermophile* (Eriksen and Goksøyr 1977) are found to produce thermostable cellulase. Other patented species also producing thermostable cellulase are *Thielavia terrestris* and *Sporotrichum cellulophilum.* The species *Sporotrichum pulverulentum* was improved genetically and was compared to the species *T. reesei* with respect to the enzymes produced. The species *Sporotrichum pulverulentum* were found to be not suitable for the process of saccharification as they have low activity on the crystalline cellulose at the temperature 50 °C.

8 Production of Cellulase

Submerged fed-batch fermentation is used for low-cost ($10–40/kg of dry protein wt.) production of cellulase by companies (e.g. Iogen, Genencor) which gives them a large amount of crude cellulase production of more than 100 g/L of broth used. Such companies use strains of *Aspergillus* and *Trichoderma* as cellulase derivative (Zhang and Zhang 2013).

Cellulase producing microorganisms use carbohydrate as their energy source as they are not capable of consuming proteins and thus degrade carbohydrates for growth and metabolism. Species like *Cellulomonas* and *Cytophaga* and other fungi have the capability to use other carbohydrates as an energy source with cellulose (Stapleton et al. 2004; Poulsen and Petersen 1988), whereas anaerobic cellulolytic species only stick to cellulose and its derivatives (hydrolytic products). Some fungi have the capability to produce high levels of cellulase and these are extensively studied. Species that can secrete extracellular proteins are *Trichoderma reesei, Humicola, Penicillium*, and *Aspergillus* which are responsible for converting cellulase to glucose (Zeikus and Ng 1982). Other common fungal species are *Humicola, Trichoderma, Aspergillus, Penicillium, Pseudomonads, Cellulomonas, Actinomycetes, Streptomyces* etc. Some fungal strains are able to secrete a cellulase enzyme complex that can hydrolyse cellulose. Species like *T. reesei, H. insolens, A. niger, Thermomonospora fusca,* and *Bacillus* are among the major fungal species which are exploited for commercial purpose (Sukumaran et al. 2005) (Table 2.2).

Table 2.2 Microorganisms involved in the production of celluloase (Sethi et al. 2013)

Strains of fungi	Substrates used
Aspergillus niger	Palm kernel cake
Aspergillus nidulans	2 mM of p-nitrophenyl-fl-o-glucoside in 50 mM acetate buffer
Aspergillus oryzae	p-nitrophenyl *β*-D-cellobioside
Fusarium solani	H_3PO_4-swollen cellulose, *O*-(carboxymethyl) cellulose, cotton fiber
Humicola insolens	CMC (Carboxymethylcellulose) for to determining activity profiles of endoglucanases (EG), cellohexaose, and cellotriose for determining the activity profiles of CBH (cellobiohydrolase I) and CBH II
Humicola grisea	*p*-nitrophenyl (PNP)-*β*-D-saccharides
Penicillium brasilianum	Azo-arabinoxylan or azo-CMCase was used as substrate for determination of endoglucanase and endoxylanase activity
Penicillium occitanis	Combination of wheat flour and wheat bran remnants with esparto grass paper pulp and cellulose substrates were shown to obtain high cellulase yields

9 Biotechnological Applications

Cellulase enzyme finds a wide range of application in agriculture, food, fermentation, etc., industries (Table 2.3).

10 Conclusion

The degradation of lignocellulosic waste by microbes is the coordinated action of several enzymes in which the cellulases are considered to be the most prominent enzyme. The researches on microbial cellulases have not only resulted in significant scientific knowledge about the enzymes but have also revealed their enormous potential in biotechnology. As a result, isolation and screening of novel cellulase-producing microorganisms that can produce profuse amounts of the enzyme at extreme conditions using cheap substrates are necessary. Although there are various species producing xylanase enzyme, species of *Trichoderma*, *Aspergillus*, and *Penicillium* are exploited for the wide variety of cellulolytic enzymes that they possess. This scenario has however shifted in the recent days to other promising strains producing xylanase that show better increased thermostability, activity, and better stability under alkaline or acidic medium. Recently, strategies are being carried out to convert cellulase into useful products, and to this end cellulase have been identified as key enzymes with diverse industrial applications. The bulk production of cellulase at an economically viable rate is the major bottleneck in commercial applications of cellulase.

Table 2.3 Industrial application of cellulase (adopted from Kuhad et al. 2011; Singh et al. 2007; Bhat 2000; Dhiman et al. 2002)

Industry	Application
Agriculture	Enzyme combinations of cellulase, pectinase, and hemicellulose improve crop growth. Also, enzymes like β-glucanases give resistance to diseases to the plant species by degrading the cell wall of the pathogenic species (Kuhad et al. 2011). Cellulase extracted from certain fungal strains can degrade the cell wall of species pathogenic to plant and help them control diseases
Bioconversion	Conversion of lignocellulosic material into useful products such as biofuels like methane and ethanol is considered to be a major industrial application of cellulase. Other products including organic acids, lipids, and single-cell proteins are also produced from cellulase. In addition, production of energy-rich animal feed with improved absorption is successfully done with the help of cellulase. Furthermore, cellulase improves performance of ruminants and helps in preservation of high-quality fodder
Detergent	Production of cellulase for manufacturing detergents that can dissolve the cellulose related stains. Cellulase imparts quality cleaning mechanism without causing damage to fibers along with enhanced color brightness and removal of dirt and rough protuberances in fabrics. In addition, it prevents ink particles from redeposition
Fermentation	Cellulase improves malting, mashing, pressing, and primary fermentation due to cellulose activity. It enhances wine aroma, stability, color extraction of grapes, and beer quality. Cellulase reduces wort viscosity which makes filterability more feasible
Food	Cellulase finds a varied application in the food processing industry as well. One such application is the production of juices extracted from fruits and vegetable. They are also used in clarification, stabilisation, and increasing the yield of the juices produced (De Carvalho et al. 2008). Enzymes such as cellulose and hemicellulose also find application in the extraction of olive oil
Pulp and paper	Cellulase acts as a co-additive in processes such as pulp bleaching and biomechanical pulping. It improves the properties in fiber brightness, strength, pulp freeness, and cleanliness as well as improving drainage in pulping and paper milling. Cellulase eases deinking process and reduces energy and chlorine demands. Production of biodegradable products such as cardboard, sanitary napkins, and paper towels is carefully done with the help of cellulase. Moreover, cellulase enzyme with xylanase enzyme is used for the purpose of deinking used paper
Textile	Cellulase is known to increase fabric quality and stability. It helps in biostoning of jeans, biopolishing, and improved absorbance of textile fibers, garment softening, color brightness restoration, and in excess dye removal from fabrics due to cellulose activity. Cellulase enzyme is used in wet processing and it improves appearances. These are used for biological polishing of cotton and other fabrics. These are also used for biological stoning of jeans. In this, the cellulase enzyme is directly placed on the fabric, which breaks off small fibers in the fabric from the surface of the yarn and thereby loosens the dye which can now be removed by washing
Soap and detergent industry	Cellulase is used with lipase and protease enzyme in detergents. The cellulase counterparts may improve the dirt-removing ability of cotton blend garments and gives the garments a more glossy, intense-colored, and brighter appearance

(continued)

Table 2.3 (continued)

Industry	Application
Animal feed industry	Pretreatment of stored fodder with cellulase improves the nutritional value of the fodder. They also help in eliminating the anti-nutritional components in the fodder like the non-starch polysaccharides like cellulose into other easily absorbed parts
Others	Cellulase improves extraction of carotenoids and enhancement of its oxidation and color stability properties. It enhanced olive oil extraction with good quality and olive paste malaxation improvement. Cellulase plays a huge role in waste management and risk reduction of cellulosic biomass by converting it to beneficial products such as enzymes, energy sources, and nutrients. Cellulases are helpful in the production of hybrid molecules like cellulosomes for improvement of cell surface properties to utilize insoluble substrates with the help of these cellulosomes. Apart from that, cellulase converts hemicellulose and is used for production of wine. It improves the clarification of must, production of wine, quality, stability, etc. Wines can be made more aromatic by the enzyme β-glucosidases

Acknowledgement The authors listed in this chapter wish to express their appreciation to the RSST Trust, Bangalore, for their continuous support and encouragement. As a corresponding author, I also express my sincere thanks to all other authors whose valuable contribution and important comments made this manuscript to this form.

Conflict of Interest: The authors listed in this chapter have no conflict of interest known best from our side. There was also no problem related to funding. All authors have contributed equally with their valuable comments which made the manuscript to this form.

References

Behera BC, Sethi BK, Mishra RR, Dutta SK, Thatoi HN (2017) Microbial cellulases—diversity & biotechnology with reference to mangrove environment: a review. J Genet Eng Biotechnol 15:197–210

Bhat MK (2000) Cellulases and related enzymes in biotechnology. Biotechnol Adv 18:355–383

Carvalho LM, Deliza R, Silva CA, Miranda RM, Maia MC (2003) Identifying the adequate process conditions by consumers for pineapple juice using membrane technology. J Food Technol 1:150–156

Chaabouni SF, Belguith H, Hassairi I, M'Rad K, Ellouz R (1995) Optimization of cellulase production by *Penicillium occitanis*. Appl Microbiol Biotechnol 43:267–269

Chawla SP, Kanatt SR, Sharma AK (2014) Chitosan. Springer, Switzerland

Coughlan MP (1985) Cellulose hydrolysis: the potential, the problems and relevant research at Galway. Biochem Soc Trans 13:405–406

Coutts AD, Smith RE (1976) Factors influencing the production of cellulases by *Sporotrichum thermophile*. Appl Environ Microbiol 31(6):819–825

De Carvalho LM, De Castro IM, Da Silva CA (2008) A study of retention of sugars in the process of clarification of pineapple juice (*Ananas comosus*, L. Merril) by micro-and ultra-filtration. J Food Eng 87:447–454

Dhiman TR, Zaman MS, Gimenez RR, Walters JL, Treacher R (2002) Performance of dairy cows fed forage treated with fibrolytic enzymes prior to feeding. Anim Feed Sci Technol 101:115–125

Durand H, Soucaille P, Tiraby G (1984) Comparative study of cellulases and hemicellulases from four fungi: mesophiles *Trichoderma reesei* and *Penicillium* sp. and thermophiles *Thielavia terrestris* and *Sporotrichum cellulophilum*. Enzym Microb Technol 6:175–180

Eriksen J, Goksøyr J (1977) Cellulases from *Chaetomium thermophile* var. dissitum. Eur J Biochem 77(3):445–450

Galbe M, Zacchi G (2012) Pretreatment: the key to efficient utilization of lignocellulosic materials. Biomass Bioenergy 46:70–78

Jorgensena H, Eriksson T, Borjesson J, Tjerneld F, Olsson L (2003) Purification and characterization of five cellulases and one xylanase from *Penicillium brasilianum* IBT 20888. Enzym Microb Technol 32:851–861

Kuhad RC, Mehta G, Gupta R, Sharma KK (2010) Fed batch enzymatic saccharification of newspaper cellulosics improves the sugar content in the hydrolysates and eventually the ethanol fermentation by *Saccharomyces cerevisiae*. Biomass Bioenergy 34:1189–1194

Kuhad RC, Gupta R, Singh A (2011) Microbial cellulases and their industrial applications. Enzyme Res 2011:1–10

Kwon KS, Hyung GK, Yung CH (1992) Purification and characterization of two extracellular β-glucosidases from *Aspergillus nidulans*. FEMS Microbiol Lett 97:149–153

Lins MAC, Daniele FC, Edelvio BG, Roberto NM, Lidia M, Melo S, Aline MC, Nei PJ (2014) Optimisation of cellulase production by *Penicillium funiculosum* in a stirred tank bioreactor using multivariate response surface analysis. Enzyme Res 2014:1–9

Liu D, Ruifu Z, Xingming Y, Hongsheng W, Dabing X, Zhu T, Qirong S (2011) Thermostable cellulase production of *Aspergillus fumigatus* Z5 under solid-state fermentation and its application in degradation of agricultural wastes. Int Biodeterior Biodegrad 65:717–725

Mai C, Kues U, Militz H (2004) Biotechnology in the wood industry. Appl Microbiol Biotechnol 63:477–494

Martin S (1997) Enzymatic properties of cellulases from *Humicola insolens*. J Biotechnol 57:71–81

Meenu K, Singh G, Vishwakarma RA (2014) Molecular mechanism of cellulase production systems in *Trichoderma*. In: Gupta VK, Schmoll M, Herrera-Estrella A, Upadhyay RS, Druzhinina I, Tuohy M (eds) Biotechnology and biology of *Trichoderma*. Elsevier, Amsterdam, pp 319–324

Ong LGA, Abd-Aziz S, Noraini S, Karim MIA, Hassan MA (2004) Enzyme production and profile by *Aspergillus niger* during solid substrate fermentation using palm kernel cake as substrate. Appl Biochem Biotechnol 118:73–79

Pandey S, Srivastava M, Shahid M, Kumar V, Singh A, Trivedi S, Srivastava YK (2015) *Trichoderma* species cellulases produced by solid state fermentation. J Data Min Genomics Proteomics 6:2–5

Poulsen OM, Petersen LW (1988) Growth of *Cellulomonas* sp. ATCC 21399 on different polysaccharides as sole carbon source induction of extracellular enzymes. Appl Microbiol Biotechnol 29:480–484

Ritter CET, Camassola M, Zampieri D, Silveira MM, Dillon AJP (2013) Cellulase and Xylanase production by *Penicillium echinulatum* in submerged media containing cellulose amended with sorbitol. Enzyme Res 2013. Article ID 240219, 9 pages. https://doi.org/10.1155/2013/240219

Sachslehner A, Haltrich D, Nidetzky B, Kulbe KD (1997) Production of hemicellulose- and cellulose-degrading enzymes by various strains of *Sclerotium rolfsii*. In: Davison BH, Wyman CE, Finkelstein M (eds) Biotechnology for fuels and chemicals. Humana Press, Totowa, NJ, pp 189–201

Sadana JC, Shewale JG, Deshpande MV (1979) Enhanced cellulase production by a mutant of *Sclerotium rolfsiit*. Appl Environ Microbiol 38:730–733

Sahin HT, Arslan MB (2008) A study on physical and chemical properties of cellulose paper immersed in various solvent mixtures. Int J Mol Sci 9:78–88

Sajith S, Priji P, Sreedevi S, Benjamin S (2016) An overview on fungal cellulases with an industrial perspective. J Nutr Food Sci 6:1–13

Sethi S, Datta A, Lal Gupta B, Gupta S (2013) Optimization of cellulase production from bacteria isolated from soil. ISRN Biotechnol 2013. Article ID 985685. 7 pages. https://doi.org/10.5402/2013/985685

Shepherd MG, Tong CC, Cole AL (1981) Substrate specificity and mode of action of the cellulases from the thermophilic fungus *Thermoascus aurantiacus*. Biochem J 193(1):67–74

Shruthi K, Yadav PS, Prasad BVS, Chandra MS (2018) Cellulase production by *Aspergillus unguis* in solid state fermentation. J Forest Res 30:205–212
Singh A, Kuhad RC, Ward OP (2007) Industrial application of microbial cellulases. In: Kuhad RC, Singh A (eds) Lignocellulose biotechnology. Int Pub House, New Delhi, India, pp 345–358
Singh A, Adsul M, Vaishnav N, Mathur A, Singhania RR (2017) Improved cellulase production by *Penicillium janthinellum* mutant. Indian J Exp Biol 55:436–440
Stapleton PC, O'Brien MM, O'Callaghan J, Dobson AD (2004) Molecular cloning of the cellobiose dehydrogenase gene from *Trametes versicolor* and expression in *Pichia pastoris*. Enzym Microb Technol 34:55–63
Sukumaran RK, Singhania RR, Pandey A (2005) Microbial cellulases-production, applications and challenges. J Sci Indus Res 64:832–844
Takashima S, Iikura H, Nakamura A, Hidaka M, Masaki H, Uozumi T (1996) Purification and characterization of cellulases from *Humicola grisea*. Biosci Biotechnol Biochem 60:77–82
Takashima S, Iikura H, Nakamura A, Hidaka M, Masaki H, Uozumi T (1998) Overproduction of recombinant *Trichoderma reesei* cellulases by *Aspergillus oryzae* and their enzymatic properties. J Biotechnol 65:163–171
Todero Ritter CE, Camassola M, Zampieri D, Silveira MM, Dillon AJP (2013) Cellulase and xylanase production by *Penicillium echinulatum* in submerged media containing cellulose amended with sorbitol. Enzyme Res 2013:1–9
Wood TM, McCrae SI (1977) Cellulase from *Fusarium solani*: purification and properties of the C1 component. Carbohydr Res 57:117–133
Zeikus JG, Ng TK (1982) Differential metabolism of cellobiose and glucose by *Clostridium thermocellum* and *Clostridium thermohydrosulfuricum*. J Bacteriol 150:1391–1399
Zhang XZ, Zhang YH (2013) Cellulases: characteristics, sources, production, and applications. In: Bioprocessing technologies in biorefinery for sustainable production of fuels, chemicals, and polymers, vol 2, pp 131–146
Zhang F, Finqing Z, Fengwu B (2018) Improvement of cellulase production in *Trichoderma reesei* rut-C30 by overexpression of a novel regulatory gene trvib-1. Bioresour Technol 247:676–683

Chapter 3
Fungal Diversity and Enzymes Involved in Lignin Degradation

Harit Jha

1 Introduction

The term lignin is derived from the Latin word *lignum* meaning wood. Lignin is an amorphous polymer formed by the free-radical-mediated polymerization of three different phenylpropane units (p-coumaryl, coniferyl and sinapyl alcohol). The amount and chemical composition of lignin vary depending on the source. Lignin in gymnosperms is prominently composed of dehydrogenation polymers of coniferyl alcohols, while angiosperm lignin contains mixed dehydrogenation polymers of coniferyl and sinapyl alcohol. The main purpose of lignin is to give the plant structural support, impermeability and resistance against microbial attack and oxidative stress. It is also not soluble in water and soluble and inactive optically; thus, lignin is difficult to degrade.

2 Structure of Lignin

The basic building unit for all lignins is the phenylpropane unit. These units are linked together by C–O–C and C–C bonds in such a way as to form a three-dimensional polymer by cross-linking (Fig. 3.1). The phenylpropanoid monomers of the lignin show similarity; however, the functional group's substitution on the phenyl ring differs. A single hydroxy or a methoxy group is present in the H-structure (4-hydroxy phenyl), two hydroxy or methoxy groups are present in the G-structure (guaiacyl) and three hydroxy or methoxy groups are present in the S-structure (syringyl). The different types of bonds found in lignin structure include carbon–carbon and carbon–oxygen bonds between various monomers. A carbon–oxygen

H. Jha (✉)
Department of Biotechnology, Guru Ghasidas Vishwavidyalaya, Bilaspur, Chhattisgarh, India

R. Naraian (ed.), *Mycodegradation of Lignocelluloses*, Fungal Biology,
https://doi.org/10.1007/978-3-030-23834-6_3

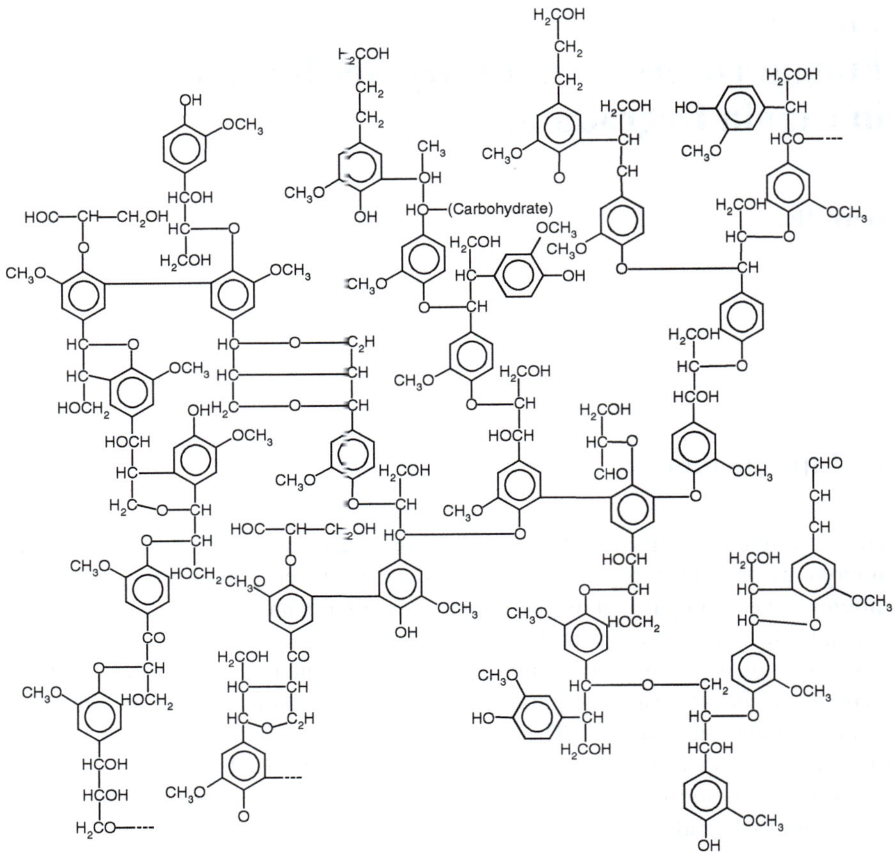

Fig. 3.1 Structure of lignin

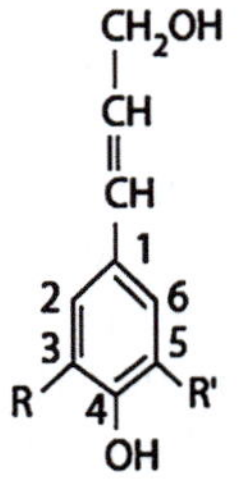

Substituents	Name	Location
R = R' = H	*p*-coumaryl alcohol	Compression wood, grasses
R = H, R' = OCH_3	coniferyl alcohol	Hardwoods and softwoods
R = R' = OCH_3	sinapyl alcohol	Hardwoods

Fig. 3.2 Lignin monomeric building blocks

link between a p-hydroxy moiety and the β- end of the propenyl group (β-O-4) is the most common bond present in lignin (Fig. 3.2) (Kukkola et al. 2004).

3 Applications of Lignin

3.1 Bioconversion of Lignocellulose into Fuel and Useful Chemicals

Technologies based on the efficient conversion of low-quality or waste lignocellulosic residues into fuel and industrially important chemical present long-term solution to the finite supplies of petroleum. Many efforts have been made to develop economically viable processes derived from the cellulose and hemicellulose and lignin components. Low-molecular-weight chemicals such as vanillin and dimethyl sulfoxide are also currently prepared on an industrial scale from lignosulfonates and kraft lignin, respectively. The range of low-molecular-weight compounds formed during lignin degradation by white-rot fungi and actinomycetes has raised the possibility of using controlled biodegradation to generate some of the high-value chemicals currently derived via the petrochemical industry (Ramasamy 1993).

Selective modification of byproducts of lignin and waste lignocellulose (demethylation, hydroxylation) in order to furnish additional reactive group has been suggested as a means of providing chemical feedstock for the manufacture of high-value polymers. Lignin degradation could improve the nutritive value of the agricultural wastes. It has been showed that presence of lignin in carbohydrate-rich residues restricts its digestibility by various enzymes of rumen fluid. Chemical and physical delignification methods have been used extensively, but with emerging and classified knowledge of lignin biodegradation, more attention is centred on delignification treatment based on lignin-degrading microorganisms particularly the basidiomycetes fungi. Digestibility of wheat straw, wood and sunflower stalk was improved using white-rot fungi pretreatment. The list of uses of lignin for which patent has been granted are as follows:

1. Drilling and additive.
2. Rubber additive.
3. Dyestuff filler.
4. Ink with lignin-based colourant.
5. Lignin derivative surfactant.
6. Cationic lignin amine surface-active agent.
7. Lignin-epoxysuccinate adduct as dispersant and metal complexing agent.
8. Lignin-reinforced polymer.
9. Polyvinyl alcohol and lignin sulfonate-containing adhesive.
10. Synthetic lignin-polyisocyanate resin.
11. Polyoxyalkylene ether polyols from lignin.
12. Isocyanate lignin products.

13. Copolymerizable colourant in polyesters.
14. Lignin use in pharmaceuticals.
15. Acid polymerized lignosulfonate as extender.
16. Polymerizable lignin derivatives.
17. Binder for glass fibre insulation.
18. Lignin-sulfonate–phenol-formaldehyde glue system.
19. Controlled released pesticides.
20. Lignin-urea composition for flame proofing.

4 Biodegradation of Lignin

Microbial degradation of lignin is an important step in carbon cycle as it enables the microorganisms to utilize the plant carbohydrates, overcoming the obstacle of lignin (Martínez et al. 2005; Kersten and Cullen 2007). Lignin removal is also a central aspect in industrial uses of cellulosic biomass, such as bioethanol production and manufacture of cellulose-based chemicals and materials, including paper cellulose pulp manufacture basically consists in breaking down (chemically or mechanically) the middle lamella in such a way that wood fibres are individualized (Sixta et al. 2006). In industrial applications, biotechnology based on lignin-degrading microbes and their enzymes can contribute to more efficient and environmentally sound use of renewable lignocellulosic feedstocks for sustainable production of materials, chemicals, biofuels and energy.

Lignin is a large polymer and the major challenges faced by a microorganism in its degradation are as follows:

1. The degrading system must be extracellular because lignin polymer (600–1000 kDa) is large and its direct uptake by microbial cells is unlikely.
2. The degradative system must be oxidative rather than hydrolytic because the structure is comprised of interunit carbon–carbon and ether bonds.
3. The ligninolytic agent must be less specific because lignin polymer is stereo-irregular.

Lignin degradation is performed by microorganisms like fungi and bacteria (Voeller et al. 2016). Lignin-degrading enzymes act to break the barrier of the plant cell wall for subsequent access to cellulose complex by cellulases and, eventually, may replace physicochemical processes, if the process becomes sufficiently rapid. Biological extraction of lignin employs wood-degrading microorganisms, including white-, brown-, and soft-rot fungi and bacteria to modify the chemical composition and/or structure of the lignocellulosic biomass so that the modified biomass was more amenable to enzyme digestion. Advantages of biological pretreatments are low energy requirement and mild operation conditions. However, long incubation time is required for enzymatic/biological treatment due to low activity of enzymes (Cardona and Sánchez 2007). Numerous studies have been carried out to find the microbes, which degrade lignin polymer. Lignin is however not biodegraded anaerobically, only minor modification in functional groups and nonlignin compounds takes place and some low-molecular-weight materials are degraded (Hatakka 1994).

The white-rot fungi are the main producers of ligninolytic enzymes. Among the several hundred species of white-rot fungi belonging to a variety of fungal families among the *Holobasidiomycetidae* are the *Agaricaceae*, *Hydnaceae*, *Corticiaceae*, *Polyporaceae* and *Thelephoraceae*. There are also a few Ascomycetes in the *Xylariaceae* that causes white -rot (Ma et al. 2016). The most efficient ligninolytic microorganisms are, however, Basidiomycetes (Kent et al. 1987). Various white-rot fungi being used to degrade lignin preferentially over cellulose or hemicelluloses are *Polyporus berkeleyi, Pycnoporus cinnabarinus, Fome sulmarius, Polyporus resinosus, Pleurotus ostreatus, Phlebia radiata* and *Merulius tremellosus* (Fritsche and Hofrichter 2000). *Phanerochaete chrysosporium* was the first fungus used for pulp delignification under conditions optimized for the mineralization of synthetic lignin.

The facts about lignin degradation derived from studies on white-rot fungi also suggest oxidative, extracellular and non-specific nature like there is substantial increase in carboxyl and carbonyl groups, and a decrease in hydrogen content in lignin residues left after decay oxidation of α-carbon of the propyl side chain to a carbonyl group hydroxylation and oxidative cleavage of the aromatic rings have been also observed. The demonstration of the insertion of an oxygen-derived hydroxyl group at beta-carbon of the side chain utilizing beta-1 lignin substructure compounds has authenticated the oxidative nature of lignin biodegradation (Shoemaker and Leisola 1990).

Bacterial lignin degradation has been most extensively studied in actinomycetes particularly *Streptomyces sp., Streptomyces viridosporus* and *Streptomyces setonii. Nocardia, Pseudomonas* and *Corynebacterium* have also been found to grow on lignin-related phenols. Lignin-related chemicals can be further processed by bacteria. For instance, the aerobic gram-negative soil bacterium *Sphingomonas paucimobilis* is able to degrade lignin-related biphenyl chemical compounds. A *Xanthomonas* strain has been reported to mineralize 30% synthetic [^{14}C] lignin in 20 days. However, generally bacteria are primarily responsible for the degradation of plant material in certain environments, but mineralization of lignin was minimal.

Wood decay and lignin degradation caused by various species of ascomycetes and fungi imperfecti include *Fusarium solani, Trichoderma strains, Trichoderma harzianum, Chaetomium piluliferum* and *Aspergillus fumigatus.* A few species of ascomycetes, e.g. species of *Xylaria, Libertella* and *Hypoxylon*, cause white-rot decay accompanied by substantial lignin loss. Lignin biodegradation by non-white-rotting basidiomycetes has been reported and it includes Gastromycete *Cyathus stercoreus, Ectomycorrhizal* fungi *Cenococcum, Amanita, Tricholoma* and *Rhizopogan*. These fungi mineralize ^{14}C-labelled lignin, albeit slowly.

The white-rot basidiomycetes degrade lignin rapidly and extensively than any other microbial groups. They invade the lumens of wood cells, where they secrete enzymes that degrade lignin and other wood components. The electron microscopic studies showed that lignin is degraded at some distance from the hyphae and is removed progressively from the lumens towards the middle lamella. During its min-

Table 3.1 Diversity of ligninolytic fungi

Class	Microorganism	Enzymes	Reference
Ascomycota	*Magnaporthe grisea, Myrothecium verrucaria*	Laccase	Dashtban et al. (2010); Su et al. (2018)
Ascomycota	*Melanocarpus albomyces*	Laccase	Hakulinen et al. (2002)
Ascomycota	*Neurospora crassa*	Laccase	Znameroski et al. (2012)
Ascomycota	*Xylaria sp.* and *Hypoxylon sp., xylariaceous*	Laccase	Dashtban et al. (2010); Pointing et al. (2005)
Basidiomycota (white-rot)	*Trametes versicolor, Trametes pubescens, Cyathus bulleri*	Aryl-alcohol oxidase, laccase	Leonowicz et al. (2001); Rana et al. (2018)
Basidiomycota (white-rot)	*Pleurotus sajor-caju, Pleurotus eryngii, Pleurotus ostreatus*	Aryl-alcohol oxidase, VP, laccase	Kalmi et al. (2008); Snajdr and Baldrian (2007)
Basidiomycota (white-rot)	*Phanerochaete chrysosporium*	MnP, aryl-alcohol oxidase, VP, laccase	Jager et al. (1985)
Basidiomycota (white-rot)	*Bjerkandera adusta, Bjerkandera fumosa*	VP	
Basidiomycota (white-rot)	*Fomes annosus*	Laccase	Papinutti and Martınez (2006); Leonowicz et al. (2001)
Basidiomycota	*Pholiota mutabilis*	Laccase	Leonowicz et al. (2001); Balaes et al. (2017)
Basidiomycota	*Coniophora puteana*	MnP, aryl-alcohol oxidase, VPs, laccase	Memić et al. (2017)
Basidiomycota	*Chaetomium thermophile*	MnP, aryl-alcohol oxidase, VPs, laccase	Chefetz et al. (1998)
Basidiomycota	*Pycnoporus coccineus, Pycnoporus sanguineus*	MnP, aryl-alcohol oxidase, VPs, laccase	Uzan et al. (2010)
Basidiomycota (white-rot)	*Pycnoporus cinnabarinus*	Laccase	Eggert et al. (1996)
Basidiomycota	*Phlebia radiata, Phlebia tremellosa*	LiP, MnP	Hatakka (1994); Fritsche and Hofrichter (2000); Maciel et al. (2010)
Basidiomycota	*Ceriporiopsis subvermispora*	MnP	Hatakka (1994); Fritsche and Hofrichter (2000)
Basidiomycota	*Nematoloma frowardii*	MnP	Sack et al. (1997); Fritsche and Hofrichter (2000)
Basidiomycota	*Stropharia rugosoannulata*	MnP	Hatakka (1994); Fritsche and Hofrichter (2000)
Basidiomycota	*Armillaria tabescens*	Laccase	He et al. (2014)
Basidiomycota	*Ganoderma lucidum*	LiP, MnP and laccase	Hariharan and Nambisan (2013)

(continued)

Table 3.1 (continued)

Class	Microorganism	Enzymes	Reference
Basidiomycota	*Agaricus bisporus*	LiP, MnP and laccase	Bonnen et al. (1994)
Basidiomycota	*Marasmius quercophilus*	Laccase	Farnet et al. (2004); Dedeyan et al. (2000)

eralization by white-rot fungi, lignin undergoes oxidative changes including aromatic ring cleavage and progressive depolymerization with release of low-molecular-weight fragments (Ward et al. 2004) (Table 3.1).

5 Enzymatic System Involved in the Lignin Biodegradation

Lignin peroxidase (LiP) was first reported from the white-rot fungus *Phanerochaete chrysosporium* with a variety of lignin-degrading reactions. There are also several other microbial enzymes that are believed to be involved in lignin biodegradation, such as LiPs (also 'ligninase', EC number 1.14.99), manganese peroxidase (MnP) (EC 1.11.1.14) and laccase (EC 1.11.1.13. Lignin degradation by fungi is an oxidative process and phenol oxidases are the key enzymes (Kuhad et al. 1997; Leonowicz et al. 1999). Of these, LiPs, MnPs, versatile peroxidases (VPs), dye-decolourizing peroxidases (DyPs) and laccases from fungi have been best studied. The oxidation of substrates by LiPs and MnPs occurs through two successive one-electron oxidation with intermediate cation radical formation. These enzymes act non-specifically through the generation of lignin free radicals, which are unstable and undergo a variety of spontaneous cleavage reactions.

Laccase has broad substrate specificity and oxidizes phenols and lignin substructures with the formation of oxygen radicals. Bacteria are relatively rich in another type of peroxidase, the so-called DyPs (EC 1.11.1.19) (Van Bloois et al. 2010; Gonzalo et al. 2016). One group of accessory enzymes are oxidases, e.g. aryl-alcohol oxidase (AAO) and glyoxal oxidase (GlO). They produce peroxide, which is essential for peroxidase activity. Other accessory enzymes include dehydrogenases that reduce lignin-derived compounds (e.g. aryl-alcohol dehydrogenase and quinone reductase) and those that modify lignin by generating hydroxyl radicals (e.g. cellobiose dehydrogenase). H_2O_2-producing enzymes and oxidoreductases are also involved in the lignin degradation processes.

The fungal peroxidases LiPs, MnPs and VPs are all heme-containing glycoproteins which require hydrogen peroxide as an oxidant and have heme (protoporphyrin IX) as their prosthetic group coordinated by two highly conserved histidine residues. The heme cofactor is located in an internal cavity connected by two access channels. The main channel is used by hydrogen peroxide and the second is the site where MnP and VP oxidize Mn^{2+} to Mn^{3+}. Typically for LiP activity, the amino acid residue needed is a tryptophan, Trp171, in the isozyme LiPA (LiP H8) of *P. chrysosporium*. Tryptophan exposed on the LiP protein surface is conserved in LiP sequences and also in VPs. It is assumed that it takes part in long-range electron transfer (LRET) from a protein radical at the surface of the enzyme, which would act as the substrate oxidizer, to the heme cofactor. This could allow the enzyme to oxidize bulky substrates such as polymeric lignin that cannot directly contact the oxidized heme in the active centre of LiP or VP.

5.1 Lignin Peroxidase (EC 1.11.1.14)

LiP (at first designated as 'ligninase') was discovered in cultures of *Phanerochaete chrysosporium* after many years of research (Glenn and Gold 1983). The glycosylated enzyme was identified and described independently by two American teams and a Japanese group in supernatant of nitrogen and carbon limited cultures of the corticoid basidiomycete *Phanerochaete chrysosporium* that were flushed with pure oxygen. It has also been reported in *Phlebia radiata, Phlebia tremellosa, Trametes versicolor*, etc. (Hatakka 1994). LiP comprises of heme in the active site. Its molecular mass varies between 38 and 47 kDa. Its catalytic cycle resembles that of horse radish peroxidases and it catalyses several oxidations in the alkyl side chains of lignin-related compounds. The major reactions involve C–C cleavages in the side chains of lignin subunit, oxidation of veratryl alcohols and related substrates to aldehydes or ketones, intradiol cleavage of phenyl glycol structures and hydroxylation of benzylic methylene groups. It is also active against highly methylated lignin (Kent et al. 1987).

LiP is also capable of cleaving aromatic rings via one-electron subtraction and subsequent incorporation of oxygen. The optimum activity of LiP is extremely low (between pH 2.5 and 3.0), while the enzyme is acidic (pI 3–5). It is also active against highly methylated lignin. The catalytic cycle of both peroxidases is similar to other peroxidase enzymes, in which the resting state of the enzyme contains ferric heme, which reacts with hydrogen peroxide to form a compound I oxo-ferryl intermediate (two-electron oxidized form), and subsequently a compound II intermediate (one-electron oxidized form). However, two aspects in their molecular structure differentiate ligninolytic peroxidases from other peroxidases: first, a heme environment that confers high redox potential to the oxo-ferryl complex due to the position of N3 of the side chain of the proximal histidine residue, increasing its electron deficiency and increasing the redox potential of theoxo-ferryl complex); secondly, the presence of binding sites for oxidation of their specific substrates, including non-phenolic aromatics in the cases of LiP and Mn^{2+} in the case of MnP.

$$\text{Ferriperoxidase} + H_2O_2 \,/\, \text{Compound I} + H_2O \tag{3.1}$$

$$\text{Compound I} + \text{AH2} \,/\, \text{Compound II} + \text{AH} \tag{3.2}$$

$$\text{Compound II} + \text{AH2} \,/\, \text{Ferriperoxidase} + \text{AH} + H_2O$$

5.2 Manganese Peroxidase (EC 1.11.1.13)

MnP was also discovered in cultures of *Phanerochaete chrysosporium* by Kuwahara et al. (1984). The enzyme resembles LiPs; it is extracellular, glycosylated and contains heme as the prosthetic group. MnP is also expressed in multiple forms with molecular weights from 40 to 48 kDa and pI is between 2.9 and 7.0. Besides *Phanerochaete chrysosporium,* MnPs have also been found in many other white-rot and soil litter decomposing basidiomycetes, e.g. *Trametes versicolor, Phlebia radiata, Ceriporiopsis subvermispora, Agaricus bisporus, Nematoloma frowardii* and *Stropharia rugosoannulata.* The catalytic cycle of MnP resembles that LiP including native ferric enzyme as well as peroxidases compound I and compound II redox states (Fritsche and Hofrichter 2000). However, in MnP during reductive reactions, Mn^{2+} is oxidized to Mn^{3+}. The stabilization of the Mn^{3+} ions to high redox potential is achieved by chelation with organic acids such as lactate, succinate, malonate, oxalate, malate or tartrate. Subsequently, oxidation of phenolic lignin structure occurs due to action of chelated Mn^{3+}acting as diffusible redox mediator. The enzyme has been successfully applied to depolymerize ^{14}C-labelled synthetic and natural lignins in vitro; as a result, low-molecular-mass, water-soluble lignin fragments and significant amounts of CO_2 have formed (Gold et al. 2000).

MnP has been shown to generate H_2O_2 in the oxidation of certain thiols (e.g. glutathione) and NAD(P)H_2. Recently, evidence was provided that the enzyme is capable of acting efficiently in the absence of external H_2O_2 by oxidizing organic acids (e.g. oxalate, malonate, malate) in 'oxidase-like' autocatalytic reactions involving the transient formation of several radical species.

MnP (EC 1.11.1.13), an extracellular glycosylated enzyme, was found to be produced by *Phanerochaete chrysosporium* and a variety of lignin-degrading white-rot fungi. MnPs have been found and purified from various organism, viz. *Dichomitus squalens, Ceriporiopsis subvermispora, Lentinus edodes, Trametes versicolor, Phlebia radiata, Pleurotus ostreatus* and *Panus tigrinus.*

5.3 Fungal Laccase (EC 1.10.3.2)

It is an enzyme secreted into medium by mycelia of basidiomycetes, ascomycetes and deuteromycetes. It could be induced by anilines, methoxy phenolic acids, lignin preparations, heat shock, 2, 5-xylidine and ferulic acid. It is shown in *Trametes versicolor, Fomes annosus, Pholiota mutabilis, Pleurotus ostreatus*, etc. (Leonowicz et al. 2001).

Initially discovered in the Japanese lacquer tree *Rhus vernicifera*, laccases have since been found in many other plants and insects. Laccases have been extensively studied in white-rot fungi, such as *Lentinus tigrinus*, *Pleurotus ostreatus* D1,*Cerrena unicolor* strain 137, *T. versicolor*, *Trametes sp.* strain AH28–2, *Trametes pubescens* and *Cyathus bulleri*, brown-rot fungi, including *Coniophora puteana*, and ascomycetes such as *Melanocarpus albomyces*, *Chaetomium thermophile*, *Magnaporthe grisea*, *Myrothecium verrucaria 24G-4* and *Neurospora crassa*. Laccase genes have been detected in different fungal species including *Pycnoporus cinnabarinus*, *Cyathus* sp., *Pycnoporus coccineus*, *Pycnoporus sanguineus* and also in *xylariaceous* ascomycetes *Xylaria* sp. and *Hypoxylon* sp. Laccases were also detected in some bacteria such as *Bacillus subtilis* and *Klebsiella aerogenes*, *Bacillus licheniformis* and *B. subtilis*, while other bacteria, like *Streptomyces griseus*, produce a laccase-like phenol oxidase (http://www.brenda-enzymes.org).

Laccase is an enzyme, which oxidizes a variety of aromatic hydrogen donors. Thus, it catalyzes the removal of an electron and a proton from phenolic hydroxyl or aromatic amino groups to form free phenoxy radicals and amino radicals, respectively (Claus 2004). Copper-containing laccase has four copper atoms all in the 2^+ oxidation state in the active site (i.e. blue oxidase) and does not only oxidize phenolic and methoxyphenolic acids but also decarboxylates them and attack their methoxyl groups through demethylation or demethoxylation. Laccase reacts with polyphenols and other lignin-derived aromatic compounds which can be polymerized and depolymerized or act as low-molecular-weight mediator.

Laccase can be used as a bleaching agent in the pulp and paper industry. It helps in removal of colour by oxidation of lignin and other chromogenic components. Laccase is also used as a stabilizer during wine processing, as a dechlorinating agent and for removal of phenolic and other aromatic pollutants from natural and industrial wastewaters. Lignin degraded by LiPs provides a substrate for laccase; both enzymes can be considered as ‘partners’ in certain biotransformation routes of lignin (Boominathan and Reddy 1992). The oxidation of phenolics generates phenoxy radicals and quinoid intermediates, which are subsequently transformed into dimers and insoluble polymers, which can be easily removed from water after sedimentation. Laccases structure is composed of four copper atoms termed Cu T1 (where the reducing substrate place) and trinuclear copper cluster T2/T3 (where oxygen binds and is reduced to water). Laccase oxidizes four substrate molecules to obtain sufficient electron (4) to reduce molecular oxygen. Thus, for complete reduction of molecular oxygen to water, oxidation of four reducing substrate molecules is necessary. The oxidation of substrate by laccase is one-electron reaction. The reaction generates free radical. This unstable free radical may either undergo second enzyme catalyzed reaction or resort to non-enzymatic reactions like hydration and polymerization. In general terms, substrate oxidation by laccase is a one-electron reaction generating a free radical. The initial product is typically unstable and may undergo a second enzyme-catalyzed oxidation or otherwise a non-enzymatic reaction such as hydration, disproportionation or polymerization. The bonds of the natural substrate, lignin, that are cleaved by laccase include Cα-oxidation, Cα–Cβ cleavage and aryl-alkyl cleavage (Madhavi and Lele 2009). Laccases can even catalyze the breakdown of non-phenolic lignin structures in the presence of redox mediators, including the cleavage of β-O-4 linkages. Mediators typically used in laccase-mediator systems include 1-hydroxybenzotriazole (1-HBT, 2,2′-azino-bis(3-ethylbenzothiazoline-6-sulfonic acid) (ABTS)) and the natural mediator (4-hydroxy-3,5-dimethoxyacetophenone) or acetosyringone.

5.4 Versatile Peroxidases (EC 1.11.1.16)

VPs have the properties of both LiPs and MnPs. It can catalyze a wide array of substrates. They can oxidize various phenolic and non-phenolic substrates and can also oxidize Mn^{2+} to Mn^{3+} like MnP. VP oxidizes typical LiP substrates, e.g. VA, methoxybenzenes and non-phenolic model lignin compounds, as well as Mn^{2+}. The structure of VP includes 11–12 helices, four disulfide bridges, two structural Ca^{2+} sites, a heme pocket and a Mn^{2+}-binding site similar to that of MnP. The substrate specificity of VPs is similar to that of LiPs, including oxidation of high and medium redox potential compounds. Yet, VP also oxidizes azo-dyes and other non-phenolic compounds with high redox potential in the absence of mediators. VP was detected in members of the genera *Pleurotus* [*Pleurotus eryngii*, *P. ostreatus*] and Bjerkandera [*Bjerkandera adusta*, *B. fumosa*]. VPs are glycoproteins secreted as several isoenzymes with a molecular mass ranging between 40 and 45 kDa with a pI ranging between 3.4 and 3.9. The basic catalytic cycle of VPs is similar to those of other peroxidases with the two intermediary compounds I and II but is more complex due to a more diversified pool of potential substrates.

5.5 Dye-Decolourizing Peroxidase (EC 1.11.1.19)

DyP-type peroxidases were first found in fungi and subsequently in bacteria. They are not included in the classical plant/microbial peroxidase superfamily as they differ in structural and functional aspects. DyPs have a protomer weight of around 40–60 kDa and various oligomeric states have been observed (Colpa et al. 2014). They belong to the peroxidase-chlorite dismutase superfamily of proteins and contain a non-covalently bound heme b cofactor (Zámocký et al. 2015). DyP-type peroxidases can degrade phenolic dyes and lignin-like compounds and act on non-phenolic lignin model compounds containing β-O-4 linkages. Most DyP substrates are too big to enter the active site and are therefore unable to interact directly with the heme cofactor; thus, similar to the typical fungal LiPs, MnPs and VPs, DyPs also seem to be able to promote lignin degradation by oxidizing redox mediators. Redox mediators as veratryl alcohol, monophenolic substrates and Mn^{2+} have been tested as DyP substrates. Some DyP-type peroxidases were shown to be active on veratryl alcohol, monophenolic substrates and Mn^{2+}.

Peroxide-Generating Enzymes The peroxidases LiP and MnP require extracellular H_2O_2 for their in vivo catalytic activity. The H_2O_2-producing enzymes are thus required for the function of peroxidases (Blanchette 1995).

Glucose Oxidase (EC 1.1.3.4) It is a FAD-dependent oxidase that can be found inside and outside fungal cells and many other organisms during the enzymatic catalysis. Glucose is oxidized first to the corresponding gluconolactone with simultaneous reduction of FAD to $FADH_2$. In the second step, $FADH_2$ is oxidized back to FAD by dioxygen (O_2) while H_2O_2 is formed (Buswell and Odier 1987).

$$\beta\ \text{D Glucose} + \text{E FAD} \rightarrow \text{D Glucose 1,5 lactone} + \text{E FADH}_2$$

$$\text{E FADH}_2 + O_2 \rightarrow \text{E FAD} + H_2O_2$$

In absence of glucose oxidase (GO) activity, *Phanerochaete chrysosporium* did not show ability to degrade lignin.

Glyoxal Oxidase (EC 1.2.3.5) GlO is a FAD-dependent, H_2O_2-generating enzyme discovered in growing cultures of *Phanerochaete chrysosporium*. Enzyme purification and characterization revealed that among others it oxidizes glyoxalic acid into oxalic acid while generating H_2O_2. In this way, GlO supports peroxidases activity and supplies chelators for Mn^{3+} (Hatakka 1994).

Aryl-Alcohol Oxidase (EC 1.1.3.7) It is an extracellular, FAD-dependent enzyme which was discovered in *Trametes versicolor* and later also found in *Phanerochaete chrysosporium, Pleurotus sajor-caju, Pleurotus eryngii*, etc. It is a glycoprotein and oxidizes a number of aromatic alcohols (aryl-α and α-β-unsaturated γ-aromatic alcohols) to the corresponding aldehydes while oxygen is reduced to H_2O_2.

6 Superoxide Dismutase (EC 1.15.1.1)

Superoxide dismutase (SOD) is an almost universal enzyme of aerobic microorganisms. It catalyzes the dismutation of the superoxide anion radicals to molecular oxygen and H_2O_2 in all oxygen-metabolizing organisms and even in certain anaerobes. SOD thus protects organisms from toxic effects of the reactive superoxide radical (O_2 SOR). All-known SODs are metal proteins containing iron (FeSOD) or manganese (MnSOD) or copper plus zinc (Cu–ZnSOD) or nickel (NiSOD) as prosthetic group.

$$2O_2^{\cdot -} + 2H \rightarrow H_2O_2 + O_2$$

In summary it appears that H_2O_2 required for a LiP and MnP may be supplied by several different oxidases (Kent et al. 1987).

The ligninolytic systems composed of H_2O_2-generating enzymes and peroxidases MnP, LiP and laccase have been shown to degrade a large number of recalcitrant compounds, xenobiotics and the organopollutants. The ligninolytic enzymes especially laccase and peroxidase are involved in ring cleavage and the breakdown of aromatic and phenolic (Leonowicz et al. 1999).

7 Conclusion

The ligninolytic enzyme system is found in mainly Basidiomycota and Ascomycota fungi. The ligninolytic system comprises of hydrogen peroxide generating system and peroxidases and oxidases. The enzymes act in tandem to degrade phenolic as well as non-phenolic substrates. This broad specificity is utilized for the bioremediation of xenobiotics, dyes and industrial pollutants. Enzymes can also be used to generate value-added products from the lignocellulosic wastes. Ligninolytic enzymes are found in fungi and in some aerobic bacteria; however, most of the reported lignin-degrading fungi are from Ascomycota and Basidiomycota. The enzymes also vary in activity and expression although the ligninolytic system is relatively nonspecific due to oxidative mediator-based system; thus, it has been extensively used for bioremediation of recalcitrant compounds. Lignin-degrading microorganisms and enzymes can be also used for targeted modification of feeder compounds under controlled condition for industrial production of array of products.

References

Balaes TP, Ungureanu CV, Mardari C, Tănase C (2017) Ligninolytic enzyme system in ecological adaptation of lignicolous macrofungi. Appl Ecol Environ Res 15(1):207–224

Blanchette RA (1995) Delignification by wood-decay fungi. Annu Rev Phytopathol 29:381–398

Bonnen AM, Anton LH, Orth AB (1994) Lignin-degrading enzymes of the commercial button mushroom, *Agaricus bisporus*. Appl Environ Microbiol 3:960–965

Boominathan K, Reddy CA (1992) Fungal degradation of lignin. In: Arora DK, Elander RP, Mukerji KG (eds) Handbook of applied mycology, vol 4: Fungal Biotechnology. Marcel Dekker, NewYork, pp 763–822

Buswell JA, Odier E (1987) Lignin biodegradation. CRC Crit Rev Biotechnol 6:1–60

Cardona CA, Sánchez OJ (2007) Fuel ethanol production: process design trends and integration opportunities. Bioresour Technol 98(12):2415–2457

Chefetz B, Chen Y, Hadar Y (1998) Purification and characterization of laccase from *Chaetomium thermophilium* and its role in humification. Appl Environ Microbiol 64(9):3175–3179

Claus H (2004) Laccases: structure, reactions, distribution. Micron 35:93–96

Colpa DI, Fraaije MW, Van Bloois E (2014) DyP-type peroxidases: a promising and versatile class of enzymes. J Ind Microbiol Biotechnol 41:1–7

Dashtban M, Schraft H, Syed TA, Qin W (2010) Fungal biodegradation and enzymatic modification of lignin. Int J Biochem Mol Biol 1(1):36–50

Dedeyan B, Klonowska A, Tagger S, Tron T, Iacazio G, Gil G, Le Petit J (2000) Biochemical and molecular characterization of a laccase from *Marasmius quercophilus*. Appl Environ Microbiol 66(3):925–929

Eggert C, Temp U, Eriksson KE (1996) The ligninolytic system of the white rot fungus *Pycnoporus cinnabarinus*: purification and characterization of the laccase. Appl Environ Microbiol 62(4):1151–1158

Farnet AM, Criquet S, Cigna M, Gil G, Ferre E (2004) Purification of a laccase from *Marasmius quercophilus* induced with ferulic acid: reactivity towards natural and xenobiotic aromatic compounds. Enzym Microb Technol 34(6):549–554

Fritsche W, Hofrichter M (2000) Aerobic degradation by microorganisms biotechnology. In: Klein J (ed) Environment processes II, vol 11b. Wiley-VCH, Weinheim, pp 145–155

Glenn JK, Gold MH (1983) Decolorization of several polymeric dyes by the lignin degrading basidiomycetes *Phanerochaete chryosporium*. Appl Environ Microbiol 45:1741–1747

Gold MH, Youngs HL, Gelpke MDS (2000) Manganese peroxidase. In: Sigel A, Sigel H (eds) Metal ions in biological systems, vol 37. Marcel Dekker, Inc, New York, pp 559–586

Gonzalo GD, Colpa DI, Habibb MHM, Fraaij MW (2016) Bacterial enzymes involved in lignin degradation. J Biotechnol 236:110–119

Hakulinen N, Kiiskinen LL, Kruus K, Saloheimo M, Paananen A, Koivula A, Rouvinen (2002) Crystal structure of a laccase from *Melanocarpus albomyces* with an intact trinuclear copper site. Nat Struct Biol 9(8):601–605

Hariharan S, Nambisan P (2013) Optimization of lignin peroxidase, manganese peroxidase, and lac production from *Ganoderma lucidum* under solid state fermentation of pineapple leaf. BioResources 8(1):250–271

Hatakka A (1994) Lignin modifying enzymes from selected white rot fungi production and role in lignin degradation. FEMS Microbiol Rev 13:120–135

He J, Ye X, Ling Q, Dong L (2014) Enhanced production of an acid-tolerant laccase by cultivation of *Armillariella tabescens*. J Chem Pharm Res 6(1):240–245

Jager A, Croan S, Kirk TK (1985) Production of ligninases and degradation of lignin in agitated submerged cultures of *Phanerochaete chrysosporium*. Appl Environ Microbiol 50(5):1274–1278

Kalmi E, Ya H, Kalyoncu F, Pazarba B, Koçyi A (2008) Ligninolytic enzyme activities in mycelium of some wild and commercial mushrooms. Afr J Biotechnol 7(23):4314–4320

Kent T, Kirk TK, Farell RL (1987) "Enzymatic combustion": the microbial degradation of lignin. Annu Rev Microbiol 41:465–505

Kersten P, Cullen D (2007) Extracellular oxidative systems of the lignin-degrading Basidiomycete *Phanerochaete chrysosporium*. Fungal Genet Biol 44(2):77–87

Kuhad RC, Singh A, Ericsson KEL (1997) Microorganisms and enzymes involved in the degradation of plant fiber cell walls. Adv Biochem Eng Biotechnol 57:45125

Kukkola EM, Koutaniemi S, Gustafsson EPM, Karhunen P, Lundell TK, Saranpaa P, Kilpela I, Teeri TH, Fagerstedt CV (2004) The dibenzodioxocin lignin substructure is abundant in the inner part of the secondary wall in Norway spruce and silver birch xylem. Planta 218:497–500

Kuwahara M, Glenn JK, Morganm MA, Gold MH (1984) Separation and characterization of two extracellular H_2O_2 dependant oxidases from ligninolytic cultures of *Phanerochaete chrysosporium*. FEBS lett 169:47–250

Leonowicz A, Matuszewska A, Luterek J, Ziegenhagen D, Wojtas-Wasilewska M, Cho NS, Hofrichter M, Rogalski J (1999) Biodegradation of lignin by white rot fungi. Fungal Genet Biol 27(2–3):175–185

Leonowicz A, Cho NM, Luterek J, Wilkolaza A, Wojtas–Wasilewska M, Matuszewska A, Hofrichter M, Wesenberg D, Rogalski J (2001) Fungal laccase: properties and activity on lignin. J Basic Microbiol 41:185–227

Ma J, Yang HY, Kun W, Liu X (2016) Structural modification of hemicelluloses and lignin based on the biorefinery process with white-rot fungal. Carbohydr Polym 153:7–13

Maciel MM, Castro Silva A, Ribeiro CTH (2010) Industrial and biotechnological applications of ligninolytic enzymes of the basidiomycota: a review. Electron J Biotechnol 13(6):14–15

Madhavi V, Lele SS (2009) Laccase: properties and applications. BioResources 4(4):1694–1717

Martínez AT, Speranza M, Ruiz-Dueñas FJ, Ferreira P, Camarero S, Guillén F, Martínez MJ, Gutiérrez A, Río JC (2005) Biodegradation of lignocellulosics: microbial, chemical, and enzymatic aspects of the fungal attack of lignin. Int Microbiol 8(3):195–204

Memić M, Vrtačnik M, Boh B, Pohleven F, Mahmutović O (2017) Biodegradation of PAHs by ligninolytic fungi *Hypoxylon fragiforme* and *Coniophora puteana*. Polycy Aroma Comp 2017. https://doi.org/10.1080/10406638.2017.1392326

Papinutti L, Martınez MJ (2006) Production and characterization of laccase and manganese peroxidase from the ligninolytic fungus *Fomes sclerodermeus*. J Chem Technol Biotechnol 81:1064–1070

Pointing SB, Pelling AL, Smith GJD, Hyde KD, Reddy AC (2005) Screening of basidiomycetes and xylariaceous fungi for lignin peroxidase and laccase gene-specific sequences. Mycol Res 109(1):115–124

Ramasamy K (1993) Proceedings of a short term training course on ligninase and its biotechnological applications held at Department of Environmental Sciences, Tamilnadu Agricultural University, Coimbatore, India

Rana R, Nanda S, Meda V, Dalai AK, Kozinski JA (2018) A review of lignin chemistry and its biorefining conversion technologies. J Biochem Eng Bioprocess Technol 1:2

Sack U, Hofrichter M, Fritsche W (1997) Degradation of polycyclic aromatic hydrocarbons by manganese peroxidase of *Nematoloma frowardii*. FEMS Microbiol Lett 152(2):227–234

Shoemaker HE, Leisola MA (1990) Degradation of lignin by *Phanerochaete chrysosporium*. J Biotechnol 13:101–109

Sixta H, Süss HU, Potthast A, Schwanninger M, Krotscheck AW (2006) Pulp bleaching: sections 7.1–7.3. 5. Handbook of pulp. pp 609–708

Snajdr J, Baldrian P (2007) Temperature affects the production, activity and stability of ligninolytic enzymes in *Pleurotus ostreatus* and *Trametes versicolor*. Folia Microbiol (Praha) 52(5):498–502

Su Y, Yu X, Sun Y, Wang G, Chen H, Chen G (2018) Evaluation of screened lignin-degrading fungi for the biological pretreatment of corn stover. Sci Rep 8:5385

Uzan E, Nousiainen P, Balland V, Sipila J, Piumi F, Navarro D, Asther M, Record E, Lomascolo A (2010) Jun high redox potential laccases from the ligninolytic fungi *Pycnoporus coccineus* and *Pycnoporus sanguineus* suitable for white biotechnology: from gene cloning to enzyme characterization and applications. J Appl Microbiol 108(6):2199–2213

Van Bloois E, Torres Pazmiño DE, Winter RT, Fraaije MW (2010) A robust and extracellular heme-containing peroxidase from *Thermobifida fusca* as prototype of a bacterial peroxidase superfamily. Appl Microbiol Biotechnol 86:1419–1430

Voeller K, Kozliak E, Kubátová A, Yao B, Ji Y, Asina F, Brzonova I (2016) Biodegradation of lignin by fungi, bacteria and laccases. Bioresour Technol 220:414–424

Ward G, Hadar Y, Dosoretz CG (2004) The biodegradation of lignocellulose by white rot fungi. In: Arora DK (ed) Fungal biotechnology in agricultural, food and environmental application. Marcel Dekkar, Inc, New York, pp 393–406

Zámocký M, Hofbauer S, Schaffner I, Gasselhuber B, Nicolussi A, Soudi M, Pirker KF, Furtmüller PG, Obinger C (2015) Independent evolution of four heme peroxidase superfamilies. Arch Biochem Biophys 574:108–119

Znameroski EA, Coradetti ST, Roche CM, Tsai JC, Iavarone AT, Cate JHD, Glass NL (2012) Induction of lignocellulose-degrading enzymes in *Neurospora crassa* by cellodextrins. PNAS 109(16):6012–6017

Chapter 4
Fungal Biodiversity Producing Xylanase Enzymes Involved in Efficient Uses of Xylanolysis

Praveen Kumar Gupta, Shreya Choudhary, C. Chandrananthi, J. Sharon Eveline, S. P. Sushmitha, Lingayya Hiremath, Ajeet Kumar Srivastava, and S. Narendra Kumar

Abbreviations

AGX	Arabino(glucurono)xylan
Araf	Arabinofuranose
AX	Arabinoxylan
DTT	Dithiothreitol
GAX	Glucurono(arabino)xylans
GlcA	Glucuronic acid
HX	Heteroxylan
MeGlcA	4-*O*-methyl-D-glucuronosyl
MLG	Mixed-linkage glucan
SbmF	Submerged fermentation
SSF	Solid-state fermentation
Wis	Water-insoluble
X	Homoxylan
Xyl	Xylanase

1 Introduction

Xylan is among the most common hemicellulose which can be found in nature and comes second among the world's most widely available biopolymer in plants (Muchlisyam et al. 2016). Xylanases are hydrolytic enzymes that depolymerize the plant cell component xylan, by cleaving the β-1,4 backbone of xylan. Fungal sources

P. K. Gupta (✉) · S. Choudhary · C. Chandrananthi · J. Sharon Eveline · S. P. Sushmitha
L. Hiremath · A. K. Srivastava · S. Narendra Kumar
Department of Biotechnology, R.V College of Engineering, Bangalore, India
e-mail: praveenkgupta@rvce.edu.in

R. Naraian (ed.), *Mycodegradation of Lignocelluloses*, Fungal Biology,
https://doi.org/10.1007/978-3-030-23834-6_4

are among the widely used producers of xylanase. But xylanases can also be produced by other species like marine algae, bacteria, yeast, snails, protozoans, crustaceans and insects (Bajpai 2014)

Many fungal species like *Aspergillus*, *Trichoderma* and *Fusarium* species are active xylanase producers. Xylanases have various biotechnological and industrial applications such as in the animal feed industry, processing of bread, food industry, cellulose pulp bleaching and textile manufacturing (Bajpai 2014).

Due to the increasing interest in the biotechnological applications of xylan, intensive research is being carried out mainly focussing on xylanolytic enzymes that are derived from microbial sources majorly mesophilic fungi. This review highlights the occurrence, structure, property of xylans and the production and application of xylanases.

2 Structure of Xylan

Hemicellulose residues were known to be first extracted from plant using dilute alkali. They are found to be of lower molecular weight than cellulose. They can have D-xylose, D-mannose, L-arabinose or D-galactose as their principal monomeric unit. Xylan contains close to or more than 90% of D-xylose as monomer and also traces of L-arabinose. Xylan homopolymer consisting of only D-xylose are very difficult to isolate and are normally found to contain 2–4 sugar monomers. In the initial broad classification, polysaccharide such as chains of D-xylose, D-galactose residues and D-mannose (either alone or in association with D-glucose) (Sunna and Antranikian 1977) can be recognized.

Xylan contains the same backbone structure of D-xylose but differences occur because of the sugar substituents present. These contain acetyl, arabinose and glucuronosyl residues. The xylan isolated from esparto grass is found to be made of the chains of (1 3 4)-P-D-xylopyranose residues. Similarly, xylan isolated from land plants are found to contain the same backbone but show the difference in the arrangement of sugar substitutes. Substitutes found were D-glucuronic acid and L-arabinose, and 4-methyl ether was also found attached (Jayme and Sätree 1942).

2.1 *Backbone (D-Xylose)*

Xylan that was isolated from plants and grass have found to contain the same backbone of β-(1 → 4)-linked xylose remnant. The main chain of xylan is composed of fl-(1 → 4)-linked β-xylopyranose residues. The presence of β-(I → 4) linkages was also found established between two xylose residues adjacent to each other (Biely 1985) (Fig. 4.1).

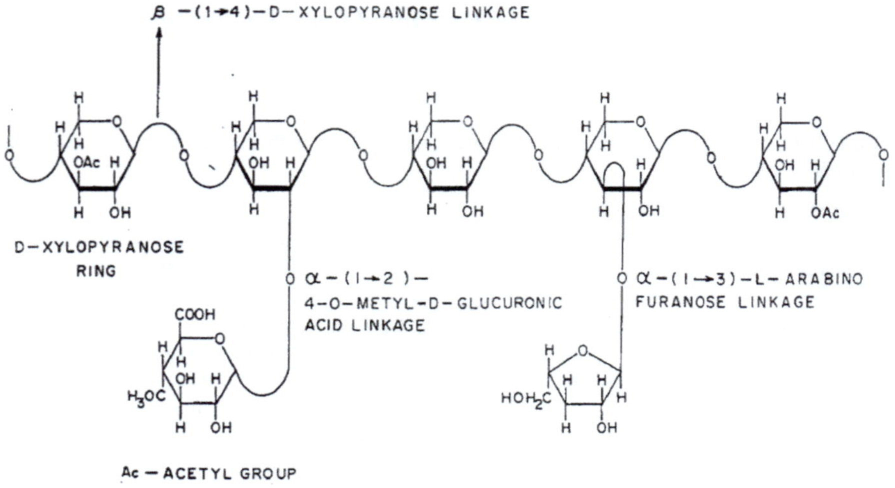

Fig. 4.1 Xylan structure with side chains (Biely 1985)

2.2 Side Chains

Common substituents like acetyl, glucuronosyl and arabinosyl residues are commonly found in the xylan backbone (Biely 1985).

3 Properties of Xylan

3.1 Solubility

Microbial and fungal sources need to secrete an enzyme complex to hydrolyze heteropolymers such as xylan to simple sugars. These enzymes, having different substrate specificity and catalytic activity, are capable of depolymerization of hemicellulose. Xylan varies insolubility. They can be either found to be soluble or non-soluble. Also, it has been shown that the solubility of the xylan structure directly depends upon the degree of substitution in the xylan structure. Another factor determining the degree of solubility is the degree of polymerization of the xylan molecule (Izydorczyk and Biliaderis 1992).

3.2 Characteristics of Xylan Extracted from Different Sources (Table 4.1)

Table 4.1 Xylanases from different microbial strains along with their distinct characteristics (Bastawde 1992)

Microorganism	Characteristics
Aspergillus fumigates	Molecular weight is 19, 8.5 kDa. Optimum temperature and pH is at 55 °C and 5.5, respectively. Its stability ranges from pH 5.0 to 9.0 and temperature of 50 °C
Aspergillus fischeri	Molecular weight is 31 kDa. Optimum temperature and pH is at 60 °C and 6.0, respectively. Its stability ranges from pH 5.0 to 9.5 and temperature of 55 °C
Streptomyces sp.	Molecular weight ranges from 42 to 44 kDa. Optimum temperature and pH is at 55 °C and 5.5, respectively. Its stability ranges from pH 4.0 to 9.0 and temperature of 50 °C
Aspergillus sojae	Molecular weight is 33, 36 kDa. Optimum temperature and pH is at 50, 60 °C and 5.0, 5.5, respectively. Its stability ranges from pH 5.0 to 9.0 and temperature of 50 °C
Aspergillus nidulans KK-99	Optimum temperature and pH is at 55 °C and 8.0, respectively. Its stability ranges from pH 4.0 to 9.5
Acrophialophora nainiana	Molecular weight is 22 kDa. Optimum temperature and pH is at 55 °C and 7.0, respectively. It's thermostable up to 55 °C
Bacillus sp	Molecular weight is 99 kDa. Optimum temperature and pH is at 75 °C and 6.0, respectively
Myceliophthora sp	Molecular weight is 53 kDa. Optimum temperature and pH is at 75 °C and 6.0, respectively. It's stable at pH 9.2 and temperature of 60 °C
Penicillium capsulatum	Molecular weight is 22 kDa. Optimum temperature and pH is at 48 °C and 3.8, respectively. It's thermostable at pH 5.0 and temperature of 50 °C
Thermomyces lanuginosus	Molecular weight is 24.7 kDa. Optimum temperature and pH is at 70 °C and in the range of 6.0–6.5, respectively. Its stability ranges from pH 5.5 to 10 and temperature of 60 °C

4 Enzyme Complex Involved in the Hydrolysis of Xylan

4.1 Endoxylanase

The enzyme 4-endoxylanase having the structure 1,4-p-o-xylan xylohydrolase is one of the enzymes responsible for xylan degradation. The substrate acts at the specific site and not randomly. The site of attack depends on the nature of the substrate and also on the presence of substituents. They are responsible for cleaving the glycosidic linkages of the xylan backbone. This converts the main products into oligosaccharides and further hydrolysis converts them into xylose, xylobiose and xylotriose (Reilly 1981).

4.2 *p-Xylosidase*

These are exoglycosidases having the formula p-D-xylosidexylo hydrolase and molecular weight in the range of 60 and 360 kDa. Short xylooligosaccharides and xylobiose are hydrolyzed to liberate xylose by this substrate (Wong et al. 1988). These are reported to be found in bacteria and fungi and also among cell associated with yeast and bacteria. The purified form of P-xylosidases are not capable of hydrolyzing xylan. However, some reports show that they attack xylan slowly producing xylose (Sunna and Antranikian 1977).

4.3 *L-Arabinofuranosidases*

This substrate plays an important role in hydrolyzing xylan despite the fact that only a few such enzymes have been isolated. The two types are only existence-exo-acting and endo-acting. Exo-acting are found to be active against branched arabinans, whereas endo-acting are found only active towards linear arabinans. Such endoarabinases are found to be present in *Bacillus subtilis*, *Clostridium felsineum* and various fungi sources. Most enzymes studied that degrade arabinan are exo-acting (Kaji and Saheki 1975; Van der Veen et al. 1994; Kaji 1984).

5 Occurrence and Isolation of Xylan

5.1 *Occurrence*

Xylanases occur in both prokaryotes and eukaryotes and are found to occur widely. They are found among higher eukaryote-like snails, insects, protozoa, etc. Among prokaryotes, bacteria and cyanobacteria are found to produce xylanase in the marine environment. Among two types of xylanase (extracellular and intracellular), intercellular xylanases are found in bacteria and protozoa from tureen sources (Dekker and Richards 1976).

Terrestrial plants and algae are known to have xylan-type polysaccharides. These are also found in different tissues of the same plant at different sites. The structural diversity in plants can be due to the fact that structural diversity is related to the functionality in the plant and hence can show the relation between the distribution of particular classes of xylan.

Xylan can also be found in the botanically oldest plant families. In green algae, Homoxylans (X) having b-(1e3) glycosidic linkages have been found to function as substitute cellulose in the cell wall architecture. Also, red seaweeds in the Palmariales and Nemaliales have cell wall comprising of homoxylans having mixed b-(1e3 and 1e4) glycosidic linkages (Painter 1983).

6 Naturally Occurring Xylans

6.1 In Plant Material

Different agricultural crops like corn stalks, sorghum, corn cobs, sugar cane, straw, hulls of fruits and dry husks of fruits and vegetable are known to be very rich sources of xylan. Other sources include forest and pulping waste products from hardwoods. Glucuronoxylan, arabinoglucuronoxylan and arabinoxylan are other structural forms of xylan produced from certain plant sources. Wheat flour xylan consists of 30–40% arabinose which are irregularly attached to the xylose backbone. A distinguishing feature of wheat bran xylan is that it has a large number of side groups attached with glucuronic acid on every 7–8 xylose positions (65% of xylose has attached arabinose). The barley husk xylan are found to contain glucuronic acid and xyloarabinose groups attached to it. Hardwood xylan are found to have 4-*O*-methylglucuronic acid on each tenth xylose. Softwood xylan along with arabinose side group contains more side groups as 4-*O*-methylglucuronic acid are found in each sixth xylose (Ebringerova and Heinze 2000) (Table 4.2).

6.2 In Cell Wall

Xylan is a hemicellulose found in the grass cell wall. The type II grass walls are found to be rich in β-(1,3/4) glucan (also called mixed-linkage glucan) and GAXs (glucurono(arabino)xylans). Grass xylans are found to contain similar general

Table 4.2 Structural features of xylans found in various plant tissues which are found within *Zea mays* and *Oryza sativa* (Ebringerova and Heinze 2000)

Plant source	Xylan type
Corn	*Mature leaves* GAX I GAX II
	Bran HX
	Coleoptile GAX I GAX II
Rice	*Cobs* wis-AGX ws-AGX
	Endosperm AX GAX
	Bran AX GAX

xylans structure (i.e. substitution with MeGlcA or α-Ara*f*, GlcA residues). In addition to it, it also contains some unique features like the presence of Xyl-arabinofuranosyl side chains and the presence of feruloyl groups at the C-5 position of Ara*f* residues. The grass xylans are of two main groups-GAX, making up to 35% of vegetative cell wall that is found to contain GlcA/MeGlcA and Ara*f* remnants; also the cell wall of cereals containing AX, lacking GlcA (Faik 2010).

7 Xylanase from Thermophilic Fungal Source

The enzymes produced from extremophilic and thermophilic microorganism show greater stability. Recently, a lot of efforts has been put in the isolation from such sources. The common thermophilic fungi involved in xylanase production includes *Chaetomium thermophile, Humicola insolens, Humicola lanuginosus, Humicola grisea, Melanocarpus albomyces, Paecilomyces variotii, Talaromyces byssochlamydoides, Talaromyces emersonii, Thermomyces lanuginosus* and *Thermoascus aurantiacus.* These xylanases are usually glycoproteins and can withstand optimum temperatures between 60 and 80 °C. They show the highest activity at an acidic pH of about 4.5–6.5 and exhibit variable molecular weights in the range of 6–38 kDa (Polizeli et al. 2005).

8 Xylanase from Mesophilic Fungal Source

The genera *Aspergillus* and *Trichoderma*, belonging to the class of mesophilic fungi, are leading in xylanase production.

8.1 Aspergillus Species

The production of xylanase from soybean residues by *Aspergillus foetidus* was optimized with the production level of 13.98 U/mL occurring at fermentation for 168 h, pH 7.0, 28 °C and 120 rpm. The specific xylanolytic activity of *Aspergillus fumigatus* was found to be 1055.6 U/g and 558.3 U/g, after 5 days of solid-state fermentation (SSF) in wheat bran and soybean, respectively, by Delabona et al. (2013). In the same work, he found that *Aspergillus niger* showed a specific xylanolytic activity of 1285.0 U/g, 484.2 U/g and 1050.0 U/g using the residue of wheat bran, soybean and wheat bran with sugarcane bagasse, respectively. De Souza Moreira et al. (2013) found that xylanases could be produced by *A. terreus* under submerged fermentation at an optimum pH and temperature of pH 6.0 and 50 °C at 120 rpm and 5.0 and 45 °C at 120 rpm, respectively. Guimaraes et al. (2013) found a xylanolytic activity of 10.50 and 11.92 U/mL, respectively, for *Aspergillus niger* and *Aspergillus flavus*,

Table 4.3 Details about the production of xylanase from different fungal sources (adopted from Delabona et al. 2013; Souza Moreira et al. 2013)

Aspergillus species	Production level	Fermentation details
Aspergillus foetidus	13.98 U/mL	Time 168 h, pH 7, 28 °C and 120 rpm
Aspergillus fumigatus	1055.6 U/g	After 5 days of SSF
Aspergillus niger	1285.0 U/g	Using wheat bran residue
Aspergillus terreus	Unknown	Under submerged fermentation at pH 6.0, 50 °C at 120 rpm
Aspergillus flavus	11.92 U/mL	Using 0.5% corn cob
Aspergillus awamori	Unknown	SSF at pH 5 and 50 °C

using wheat bran 0.5% and corn cob 0.5% as residues. Based on the data obtained from different literature, it was found that the xylanase from *A. foetidus* was more effective in comparison to xylanase from *A. niger, A. flavus and A. fumigatus.* (Cunha et al. 2018). The SSF of tomato pomace by *Aspergillus awamori* was known to produce hydrolytic enzymes like xylanase. At pH 5, the enzyme shows optimum activity and temperature of 50 °C. Hg^{2+} and Cu^{2+} could strongly inhibit the enzyme, whereas the enzyme can be activated by Mg^{2+}. The enzymatic activity was observed to be quite high when the extract was preserved at a pH of 3–10 and a temperature range of 30–40 °C (Umsza-Guez et al. 2011).

Xylanase can be produced by *Aspergillus* species on fermentation of soybean residues, wheat bran, residues of wheat bran with sugarcane bagasse, corn cob and tomato pomace. Based on the data obtained from different literature, it was found that the xylanase from *A. foetidus* was more effective in comparison to xylanase from *A. niger, A. flavus* and *A. fumigatus* (Cunha et al. 2018). Hg^{2+} and Cu^{2+} could strongly inhibit the enzyme, whereas the enzyme can be activated by Mg^{2+} (Table 4.3).

8.2 *Trichoderma Species*

Ascomycetes filamentous fungi *Trichoderma reesei* are known to produce alkaline xylanase (Mewada et al. 2017). The cultivation of the crude extracellular extract of the strain *Trichoderma inhamatum*, Xyl I and Xyl II were stable at a temperature of 40 °C and at pH 4.5–6.5 for Xyl I and 4.0–8.0 for Xyl II. These xylanases were stable at a temperature of 40 °C and at pH 4.5–6.5 for Xyl I and 4.0–8.0 for Xyl II. The activity of enzyme was strongly reduced by ion Hg^{2+} and the detergent SDS, whereas 1,4-dithiothreitol was found to stimulate both Xyl I and Xyl II enzymes (Silva et al. 2015). *Trichoderma viride* VKF-3 produces xylanase using coconut oil cake as substrate with an activity of 3.045 IU/mL (Nathan et al. 2017). Rifai, a strain of *Trichoderma harzianum,* isolated from decaying *Aspidosperma sp.* (peroba) wood was used to produce xylanase using sugarcane bagasse as substrate. The highest xylanase activity was observed to be 288 U/mL on the seventh day using this method (Rezende et al. 2002).

8.3 Fusarium Species

Xylanase can be purified from *Fusarium heterosporum,* using barley remnants by SSF. The molecular mass of xylanase obtained by this method was found to be 19.5 kDa. The optimum pH for the xylanase was 5.0. Xylanase activity can be enhanced by Ba^{2+}, Ca^{2+}, Mg^{2+}, dithiothreitol (DTT) and β-mercaptoethanol, whereas Hg^{2+}, Pb^{2+} and Zn^{2+} strongly inhibited the enzyme activity (Heinen et al. 2014). High amounts of xylanase can be produced on birchwood xylan and agricultural residues like wheat bran and peptone by using a mutant strain of *Fusarium oxysporum* (NTG-19) (Kuhad et al. 1998). *Fusarium sp.* BVKT R2 are known producers of xylanase. For maximum production, a temperature of 32.5 °C and a pH of 5 are required along with 1.5% of yeast and sorbitol extract. This should also be kept under agitation at 175 rpm (Ramanjaneyulu and Reddy 2016).

8.4 Penicillium Species

Using cheap substances like wood, rice, wheat bran and sesame oil, the production of xylanase from *Penicillium oxalicum* under SSF was optimized. Xylanase can also be produced by the fungus *Penicillium sclerotiorum* under submerse cultivation. The best inducers of xylanase activity were oat spelts xylan and wheat bran. The conditions for optimum production of xylanase were found to be at a temperature of 30 °C and a pH of 6.5 and under stationary condition for 5 days using liquid Vogel medium. The optimum activity was found at a temperature of 50 °C and at pH 4.5 (Knob and Carmona 2008). Xylanase can also be produced from submerged cultures of *Penicillium echinulatum* 9A02S1. The highest activity of xylanases was observed in the medium containing 0.25% (w/v) cellulose and 0.75% (w/v) sorbitol added after 36 h of cultivation (Todero Ritter et al. 2013). Xylanase can be characterized and purified by the usage of agro-industrial waste which was proved when *Penicillium glabrum* was found to produce xylanase enzyme by the use of brewer's spent grain as a substrate. For optimum production of Xylanase by *P. glabrum*, it was grown at a pH of 5.5 and a temperature of 25 °C in a liquid medium kept under stationary condition for straight 6 days. Xylanase was purified by *P. glabrum* by using inexpensive procedures like molecular exclusion chromatography and also ammonium sulfate fraction at ion. The optimum activity of xylanase was found at a pH of 3.0 and temperature of 60 °C. The xylanase activity can be enhanced by the use of Mn^{2+} ion and by DTT and *β*-mercaptoethanol, both of which are reducing agents, whereas detergent SDS, the ions Hg^{2+}, Zn^{2+} and Cu^{2+}, were found to be strongly inhibiting the enzymes (Knob et al. 2013). *Penicillium janthinellum* FS22A are known to produce cellulolytic-xylanolytic enzymes (Okeke et al. 2015) (Table 4.4).

Table 4.4 Table showing characteristics of some xylanases produced by different microorganisms (Polizeli et al. 2005)

Microbial species	Optimal temperature (°C)	Optimal pH
Aspergillus aculeatus	50, 50, 70	4.0, 4.0, 5.0
Aspergillus oryzae	60	5.0
Aspergillus sydowii	50	4.0
Aspergillus kawachii	60, 55, 50	5.5, 4.5, 2.0
Aspergillus fischeri	60	6.0
Aspergillus sojae	60, 50	5.0, 5.5
Acrophialophora nainiana	22 55	7.0
Myceliophthora sp.	75	6.0
Cryptococcus sp.	40	2.0
Chaetomium cellulolyticum	50	2.0
Thermomyces lanuginosus	50	5.0–7.0
Penicillium capsulatum	48	3.8
Penicillium sp.	70	6.0–6.5

9 Applications of Xylanases

Xylanases have great potential application in many industrial processes. About 20% of the world enzyme market is occupied by xylanase, cellulase and pectinase. Due to the potential effectiveness of xylanolytic enzymes in bread making, they are used in baking and brewing industries. Xylanases are used as animal feed, where arabinoxylans which are present in feed ingredients are broken down by them reducing the viscosity of the raw material (Harris and Ramalingam 2010). Xylan and xylanases are also used in pharmaceutical industries. Xylanases along with enzymes like proteases and hemicellulase are used for treating indigestion. The hydrolytic products of xylan can be used to make artificial sweeteners, solvents, ethanol etc. Xylanolytic complex also finds varied application in textile industries and is also used for making plant fibres like linen or hessian. Xylanases are also used in bleaching of cellulose pulp. Xylanases are also used to enhance sugar recovery from agricultural residues (Chakdar et al 2016) (Table 4.5).

10 Comparison of Xylanase Production from Bacterial and Fungal Sources

Xylanases are known to be produced by organisms like bacteria, snails, fungi, yeast, protozoans, fruit and vegetable seeds and insect. Bacterial genera like *Bacillus, Pseudoxanthomonas, Rhodothermus, Paenibacillus, Cellulomonas, Staphylococcus* and *Microbacterium* are known to produce xylanase enzyme. Also, thermostable

Table 4.5 Table showing commercial xylanases produced by microorganisms

Microorganisms	Optimal pH	Optimal temperature (°C)	Application
Aspergillus niger	5.3	5.3	Used in the improvement of animal feed
Humicola	n.c.	n.c.	
A. niger	n.c.	n.c.	Used in the preparation of feed for pigs and birds
Bacillus sp	9.5	50	Used in paper-making industry
Trichoderma reesei	5.0–6.0	5.0–6.0	Cellulose pulp bleaching
T. longibrachiatum	5.0–5.5	5.0–5.5	Used in food industry
Trichoderma koningii	5	5	Used in the production manufacture of mushrooms and vegetable extracts, peeling of cereals through enzymatic methods, making of bread, preparation of animal feed
Trichoderma sp	n.c.	n.c.	Utilized for structural studies of carbohydrates

SbmF submerged fermentation, *SSF* solid-substrate fermentation, *n.c.* not cited

xylanases that are active at 60–70 °C are produced by *Thermotoga* sp., *Streptomyces* sp., *Clostridium thermocellum, Bacillus* spp., *Rhodothermus marinus* and *Stenotrophomonas maltophilia. Flavobacterium frigidarium, Penicillium Strain* and *Clostridium* sp. produce cold-adaptive xylanases (Chakdar et al. 2016).

Different organisms produce xylanase in different conditions. For example, bacterial species produce xylanase neutral/alkaline pH and xylanase are produced in acidic range by fungal species. Fungal xylanases are less attractive as the low pH requirement for growth of fungi and production of fungal xylanases necessitates additional steps in the subsequent stages. The secondary structures especially at the loop areas are different in bacterial and fungal xylanases. The nucleophile and proton donor in bacterial and fungal xylanases are always the glutamic acid, though their position may change. Fungal xylanases are produced with cellulose and hence there is increase in the length of downstream processing. In bacteria, xylanases are mostly produced alone and hence reduces the downstream process time (Chakdar et al. 2016).

11 Conclusion

Xylanases are hydrolases which are capable depolymerizing xylan polysaccharide. Xylan being a cell component is also known to be the second most widely available polysaccharide. Xylanases are known to be produced by insect, snails, bacteria, fungi, algae (marine), crustaceans, hulls and husks, seeds, protozoans, etc., but the filamentous fungi are the principal commercial sources of xylanase. These enzymes are synthesized by extremophiles, mesophiles and thermophiles microorganisms as the enzyme produced by them is found to be more stable. In order to consolidate the entire economics of the pre-treatment of lignocellulosic material, strategies are being carried out to convert xylan into useful products, and to this end xylanases have been identified as key enzymes with diverse industrial applications. The bulk production of xylanases at an economically viable rate is the major bottleneck in commercial applications of xylanase-based enzymatic processes. The study conducted by authors of this chapter is focussed on the sources, production and applications of fungal xylanolytic enzymes in the context of their biotechnological potential.

Acknowledgement The authors listed in this chapter wish to express their appreciation to the RSST Trust, Bangalore, for their continuous support and encouragement. As a corresponding author, I also express my sincere thanks to all other authors whose valuable contribution and important comments made this manuscript to this form.

Conflict of Interest: The authors listed in this chapter have no conflict of interest known best from our side. There was also no problem related to funding. All authors have contributed equally with their valuable comments which made the manuscript to this form.

References

Bajpai P (2014) Xylanolytic enzymes. Academic Press, Cambridge, p 120

Bastawde KB (1992) Xylan structure, microbial xylanases, and their mode of action. World J Microbiol Biotechnol 8(4):353–368

Biely P (1985) Microbial xylanolytic system. Trends Biotechnol 3:286–290

Chakdar H, Kumar M, Pandiyan K, Singh A, Nanjappan K, Kashyap PL, Srivastava AK (2016) Bacterial xylanases: biology to biotechnology. 3 Biotech 6(2):1–15

Cunha L, Martarello R, de Souza PM, de Freitas MM, Barros KVG, Filho EXF, de-Mello MH, Magalhães PO (2018) Optimization of xylanase production from *Aspergillus foetidus* in soybean residue. Enzyme Res 2018:1–7

Dekker RF, Richards GN (1976) Hemicellulases: their occurrence, purification, properties, and mode of action. Adv Carbohydr Chem Biochem 32:277–352

Delabona PDS, Pirota RDPB, Codima CA, Tremacoldi CR, Rodrigues A, Farinas CS (2013) Effect of initial moisture content on two Amazon rainforest *Aspergillus* strains cultivated on agro-industrial residues: biomass-degrading enzymes production and characterization. Ind Crop Prod 42(1):236–242

Ebringerova A, Heinze T (2000) Xylan and xylan derivatives–biopolymers with valuable properties, 1. Naturally occurring xylans structures, isolation procedures and properties. Macromol Rapid Commun 21(9):542–556

Faik A (2010) Xylan biosynthesis: news from the grass. Plant Physiol 153(2):396–402

Guimaraes NCA, Sorgatto M, Peixoto-Nogueira SC, Betini JHA, Zanoelo FF, Marques MR, de Lourdes M, Polizeli TM, Giannesi GC (2013) Bioprocess and biotechnology: effect of xylanase from *Aspergillus niger* and *Aspergillus flavus* on pulp biobleaching and enzyme production using agroindustrial residues as substract. Springerplus 2(1):1–7

Harris AD, Ramalingam C (2010) Xylanases and its application in food industry: a review. J Exp Sci 1(7):1–11

Heinen PR, Henn C, Peralta RM, Bracht A, Simão RCG, Silva JLC, Polizeli MLTM, Kadowaki MK (2014) Xylanase from *Fusarium heterosporum*: properties and influence of thiol compounds on xylanase activity. Afr J Biotechnol 13(9):1047–1055

Izydorczyk MS, Biliaderis CG (1992) Influence of structure on the physicochemical properties of wheat arabinoxylan. Carbohydr Polym 17:237–247

Jayme G, Sätree M (1942) Oxidation of xylan with periodic acid I. Use of buttered periodic acid. Berichte der Deutschen Chemischen Gesellschaft B 75:1840–1850

Kaji A (1984) L-Arabinosidases. Adv Carbohydr Chem Biochem 42:382–394

Kaji A, Saheki T (1975) Endo-arabinase from *Bacillus subtilis* F-1 1. Biochim Biophys Acta (BBA) Enzymol 410:354–360

Knob A, Carmona EC (2008) Xylanase production by *Penicillium sclerotiorum* and its characterization. World Appl Sci J 4(2):277–283

Knob A, Beitel SM, Fortkamp D, Terrasan CRF, de Almeida AF (2013) Production, purification, and characterization of a major *Penicillium glabrum* xylanase using Brewer's spent grain as substrate. Biomed Res Int 2013:1–8

Kuhad C, Manchanda M, Singh A (1998) Optimization of xylanase production by a hyperxylanolytic mutant strain of *Fusarium oxysporum*. Process Biochem 33(6):641–647

Mewada M, Albert S, Pandya B (2017) Enhancement of ligninolytic & xylanolytic enzyme activities in *Trichoderma reesei* co-cultured with two white rot fungi. Int J Biotech Biochem 13(4):429–439

Moreira LRS, Campos CM, de Siqueira PH, Silva LP, Ricart CA, Martins PA, Queiroz RM, Filho EX (2013) Two β-xylanases from *Aspergillus terreus*: characterization and influence of phenolic compounds on xylanase activity. Fungal Genet Biol 60:46–52

Muchlisyam, Silalahi J, Harahap U (2016) Hemicellulose: isolation and its application in pharmacy. In: Thakur VK, Thakur MK (eds) Handbook of sustainable polymers: processing and applications. Pan Stanford Publishing Pte. Ltd, Singapore, pp 305–339

Nathan VK, Esther Rani M, Rathinasamy G, Dhiraviam NK (2017) Low molecular weight xylanase from *Trichoderma viride* VKF3 for bio-bleaching of newspaper pulp. BioResources 12(3):5264–5278

Okeke BC, Hall RW, Nanjundaswamy A, Thomson MS, Deravi Y, Sawyer L, Prescott A (2015) Selection and molecular characterisation of cellulolytic-xylanolytic fungi from surface soil-biomass mixtures from Black Belt sites. Microbiol Res 175:24–33

Painter TJ (1983) The polysaccharides. In: Aspinall GO (ed) Academic, San Diego, pp 196–285

Polizeli MLTM, Rizzatti ACS, Monti R, Terenzi HF, Jorge JA, Amori DS (2005) Xylanases from fungi: properties and industrial applications. Appl Microbiol Biotechnol 67(5):577–591. https://doi.org/10.1007/s00253-005-1904-7

Ramanjaneyulu G, Reddy BR (2016) Optimization of xylanase production through response surface methodology by *Fusarium* sp. BVKT R2 isolated from forest soil and its application in saccharification. Front Microbiol 22(7):1450–1466

Reilly PJ (1981) Xylanases: structure and function. In: Trends Biol Fernent Fuels Chem, pp 111–129

Rezende MI, Barbosa AM, Vasconcelos AFD, Endo AS (2002) Xylanase production by *Trichoderma harzianum* rifai by solid state fermentation on sugarcane bagasse. Braz J Microbiol 33(1):67–72

Silva LAO, Terrasan CRF, Carmona EC (2015) Purification and characterization of xylanases from *Trichoderma inhamatum*. Electron J Biotechnol 18(4):307–313

Sunna A, Antranikian G (1977) Xylanolytic enzymes from fungi and bacteria. Crit Rev Biotechnol 17(1):39–67

Todero Ritter CE, Camassola M, Zampieri D, Silveira MM, Dillon AJ (2013) Cellulase and xylanase production by *Penicillium echinulatum* in submerged media containing cellulose amended with sorbitol. Enzyme Res 2013:240219. https://doi.org/10.1155/2013/240219

Umsza-Guez MA, Díaz AB, de Ory I, Blandino A, Gomes E, Caro I (2011) Xylanase production by *Aspergillus awamori* under solid state fermentation conditions on tomato pomace. Braz J Microbiol 42(4):1585–1597. https://doi.org/10.1590/S1517-838220110004000046

Van der Veen, Flipphi MJ, Visser J, de Graaff PLH (1994) Arabinase gene expression in *Aspergillus niger*: indications for coordinated regulation. Microbiology 140(Pt 10):2673–2682. https://doi.org/10.1099/00221287-140-10-2673

Wong KK, Tan LU, Saddler JN (1988) Multiplicity of beta-1, 4-xylanase in microorganisms: functions and applications. Microbiol Rev 52(3):305–317

Chapter 5
Biodegradation of Lignocellulosic Biomass and Production of Ethanol Using Potential Microorganisms

Ajeet Kumar Srivastava, Praveen Kumar Gupta, Lingayya Hiremath, S. Narendra Kumar, and A. V. Narayan

1 Introduction

The consumption of world petroleum is at a rapid rate. This has led to resurgence in the development of alternative energy sources that can displace fossil transportation fuel (Alper and Stephanopoulos 2009; Gupta et al. 2009). Ethanol as biofuel has several advantages such as low toxicity, biodegradable nature, generation of fewer airborne pollutants than petroleum fuel, and easy integration in the existing vehicle (McCarthy and Tiemenn 1998). Lignocellulosic biomass represents an abundant carbon-neutral renewable resource for the production of bioethanol (Ragauskas et al. 2006). Lignocelluloses are mainly comprised of cellulose, hemicellulose, and lignin (Kuhad et al. 1997). The bioconversion process mainly consists of three major steps, i.e. pretreatment, hydrolysis, and fermentation. Pretreatment is required to remove lignin and hemicellulose fraction, increase the porosity of the materials, and reduce cellulose crystallinity.

Researchers have displayed that biological pretreatment by microorganisms like bacteria and fungus (white-, brown-, and soft-rot fungi) enhances the hydrolysis productivity as the inhibitors generation is low and the energy utilized is also low (Wang et al. 2013; Taha et al. 2015; Hatakka 2005). Biological pretreatment is also considered reasonably cost-effective compared to other pretreatment processes. Such biological pretreatments have been less investigated (Chaturvedi and Verma 2013). The highest efficiency among the pretreatment methods has been achieved by lignin-degrading white-rot fungi (WRF) for the soft and brown fungi only attack cellulose (Kang et al. 2013; Fernandez-Fueyo et al. 2012; Suhara et al. 2012). Moreover, using white rot fungi consumes less environmental damage and less energy consumption (Narayanaswamy et al. 2013; Kumar et al. 2015). The ability of biological pretreat-

A. K. Srivastava (✉) · P. K. Gupta · L. Hiremath · S. N. Kumar · A. V. Narayan
Department of Biotechnology, R.V College of Engineering, Bangalore, India
e-mail: ajeeth@rvce.edu.in

R. Naraian (ed.), *Mycodegradation of Lignocelluloses*, Fungal Biology,
https://doi.org/10.1007/978-3-030-23834-6_5

ment to degrade lignocellulosic materials is due to their highly efficient enzymatic system. The effectiveness of the white-rot fungi (WRF) is studied here for the biological pretreatment and further fermentation to produce bioethanol.

2 History of Ethanol as Biofuel

Ethanol as a fuel has been used throughout technologically advanced man's modern history. Even the invention of ignition engines was with bioethanol. Ethanol was one of the most popular lamp illuminants used in the 1850s, and approximately 90 million gallons of ethanol was produced in the United States. The tax imposition on ethanol to assist the finance of the civil war brought changes in its established status. Kerosene became cheaper, and it quickly replaced ethanol as the premier illuminant in the 1860s (Morris 1993). In 1906, alcohol tax was lifted, which renewed the interest in ethanol, and in 1908, Henry Ford designed the automobile car Model T to run on ethanol. By 1914, the production of ethanol had come back and could reach ten million gallons. But in 1919, the emergence of petroleum as fuel made the use of ethanol as fuel decreased again. By the early 1940s, the production of ethanol reversed again when it was used during World War II for fuel and to make synthetic rubber. During this period, the United States produced about 600 million gallons of ethanol annually (Morris 1993). At the end of World War II, the demand for ethanol reduced and continued to decline for the next two decades, mostly due to cheap petroleum imports. But again, the oil embargo by Arab countries in 1973 created petroleum shortages which resulted in significant increase in gasoline price (Campbell and Laherrere 1998). Since the 1970s, the gasoline shortage had accelerated the concerns about the rising prices for the crude oil and increasing political instability which made the use of ethanol as biofuel is again under consideration worldwide.

3 Bioethanol: First and Second Generations

3.1 First-Generation Bioethanol

The bioethanol produced by fermentation of sugar (sugarcane juice, molasses, sugar beet juice, fruit juice) and starchy feedstocks (wheat, corn, potato) is commonly known as first-generation bioethanol (Antoni et al. 2007). The ethanol production methods used are enzymatic digestion (to release sugars from stored starches), fermentation of the sugars, distillation, and drying. According to the US Energy Information Administration, the first-generation bioethanol have played an important role in establishing the infrastructure and policy drivers required to support renewable transport fuels in the international marketplace. However, the International

Energy Agency (IEA 2008) has a number of concerns about the potential drawbacks of first-generation bioethanol, such as:

(a) **Competition between food and fuel**: It is clear that the development of bioenergy options, particularly food-based bioethanol, may adversely affect food demands. It is an obvious fact that the higher food prices will have adverse effects on the developing world, where disposable incomes are lower. This alarm to use food resources led to a search for alternative biofuels.
(b) **Deforestation**: The constant production of first-generation biofuels might lead to major deforestation, and land thus available may be changed from permanent forest cover to agriculture.
(c) **Multi-feedstock flexibility**: For commercial viability, technologies and plant designs which are able to process a number of different feedstock in a flexible way are preferable. The storage of crops raw materials may accelerate the cost of biofuel production if many single food crops used for biofuels are seasonal for the plant to operate around the year, and the cost of biofuel production may be accelerated due to the storage of raw materials.

3.2 Second-Generation Bioethanol

Now, it is clearly understood that the production of first-generation bioethanol is an unsustainable approach, and the increasing criticisms have highlighted for the use of non-food crops for the production of second-generation bioethanol. The production of second-generation bioethanol is from lignocellulosic biomass comprised of the residual non-food parts of the food crops and other crops that are not used for food purposes along with municipal, industrial and construction waste. Second-generation biofuels are expected to reduce net carbon emission, increase energy efficiency and reduce energy dependency, potentially overcoming the limitations of first-generation biofuels (Antizar-Ladislao and Turrion-Gomez 2008). In addition to these, the renewable nature of ethanol, its long-term sustainability, low net carbon emission, high energy efficiency, and low energy dependency are the other major benefits of switching to cellulosic ethanol (IEA 2008). However, there is still much work to be done in terms of improving second-generation biofuel technology pathways, to reduce costs and to improve the performance and reliability of the conversion process.

4 Current Status of Bioethanol

4.1 Status of Bioethanol Production Worldwide

Bioethanol production worldwide has increased considerably since the oil crisis in 1970 (Campbell and Laherrere 1998). Its market grew from less than a billion litres in 1975 to more than 65 billion litres in 2008 and is expected to reach 100 billion litres in 2015 (Licht 2005). According to the IEA (2008), a projection of 1% rise per year has been made for worldwide demand for oil. This is mostly due to increasing demand in energy market of developing countries, especially India (3.9%/year) and China (3.5%/year). With regard to bioethanol, the share of the USA in global production is 50% and Brazil provides 39% of the total global supply. The share of the Organisation for Economic Cooperation and Development (OECD), Europe, is 5% (Gnansounou 2010). Since Brazil is one of the most developed nations in ethanol production, almost all the Brazilian vehicles use either pure ethanol or the blend of gasoline and ethanol (75:25) (Mussatto et al. 2010). The high percentage in which ethanol is added to gasoline in Brazil is also an effort on part of the government to reduce the imports of oil (Prasad et al. 2007). As a result of these efforts, ethanol production in Brazil has substantially risen from 555 million litres (1975/1976) to 16 billion litres (2005/2006) (Orellana and Bonalume 2006), but a major reason for this is sugarcane juice. The innovations introduced by the automobile industry with flex-fuel cars, which may be fuelled with ethanol and/or gasoline in any proportion, increased the market for ethanol.

It is noteworthy that the United States, the largest consumer of petroleum products (2.42 billion litres/day or 20.7 million barrels/day in 2007), meets its demand by importing about 58%, i.e. 1.4 billion litres or 12 million barrels/day. It is predicted that gasoline consumption will rise further along with the rising population, as gasoline is a primary energy source that meets non-commercial transportation demands. Similar to Brazil, the USA is also a big investor in bioethanol research (Solomon et al. 2007) and has increased the ethanol production from 6.16 billion litres or 1.63 billion gallons in 2000 to 39.3 billion litres or 10.4 billion gallons in 2009, representing a sevenfold increase (Petrova and Ivanova 2010). At present, more than 95% of ethanol production in the United States is from corn. The rest is made from wheat, barley, cheese whey and beverage residues (Solomon et al. 2007). However, it is expected that about 1.53 billion litres or 405 million gallons of cellulosic ethanol will be produced by the end of 2012 (Solomon et al. 2007).

In Europe, the maximum amount of ethanol is produced from wheat and sugar beet, and France, Germany and Spain are the European countries more strongly committed to ethanol production (Prieur-Vernat and His 2006). The European Union strategy for biofuels is to decrease their dependence on oil and reduce the negative impact caused to the environment. The share of OECD countries in global oil demand is also expected to decrease from 57% in 2007 to 43% in 2030 (IEA 2008).

China has also invested much in the production of ethanol, and since it is the world's largest auto market, it imported about 52% of the total transportation oil consumed in 2008 (Fang et al. 2010). The production of ethanol for biofuels began in China in 2001, using corn as a raw material, and by 2007 four grain-based ethanol plants with the production of about 1.75 billion litres or 1.4 million metric tons (MMT) ethanol have been developed. However, the Chinese government supported the production of "non-food ethanol" (ethanol made from non-food crops) to restrict the competition for ethanol and food applications. The projects on ethanol based on grains were restricted by the Chinese government (Fang et al. 2010). Many technologies of ethanol production based on non-food crops, such as cassava, sweet sorghum, sweet potato, Jerusalem artichoke and kudzu vine root, are being developed (Li and Chan-Halbrendt 2009). Till now, the exclusive application of gasoline containing 10% ethanol to motor vehicles has been enforced in all areas of Heilongjiang, Jilin, Liaoning, Henan, Anhui, Guangxi and selected areas of Hebei, Shandong, Jiangsu and Hubei provinces (Fang et al. 2010). Similarly, Thailand has also invested in the production of ethanol. In 2007, there were seven ethanol plants with a total capacity of 955 thousand litres/day, comprising 130 thousand litres/day cassava ethanol and 825 thousand litres/day molasses ethanol, and as a result of government promotions, 12 new plants with a total installed capacity of 1.97 million litres/day are being constructed (Silalertruksa and Gheewala 2009).

Other countries like Japan and Korea are also in the race. An indigenous and affordable source of energy has become a high priority for energy security and sustainability. The production of biofuels in Japan started in 2003, and by 2007 the total amount of bioethanol production reached approximately 30 thousand litres per year (Matsumoto et al. 2009). Similarly, Korea is also very concerned about its high CO_2 emission and dependence on imported crude oil (Kim et al. 2010). Korea consumes about ten billion litres of gasoline annually and three or five million litres of ethanol would be needed in order to implement 3% ethanol-blending (E3) or 5% ethanol-blended gasoline (E5) countrywide (Kim et al. 2010). Therefore, the Korean government announced its plan to increase the supply of transportation biofuels from 0.2 billion litres (2008) to 5 billion litres by 2030.

4.2 Status of Bioethanol Production in India

In the year 2003, the Planning Commission of the Government of India brought out an extensive report on the development of biofuel, and bioethanol was identified as a principal biofuel. The ethanol blend in gasoline was proposed to 10% by 2011–2012 in India, and 5% ethanol blend in gasoline was made mandatory in 11 states and 3 union territories of the nation (Sukumaran et al. 2010). In 2006, the demand for ethanol for 5% gasoline doping/blending level was 0.64 billion litres. The estimated current demand for 10% blending is projected to be 2.2 billion litres in 2017 (Sukumaran et al. 2010). According to a 2006 estimate, the actual production of ethanol was only 0.39 billion litres which was not sufficient to meet the fuel demand if the entire gasoline had to be doped at 5% level.

In India, ethanol is mainly produced from sugarcane molasses, but the substrate has to compete with the food demand and therefore cannot supply the required amount of ethanol. Therefore, the nation needs to develop bioethanol technologies, which use biomass feedstock that does not have food or feed value. The most appropriate bioethanol technology for the nation would be to produce it from lignocellulosic biomass, such as rice straw, rice husk, wheat straw, sugarcane tops and bagasse, municipal waste and forest waste (Sukumaran et al. 2010). According to Kim and Dale (2004), the total bioethanol production from plant biomass is estimated to be 491 gigalitres (GL)/year globally.

India alone has the capacity to produce 25%, i.e. 123 GL/year, of the total world ethanol production, if the entire lignocellulosic residues available are used for ethanol production. Hence, to contemplate a bioethanol production plant, the lignocellulosic biomass assessment with geographical distribution and accurate information on availability of biomass in different parts of the country is a prerequisite. With this in view during the Ninth Plan, the Ministry had sponsored 500 taluka-level biomass assessment studies in 23 states to compile data on availability of lignocellulosic biomass.

As an extension of this effort, a project for preparation of "Biomass Resource Atlas of India" has been jointly sponsored to the Indian Institute of Science (IISc), Bangalore, and Regional Remote Sensing Service Centre (RRSSC), Bangalore, which aims at the integration of the data on biomass availability obtained from taluka-level studies and from other reliable sources, with information on crop distribution pattern derived from GIS-based maps provided by RRSSC.

4.3 Plant Biomass

Plant biomass is the most abundantly available and renewable natural resource on earth. Lignocellulose is comprised of three main polymers: cellulose, hemicellulose and lignin (Fig. 5.1), collectively termed as lignocellulose. The chemical properties of the components of lignocellulose make them a substrate for enormous biotechnological products (Kuhad and Singh 1993; Kuhad et al. 1997, 2007).

4.4 Cellulose

Cellulose is a glucan polymer of D-glucopyranose units, which are linked with each other by β-1, 4-glucosidic bonds. The wood cellulose has an average degree of polymerization (DP) of at least 9000–10,000 and possibly as high as 15,000. An average DP of 10,000 corresponds to the linear chain length of approximately 5 μm in wood. An approximate molecular weight for cellulose ranges from about 10,000 to 150,000 Da (Goring and Timell 1962).

Wood has many types of cellulose: crystalline and non-crystalline and accessible and inaccessible. Most wood-derived cellulose is highly crystalline which may contain as much as 65% crystalline regions. The remaining portion is referred to as amorphous cellulose which has a lower packing density. Accessible and nonaccessible refer to the availability of the cellulose to water, microorganisms, etc. The surfaces of crystalline cellulose are accessible but the rest of the crystalline cellulose is nonaccessible, whereas most of the non-crystalline cellulose is accessible. Part of the non-crystalline cellulose is covered with both hemicelluloses and lignin which makes it nonaccessible (Rowell et al. 2005; Kuhad et al. 2011). Concepts of accessible and nonaccessible cellulose are of significance in moisture adsorption,

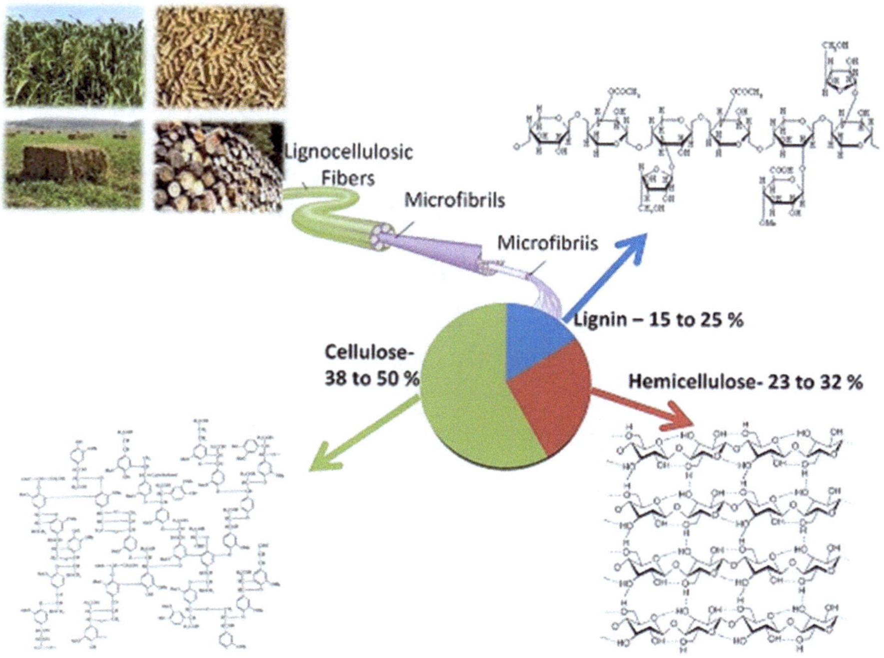

Fig. 5.1 Different components of lignocellulosic biomass and structures

pulping, extractions, chemical modification and interactions with microorganisms. Unlike cellulose, hemicelluloses are not chemically homogenous (Eriksson et al. 1990; Kuhad et al. 1997, 2011). The hemicelluloses are comprised of both linear and branched heteropolymers of D-xylose, L-arabinose, D-mannose, D-glucose, D-galactose and D-glucuronic acid (Fig. 5.2a). In general, the hemicellulose fraction of woods consists of a collection of polysaccharide polymers with a lower DP than cellulose (100–200), containing mainly the sugars D-xylopyranose, D-glucopyranose, D-galactopyranose, L-arabinofuranose, D-mannopyranose, D-glucopyranosyluronic acid and D-galactopyranosyluronic acid with lower amounts of other sugars. It usually has one repeating sugar unit linked β-(1 → 4) with branch points (1 → 2), (1 → 3) and/or (1 → 6).

4.5 *Hemicelluloses*

Hemicelluloses usually consist of more than one type of sugar unit and are called according to their components, e.g. galactoglucomannan, arabinoglucuronoxylan, arabinogalactan, glucuronoxylan and glucomannan. The hemicelluloses also contain acetyl- and methyl-substituted groups (Rowell et al. 2005). The hemicelluloses

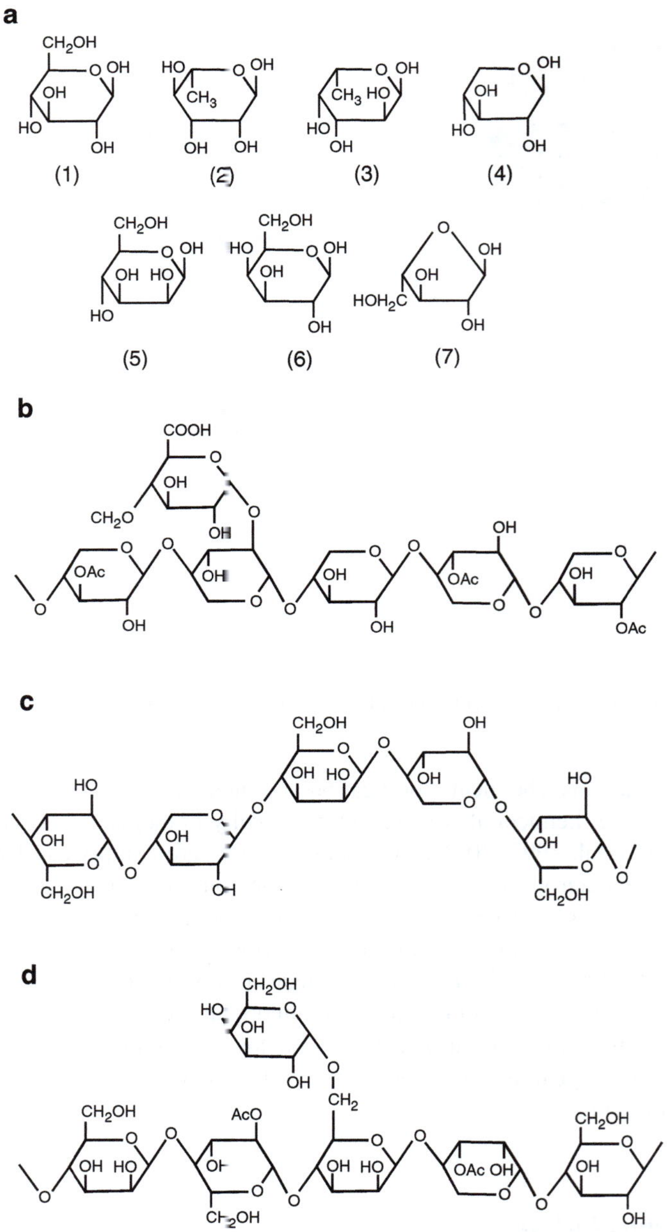

Fig. 5.2 (**a**) Structure of monosaccharides commonly present in xylan backbone (where (1) β-D-glucopyranose; (2) α-L-rhamnopyranose; (3) α-L-fucopyranose; (4) β-D-xylopyranose; (5) β-D-mannopyranose; (6) β-D-galactopyranose and (7) α-L-arabinofuranose). (**b**) Polymeric unit structures of hardwoods (o-acetyl 1-4-*O*-methylglucoronoxylan). (**c**) Polymeric unit structures of hardwoods (glucomannans). (**d**) Polymeric unit structure of softwoods (cetyl-galactoglucomannan)

from agricultural residues and hardwood are rich in xylan. Softwood contains less xylan and more mannan (Kuhad et al. 1997, 2011).

4.5.1 Softwood Hemicelluloses

Galactomannans are the principal hemicelluloses in softwood. Their backbone is a linear chain built up by 1, 4-linked β-D-glucopyranose and β-D-mannopyranose units (Fig. 5.2a). The mannose and glucose units in the backbone are partially substituted at C-2 and C-3 positions by acetyl groups, approximately 1 per 3-4 hexose units as reviewed earlier (Kuhad et al. 1997). Arabino-glucuronoxylan is another major hemicellulosic sugar and is composed of 1,4-linked-β-D-xylopyranose units. This chain is substituted at C-2 by 4-*o*-methyl-α-D-glucuronic acid group with approximately two such units per ten xylose units. The xylose backbone is also substituted by α-L-arabinopyranose units, on the average 1.3 residues per ten xylose units (Kuhad et al. 1997, 2007, 2011). Arabinogalactan is a minor component in hardwoods and softwoods. The backbone of this galactan is built up by 1, 3-linked α-D-galactopyranose units, and almost every galactose unit is substituted at the C-6 position.

4.5.2 Hardwood Hemicelluloses

The o-acetyl-4-*o*-methyl-glucurono-β-xylan (commonly known as glucuronoxylan) is the major component of hardwood hemicelluloses (Kuhad et al. 1997) (Fig. 5.2b–d). The xylan content varies between 15 and 30% in different hardwood species. The backbone of xylan consists of β-D-xylopyranose units linked by 1,4-bonds. Seven of ten xylose units are substituted by acetyl group at C-2 or C-3 position and in one of ten xylose units, the 4-*o*-methyl-α-D-glucuronic acid residue unit is linked at C-1 and C-2 positions to the hemicellulose backbone (Kuhad et al. 1997). Glucomannan is another hemicellulose in hardwoods. This polysaccharide comprises 2–5% of the wood and is composed of β-D-glucopyranose and β-D-mannopyranose units by 1,4-bonds. Depending on the wood species, the glucose–mannose ratio varies between 1:1 and 1:2 (Kuhad et al. 1997; Rowell et al. 2005).

4.6 Lignin

Lignin is an amorphous and highly complex polymer of phenylpropanoid units and is considered to be an encrusting substance in the plant biomass (Kuhad et al. 1997). The precursors of lignin biosynthesis are coniferyl alcohol, *p*-coumaryl alcohol and sinapyl alcohol. *p*-Coumaryl alcohol is a minor precursor of hardwood and softwood lignins. The predominate precursor of softwood lignin is coniferyl alcohol.

Coniferyl and sinapyl alcohols are both precursors of hardwood lignin. Softwood lignin has methoxyl content of ~15–16%. Hardwood lignin has a methoxyl content of ~21%. Moreover, ecological factors such as age of the wood, climate, plant sustenance and amount of sunlight also affect the chemical structure of lignin (Kuhad and Singh 1993; Kuhad et al. 1997). Lignin does not have a single repeating unit like cellulose does but instead consists of a complex arrangement of substituted phenolic units. Lignins are usually divided according to their structural elements. Wood lignins consist of three basic building blocks which are guaiacyl, syringyl and *p*-hydroxyphenyl moieties; other aromatic units also exist in different types of woods. There is a wide variation of structures in different wood species. The lignin content of softwoods is usually in the range of 18–25%, whereas the lignin content of hardwoods varies between 25 and 35%. The phenylpropane can be substituted at the α, β or γ positions into various combinations linked together both by ether (C-O-C) and carbon to carbon (C-C) linkages. Lignins from softwoods are called guaiacyl lignin which is mainly a polymerization product of coniferyl alcohol. Hardwood lignins are mainly syringyl–guaiacyl lignin, because they are a copolymer of coniferyl and sinapyl alcohols. The ratio of the two types differ from one lignin to others from about 4:1 to 1:2.

4.7 *Bioconversion of Lignocellulosic Biomass to Ethanol*

The bioconversion of lignocelluloses to ethanol consists of two main processes: hydrolysis of lignocellulosic carbohydrate to fermentable reducing sugars and fermentation of the sugars to ethanol (Fig. 5.3). The hydrolysis is usually catalyzed by cellulase enzymes, and the fermentation is carried out by bacteria or yeasts. The factors which have been identified to affect the hydrolysis of cellulose include porosity (accessible surface area) of the waste materials, cellulose fibre crystallinity and lignin and hemicellulose content (Mosier et al. 2005; Kuhad et al. 2011). The presence of hemicellulose and lignin in lignocellulosic materials makes the access of cellulase enzymes difficult, thus reducing the efficiency of the hydrolysis (Himmel et al. 2007). Pretreatment of lignocellulosic biomass prior to hydrolysis significantly improves the efficiency of hydrolysis due to the removal of lignin and hemicellulose, reduction of cellulose crystallinity and increase of porosity (McMillan 1994; Mosier et al. 2005; Kuhad et al. 2011).

4.7.1 Pretreatment of Lignocellulosic Biomass

The effect of pretreatment strategies of lignocellulosic materials has been recognized well for a long time (Mosier et al. 2005; Sanchez and Cardona 2008; Kuhad et al. 2011). The pretreatment strategies must meet the following requirements: (1) improve the formation of sugars or the ability to subsequently form sugars by enzymatic hydrolysis; (2) avoid the degradation or loss of carbohydrate; (3) avoid

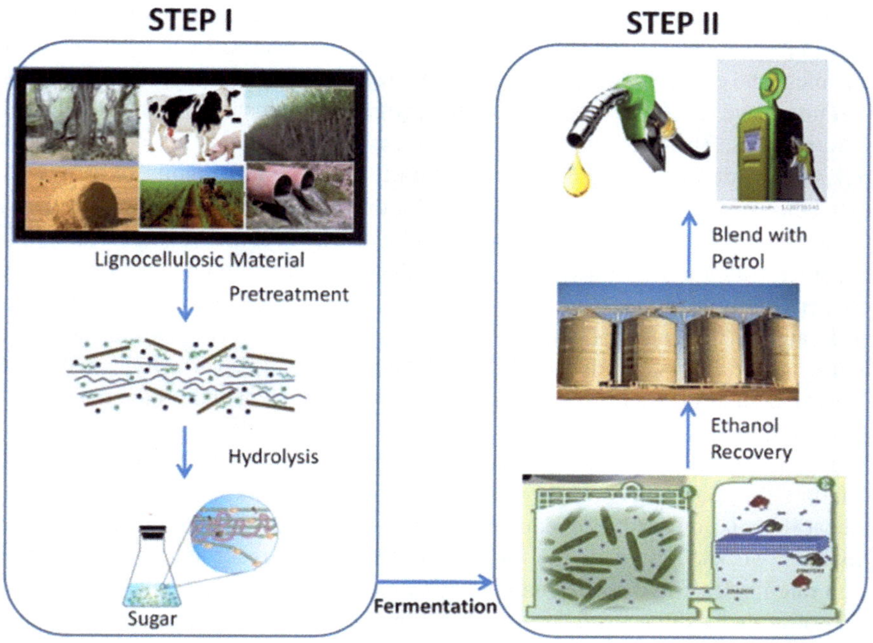

Fig. 5.3 Bioethanol production from lignocellulosic biomass

the formation of hydrolysis and fermentation inhibitory byproducts; and (4) cost-effectiveness of the process. Various physical, biological, physico-chemical and chemical processes have been used for pretreatment of lignocellulosic materials.

4.7.2 Bioconversion of Lignocellulosic Materials to Ethanol

The pretreatment has become a necessity to maximize the hydrolysis of lignocelluloses and eventually the production of ethanol. The advantages of biological delignification of plant material over chemical and mechanical pretreatment methods include (1) mild reaction conditions, (2) avoidance of the use of toxic and corrosive chemicals, (3) higher product yield, (4) fewer side reactions, (5) less energy demand and (6) less reactor resistance to pressure and corrosion (Lee 1997; Kuhar et al. 2008; Sanchez 2009). In situ microbial delignification appears to be a feasible strategy for achieving improved depolymerization of hemicellulose and cellulose.

The white-rot fungi (WRF) are the most effective microorganisms for biological pretreatment as they degrade lignin more extensively and rapidly than any other known group of organisms (Eriksson 1993; Kuhad et al. 1997, 2007; Keller et al. 2003). Some WRF have been reported to degrade lignin selectively and this capability of selected WRF can be exploited for delignification of plant materials without affecting much of cellulose (Kuhar et al. 2008; Gupta et al. 2011). Thus, selected lignin-degrading WRF with comparatively low cellulase and xylanase activities could be advantageous for efficient delignification and eventually in the reduction of chemical and energy inputs for chemical or enzymatic hydrolysis of the substrate(s).

Fungi have two types of extracellular enzymatic systems: hydrolytic and ligninolytic systems. The hydrolytic system produces hydrolases that are responsible for polysaccharide degradation and production of sugars. The unique oxidative and extracellular ligninolytic system degrades lignin component and opens phenyl rings (Lundell et al. 2010). WRF is known to produce one or more of four such cellular enzymes which are of importance for the degradation of lignin. Laccase contains copper (Lac), heme contains manganese peroxidase (MnP) and lignin peroxidase (LiP) and also aryl alcohol oxidase (AAO). Major hydrolytic enzymes like amylases and xylanase are also produced by them (Rosales et al. 2002; Elisashvili et al. 2006) These enzymes of WRF are very significant for the efficient and productive bioconversion of plant residues (Radhika et al. 2013; Songulashvili et al. 2006).

Few studies have been reported on the pretreatment of plant biomass with WRF for its effect on cellulose hydrolysis. According to Hatakka (1983), 35% of the wheat straw is convertible to reducing sugars when pretreated with *Pleurotus ostreatus* for 5 weeks. Taniguchi and co-workers (2005) also recorded a similar conversion rate in rice straw pretreated with *P. ostreatus* for 60 days. Keller et al. (2003) also observed a three- to fivefold improvement in the enzymatic cellulose digestibility in corn stover pretreated with *Coriolus versicolor* in over 30 days. Thus, most of these fungal pretreatments have suffered because of long incubation periods. Therefore, to economize microbial pretreatment of lignocellulosic materials, to improve the hydrolysis of carbohydrates to reducing sugars and to eventually improve ethanol yield, there is a need to test more and more basidiomycetous fungi for their ability to delignify the plant material quickly and efficiently (Kuhad et al. 2011).

Biological pretreatment in combination with other pretreatment technologies has also been studied (Itoh et al. 2003; Balan et al. 2008). Itoh et al. (2003) reported production of ethanol by simultaneous saccharification and fermentation (SSF) from beech wood chips after bio-organosolvation pretreatments by ethanolysis and white-rot fungi, *Ceriporiopsis subvermispora*, *Dichomitus squalens*, *Pleurotus ostreatus* and *Coriolus versicolor*. The yield of ethanol obtained was 0.294 g/g of ethanolysis pulp and 0.176 g/g of beech wood chips after pretreatment with *C. subvermispora* for 8 weeks. The yield was 1.6 times higher than when obtained without the fungal treatments. The combined process allowed the separation of lignin, cellulose and hemicelluloses with the use of only water, ethanol and wood-rot fungi. The biological pretreatments saved 15% of the electricity which was needed for ethanolysis. In another approach, Balan et al. (2008) studied the effect of fungal treatment of rice straw followed by AFEX pretreatment and enzymatic hydrolysis. They reported that treating rice straw with white-rot fungus, followed by AFEX gave significantly higher glucan and xylan conversions. Recently, it was found that the SSF of *Prosopis juliflora* and *Lantana camara* followed by acid hydrolysis and observed that fungal treatment significantly reduce the number of inhibitors generated, and eventually the requirement of the detoxifying agent was also reduced

(Gupta et al. 2011). The pretreated cellulose-rich biomass is then hydrolysed with a consortium of cellulolytic enzymes, i.e. endocellulase, exocellulase and β-glucosidase. These enzymes act in synergy and hydrolyse the cellulosic polymer to simple sugars (glucose) (Zhang et al. 2006; Kuhad et al. 2011). Enzymatic hydrolysis has demonstrated better results for the subsequent fermentation because no degradation components of glucose are formed; however, the process is costlier (Sanchez and Cardona 2008). Various improvements have been attempted in the last three decades that would lower the effective enzyme cost, including enzyme reusage, higher enzyme production or using genetically engineered systems (Zhang et al. 2006; Kuhad et al. 2011); but there is still a need to search for more competent solutions.

Since the lignocellulose contains both hexose and pentose sugar, the efficient fermentation of both the sugars has become a prerequisite for the cost-effective production of bioethanol. A variety of microorganisms are known to ferment hexose sugars, but the microbes for pentose fermentation are limited in number. The most common microbes tried for pentose fermentation are *Candida shehatae*, *Pichia stipitis* and *Pachysolen tannophilus* (Abbi et al. 1996a, b; Kuhad et al. 2011). However, none of these yeasts have been found to be very promising. Intensive efforts have been made towards the utilization of all the sugars present in hydrolysates by employing both genetic manipulations as well as process improvement approaches. Currently, most commercial cellulases are produced from *Phanerochaete* sp., *Aspergillus* sp. and *Trichoderma reesei* (Adney et al. 2003), usually used to describe a mixture of cellulolytic enzymes whose synergistic action is required for effective breakdown of substrate to its monomeric units.

The action of cellulases involves the concerted action of endoglucanases (endo-1, 4-β-glucanases, EGs), which can hydrolyse internal bonds preferably in cellulose amorphous regions releasing new terminal ends, which randomly attacks the internal, β1,4-linkages cellobiohydrolase (exo-1,4 β-glucanases, Cbhs), which acts on the existing or endoglucanase generated chain ends. Both enzymes can degrade amorphous cellulose but, with some exceptions, Cbhs are the only enzymes that efficiently degrade crystalline cellulose. Cbhs and EGs release cellobiose molecules which cleaves off cellobiose units from the non-reducing ends of the glucan and β3-glucosidase, which hydrolyses cellobiose to glucose. The pretreated substrate can be then hydrolysed with microorganisms and subsequently be fermented to ethanol, which will serve as biofuel. Overall, the use of biomass to produce bioethanol holds much promise for providing a renewable, indigenously produced liquid energy source that can be a viable alternative to petroleum-based fuels.

5 Conclusion

Plant cells are mainly composed of lignocellulosic material, which includes cellulose, hemicellulose and lignin. Pretreatment of lignocellulosic biomass prior to hydrolysis significantly improves the efficiency of hydrolysis due to the removal of lignin and hemicellulose and reduction of cellulose. Microorganisms, especially fungi, can degrade the plant cell wall through a set of enzymes. Fungi have two types of extracellular enzymatic systems: hydrolytic and ligninolytic systems. The hydrolytic system produces hydrolases and the ligninolytic system degrades lignin component and opens phenyl rings. The white-rot fungi (WRF) are effective for biological pretreatment as they degrade lignin more extensively and rapidly than other microorganisms. The selected lignin-degrading WRF with comparatively low cellulase and xylanase activities could be advantageous for efficient delignification and eventually in the reduction of chemical and energy inputs for chemical or enzymatic hydrolysis of the substrate.

References

Abbi M, Kuhad RC, Singh A (1996a) Bioconversion of pentose sugars to ethanol by free and immobilized cells of *Candida shehatae* NCL-3501: fermentation behaviour. Process Biochem 31:555–560

Abbi M, Kuhad RC, Singh A (1996b) Fermentation of xylose and rice straw hydrolysate to ethanol by *Candida shehatae* NCL-3501. J Ind Microbiol 17:20–23

Adney WS, Chou YC, Decker SR, Ding SY, Baker JO, Kunkel G, Vinzant TB, Himmel ME (2003) Heterologous expression of *Trichoderma reesei* 1,4-beta-D-glucan cellobiohydrolase (Cel 7A). In: Applications of enzymes to lignocellulosics, vol 855. Amer Chemical Soc, Washington, pp 403–437

Alper H, Stephanopoulos G (2009) Engineering for biofuels: exploiting innate microbial capacity or importing biosynthetic potential? Nat Rev Microbiol 7:715–723

Antizar-Ladislao B, Turrion-Gomez JL (2008) Second-generation biofuels and local bioenergy systems. Biofuels Bioprod Biorefin 2:455–469

Antoni D, Zverlov VV, Schwarz WH (2007) Biofuels from microbes. Appl Microbiol Biotechnol 77:23–35

Balan V, Souca LDC, Chundawat SS, Vismeh R, Jones AD, Dale BE (2008) Mushroom spent straw: a potential substrate for an ethanol-based biorefinery. J Ind Microbiol Biotechnol 35:293–301

Campbell CJ, Laherrere JH (1998) The end of cheap oil. Science 3:78–83

Chaturvedi V, Verma P (2013) An overview of key pretreatment processes employed for bioconversion of lignocellulosic biomass into biofuels and value added products. Biotech 3:415–431

Elisashvili V, Penninckx M, Kachlishvili E, Asatiani M, Kvesitadze G (2006) Use of *Pleurotus dryinus* for lignocellulolytic enzymes production in submerged fermentation of mandarin peels and tree leaves. Enzym Microb Technol 38:998–1004

Eriksson KEL (1993) Lignin biodegradation and practical utilization. J Biotechnol 30:149–158

Eriksson KL, Blanchette RA, Ander P (1990) Microbial and enzymatic degradation of wood and wood components. Springer, Berlin

Fang X, Shen Y, Zhao J, Bao X, Qu Y (2010) Status and prospects of lignocellulosic bioethanol production in China. Bioresour Technol 101:4814–4819

Fernandez-Fueyo E, Ruiz-Dueñas FJ, Ferreira P, Floudas D, Hibbett DS (2012) Comparative genomics of *Ceriporiopsis subvermispora* and *Phanerochaete chrysosporium* provide insight into selective ligninolysis. Proc Natl Acad Sci U S A 109:5458–5463

Gnansounou E (2010) Production and use of lignocellulosic bioethanol in Europe: current situation and perspectives. Bioresour Technol 101:4842–4850

Goring DAI, Timell TE (1962) Molecular weight of native celluloses. Tappi 45:454–460

Gupta R, Sharma KK, Kuhad RC (2009) Separate hydrolysis and fermentation (SHF) of *Prosopis juliflora*, a woody substrate, for the production of cellulosic ethanol by *Saccharomyces cerevisiae* and *Pichia stipitis*-NCIM 3498. Bioresour Technol 100:1214–1220

Gupta R, Mehta G, Khasa YP, Kuhad RC (2011) Fungal delignification of lignocellulosic biomass improves the saccharification of cellulosics. Biodegradation 22:797–804

Hatakka AI (1983) Pretreatment of wheat straw by white-rot fungi for enzymatic saccharification of cellulose. Eur J Appl Microbiol Biotechnol 18:350–357

Hatakka A (2005) Biodegradation of lignin. In: Biopolymer online. p 1

Himmel ME, Ding SY, Johnson DK, Adney WS, Nimlos MR, Brady JW, Foust TD (2007) Biomass recalcitrance: engineering plants and enzymes for biofuels production. Science 315:804–807

IEA (2008) World energy outlook

Itoh H, Wada M, Honda Y, Kuwahara M, Watanabe T (2003) Bioorganosolve pretreatments for simultaneous saccharification and fermentation of beech wood by ethanolysis and white rot fungi. J Biotechnol 103:273–280

Kang S, Li X, Fan J, Chang J (2013) Hydrothermal conversion of lignin: a review. Renew Sustain Energy Rev 27:546–558

Keller FA, Hamillton TE, Nguyon QA (2003) Microbial pretreatment of biomass potential for reducing severity of thermo-chemical biomass pretreatment. Appl Biochem Biotechnol 105:27–41

Kim S, Dale EB (2004) Global potential bioethanol production from wasted crops and crop residues. Biomass Bioenergy 26:361–375

Kim JS, Park SC, Kim JW, Park JC, Park SM, Lee JS (2010) Production of bioethanol from lignocellulose: status and perspectives in Korea. Bioresour Technol 101:4801–4805

Kuhad RC, Singh A (1993) Lignocellulose biotechnology: current and future prospects. Curr Rev Biotechnol 13:151–172

Kuhad RC, Singh A, Eriksson KEL (1997) Microorganisms and enzymes involved in the degradation of plant fiber cell wall. Adv Biochem Eng Biotechnol 57:47–125

Kuhad RC, Kuhar S, Kapoor M, Sharma KK, Singh A (2007) Lignocellulolytic microorganisms, their enzymes and possible biotechnologies based on lignocellulolytic microorganisms and their enzymes pp 3-32 In: Lignocellulose biotechnology: future prospects (edited by) R. C. Kuhad and A. Singh. IK International, New Delhi.

Kuhad RC, Gupta R, Khasa YP (2011) Bioethanol production from lignocellulosics: an overview. In: Lal B, Sharma PM (eds) Wealth from waste, 3rd edn. TERI Press, New Delhi, India

Kuhar S, Nair LM, Kuhad RC (2008) Pretreatment of lignocellulosic material with fungi capable of higher lignin degradation and lower carbohydrate degradation improves substrate acid hydrolysis and the eventual conversion to ethanol. Can J Microbiol 54:305–313

Kumar G, Bakonyi P, Periyasamy S, Kim SH, Nemestóthy N (2015) Lignocellulose biohydrogen: practical challenges and recent progress. Renew Sustain Energy Rev 44:728–737

Lee J (1997) Biological conversion of lignocellulosic biomass to ethanol. J Biotechnol 56:1–24

Li SZ, Chan-Halbrendt C (2009) Ethanol production in China: potential and technologies. Appl Energy 4:4–9

Licht FO (2005) World ethanol and biofuels report 3:15

Lundell TK, Mäkelä MR, Hildén K (2010) Lignin-modifying enzymes in filamentous basidiomycetes: ecological, functional and phylogenetic review. J Basic Microbiol 50:1–16

Matsumoto N, Sano D, Elder M (2009) Biofuel initiatives in Japan: strategies, policies and future potential. Appl Energy 86:569–576

McCarthy JE, Tiemenn M (1998) CRS report for congress MTBE in gasoline clean air and drinking water. http://www.epa.gov/otaq/consumer/fuels/mtbeMTBE.pdf

McMillan JD (1994) Pretreatment of lignocellulosic biomass. ACS Symp Ser 566:292–324

Morris D (1993) Ethanol: a 150 year struggle toward a renewable future. Institute for Local Self-Reliance, Washington

Mosier NS, Wyman C, Dale B, Elander R, Lee YY, Holtzapple M, Ladisch MR (2005) Features of promising technologies for pretreatment of lignocellulosic biomass. Bioresour Technol 96:673–686

Mussatto SI, Dragone G, Guimarães PMR, Silva JPA, Carneiro LM, Roberto IC, Vicente A, Domingues L, Teixeira JA (2010) Technological trends, global market and challenges of bioethanol production R1. Biotechnol Adv 28:817–830

Narayanaswamy N, Dheeran P, Verma S, Kumar S (2013) Biological pre-treatment of lignocellulosic biomass for enzymatic saccharification. In: Fang Z (ed) Pre-treatment techniques for biofuels and biorefineries. Springer, Berlin

Orellana C, Bonalume NR (2006) Brazil and Japan give fuel to ethanol market. Nat Biotechnol 24:232

Petrova P, Ivanova V (2010) Perspectives for the production of bioethanol from lignocellulosic materials. Biotechnol Bioeng 24:529–546

Prasad S, Singh A, Joshi HC (2007) Ethanol as an alternative fuel from agricultural, industrial and urban residues. Resour Conserv Recycl 50(1):1–39

Prieur-Vernat A, His S (2006) Biofuels in Europe, panorama. IFP, Paris

Radhika R, Jebapriya GR, Gnanadoss JJ (2013) Production of cellulase and laccase using *Pleurotus* sp. under submerged and solid-state fermentation. Int J Curr Sci 6E:7–13

Ragauskas AJ, Williams CK, Davison BH, Britovsek G, Cairney J, Eckert CA, Frederick WJ Jr, Hallett JP, Leak DJ, Liotta CL, Mielenz JR, Murphy R, Templer R, Tschaplinski T (2006) The path forward for biofuels and biomaterials. Science 311:484–489

Rosales E, Rodriguez Couto S, Sanroman MA (2002) A new uses of food wastes: application to laccase production by *Trametes hirsute*. Biotechnol Lett 24:701–704

Rowell RM, Pettersen R, Han JS, Rowell JS, Tshabalala MA (2005) Cell wall chemistry. In: Rowell RM (ed) Handbook of wood chemistry and wood composites. CRC Press LLC, Boca Raton, pp 35–62

Sanchez C (2009) Lignocellulosic residues: biodegradation and bioconversion by fungi. Biotechnol Adv 27:185–194

Sanchez OJ, Cardona CA (2008) Trends in biotechnological production of fuel ethanol from different feedstocks. Bioresour Technol 99:5270–5295

Silalertruksa T, Gheewala SH (2009) Environmental sustainability of bio-ethanol production in Thailand. Energy 34:1933–1946

Solomon BD, Barnes JR, lvorsen KE (2007) Grain and cellulosic ethanol: history, economics and energy policy. Biomass Bioenergy 31:416–425

Songulashvili G, Elisashvili V, Wasser S, Nevo E, Hadar Y (2006) Laccase and manganese peroxidases activities of *Phellinus robustus* and *Ganoderma adspersum* grown on food industry wastes in submerged fermentation. Biotechnol Lett 28:1425–1429

Suhara H, Kodama S, Kamei I, Maekawa N, Meguro S (2012) Screening of selective lignin-degrading basidiomycetes and biological pretreatment for enzymatic hydrolysis of bamboo culms. Int Biodeterior Biodegrad 75:176–180

Sukumaran RK, Surender VJ, Sindhu R, Binod P, Janu KU, Sajna KV, Rajasree KP, Pandey A (2010) Lignocellulosic ethanol in India: prospects, challenges and feedstock availability. Bioresour Technol 101:4826–4833

Taha M, Shahsavari E, Al-Hothaly K, Mouradov A, Smith AT (2015) Enhanced biological straw saccharification through coculturing of lignocellulose-degrading microorganisms. Appl Biochem Biotechnol 175:3709–3728

Taniguchi M, Suzuki H, Watanabe D, Sakai K, Hoshino K, Tanaka T (2005) Evaluation of pretreatment with *Pleurotus ostreatus* for enzymatic hydrolysis of rice straw. J Biosci Bioeng 100:637–643

Wang W, Yuan T, Cui B, Dai Y (2013) Investigating lignin and hemicellulose in white rot fungus-pretreated wood that affect enzymatic hydrolysis. Bioresour Technol 134:381–385

Zhang YHP, Himmel ME, Mielenz JR (2006) Outlook for cellulase improvement: screening and selection strategies. Biotechnol Adv 24:452–481

Chapter 6
Myco-Degradation of Lignocellulose: An Update on the Reaction Mechanism and Production of Lignocellulolytic Enzymes by Fungi

Arvind Kumar

1 Introduction

Lignocellulosic biomass is the most abundant renewable feedstock for second-generation biofuel and other bio-based valuable products. It is constantly produced as a by-product of agro-industry, forestry, and solid waste of municipality. Lignocellulosic biomass represents approximately half of the global biomass production. Dry lignocellulosic biomass production was estimated 10–50 tons per year worldwide (Galbe and Zacchi 2002). Lignocellulose is majorly composed of cellulose, hemicellulose, and lignin. Their proportions are dependent on the source of biomass. For example, agricultural residues contain approximately 30–40% cellulose, 20–30% hemicellulose, and 15–30% lignin, whereas angiosperm and gymnosperm woods (hardwood and softwood) contain 40–42% cellulose, 20–30% hemicellulose, and 25–32% lignin (Bornscheuer et al. 2014). In general scenario, it has been proposed that lignocellulosic materials hold 40–60% cellulose, 20–40% hemicellulose, and 10–25% lignin (Jaramillo et al. 2015).

Cellulose is a linear polymer of D-glucopyranose subunits joined together by β-1,4-glycosidic linkage (Bornscheuer et al. 2014). Each glucopyranosyl subunit is rotated 180° in relation to its neighbor molecule forming the repeat unit of cellulose known as cellobiose. The chain length of monomeric subunits in a cellulose molecule depends on the source, and it varies from 250 residues to well over 10,000 residues. These molecules form intra- and intermolecular hydrogen bonds and are compactly packed at the distance of van der Waals' radii due to the absence of side groups which accounts for the rigid and insoluble (crystalline) nature of cellulose microfibrils. Cellulose molecules also contain amorphous regions in addition to

A. Kumar (✉)
Department of Biotechnology, Guru Ghasidas Vishwavidyalaya (A Central University), Bilaspur, Chhattisgarh, India

R. Naraian (ed.), *Mycodegradation of Lignocelluloses*, Fungal Biology,
https://doi.org/10.1007/978-3-030-23834-6_6

crystalline regions, and generally a cellulose microfibril is composed of 36 chains (Guerriero et al. 2016).

Hemicelluloses are a group of linear and branched complex carbohydrate structures arranged in different proportions with different substituents. Hemicelluloses are mainly composed of xylose, arabinose, glucose, mannose, and galactose residues. Thus, they are the polymers of pentoses (xylan and araban), hexoses (mannan, glucan, glucomannan, and galactan), and pectin (Bornscheuer et al. 2014; Johansen 2016). Among the hemicelluloses found in the plant cell walls, xylan is characterized as polymer of β-1,4 linked xylosyl residues substituted with acetyl, feruloyl, glucuronic acid, 4-*O*-methylglucuronic acid, arabinose, and galactose side chains (Guerriero et al. 2016). Hemicelluloses of plant origin are also known as acidic and neutral hemicelluloses based on the glucuronic and 4-*O*-methylglucuronic acid content. Similarly, araban, mannan and galactan are polymers of arabinose, mannose, and galactose, whereas glucomannan is composed of mannose chains interrupted by glucose residues. Hemicelluloses are often joined with lignin which provides additional strength for structural stability.

Lignin is a complex heterogeneous polymer of phenylpropanoid alcohols known as monolignols. These monolignols are *p*-coumaryl, coniferyl, and sinapyl alcohols polymerized by ether and C-C bonds via free radical reaction (Guerriero et al. 2016). Gymnosperm lignin is almost entirely composed of coniferyl alcohols, whereas angiosperm dicot lignin is a mixture of coniferyl and sinapyl monomers. All the three kinds of monolignols are constituents of monocotyledonous lignin (Bornscheuer et al. 2014). The ratios of constituent monomers vary in different plants, woods tissues, and cell wall layers (Johansen 2016).

Plant-derived biomass is a natural resource of carbohydrate polymers. Its utilization for value and revalue has received considerable interest in the recent years with ever-increasing scientific knowledge, worldwide economic and environmental consciousness, demands of legislative authorities, and the manufacture, use, and removal of petrochemical-based by-products (Arevalo-Gallegos et al. 2017). Since a lignocellulosic material consists of 70% sugars, therefore, it is considered one of the valuable biomass for future bio-economy and alternate energy source (Horn et al. 2012; Arevalo-Gallegos et al. 2017). Such biomass offers a great potential for biofuel and many other valuable products without any competing interest for food production and also provides an opportunity for reduced greenhouse gas emission. Thus, biorefinery approach is emerging as a promising sector for the capitalization of lignocellulosic biomass into a variety of bio-based useful products such as biodegradable plastic, biochemicals, and advanced biofuels (Arevalo-Gallegos et al. 2017). The term biorefinery refers to the sustainable conversion of biomass into a variety of valued chemicals and energy in an economical and environment-friendly manner (Arevalo-Gallegos et al. 2017). The important biorefinery-based products/chemicals selected by the US Department of Energy from lignocellulose are ethanol, furans (furfural and 5-hydroxymethyl furfural), glycerol and its derivatives (propanediol, glycerol carbonate, and epichlorohydrin), hydrocarbons, lactic acid, succinic acid, levulinic acid, sorbitol, and xylitol (Chandel et al. 2018). Excluding biofuels, the production of bio-based other valued chemicals is estimated to be 50 billion kilos per year with a 3–4% annual growth (de Jong 2011).

Recalcitrant nature is the major obstacle in the saccharification and bioconversion of lignocellulosic material into valued end products. This recalcitrance arises due to an intricate arrangement between cell wall polysaccharides (cellulose and hemicellulose) and lignin (Jaramillo et al. 2015). The paracrystallinity of cellulose, complex hemicellulose coating of cellulose microfibrils, and interpenetration and encapsulation of polysaccharides by lignin impede the accessibility of lignocellulosic materials for carbohydrases, ligninases, and accessory proteins (Cragg et al. 2015). These barriers are overcome by harsh physical, chemical, and physicochemical pretreatments in the industries which aid production cost. High cost of biofuel production is the major constraint of commercialization. It has been estimated that cellulosic ethanol production cost ($0.5 per liter) is almost double in comparison with starch-based ethanol (Bornscheuer et al. 2014). A considerable cost in the biomass-based production of biofuel is related with enzymes. Thus, it necessitates the development of an efficient and economical lignocellulolytic processing for viable commercialization. Microbial sources of enzymes for lignocellulose biomass degradation offer opportunities to overcome the high cost barrier. Microorganisms degrade lignocellulosic biomass under physiologically tolerable conditions, rendering them to be the suitable host for the production of efficient lignocellulolytic enzymes. Among the microorganisms, saprophytic fungi express higher levels of lignocellulose biomass degrading enzymes and are still considered the most promising hosts to deal with the challenging demand of lignocellulose utilization (Mitrovic and Glieder 2015; Li et al. 2017). There has been intensive progress in the screening of hyper lignocellulolytic fungi and development of efficient cellulolytic platforms in the recent years. Thus, this chapter discusses lignocellulose-degrading enzymes, mechanisms of enzyme actions, and some recent developments in the production of lignocellulolytic enzymes by fungal strains.

2 Lignocellulose-Degrading Enzymes

Enzymes for deconstruction of lignocellulosic material are broadly categorized into two classes known as glycosyl hydrolases and oxidative enzymes. Glycosyl hydrolases degrade the Earth's most abundant cellulose and hemicellulose carbon polymers, whereas oxidative enzymes decompose the most recalcitrant lignin biomolecule as well as cellulose (Hallberg et al. 2003; Mitrovic and Glieder 2015; Beeson et al. 2015). Efficient degradation of lignocellulosic material requires a plethora of enzymes collectively known as enzyme cocktail. Enzymatic cocktail is composed of cellulases, hemicellulases, ligninases, auxiliary activity, and accessory proteins which are discussed in the later sections of the text.

2.1 Cellulases

Cellulases play an important role in the development of an efficient lignocellulolytic process because cellulose content of the lignocellulosic materials is up to 60% (Bilal et al. 2017). Cellulases include a consortium of β-1,4-endoglucanases (EC 3.2.1.4), cellobiohydrolases (EC 3.2.1.91), and β-glucosidases (EC 3.2.1.21) which hydrolyze polymeric cellulose into fermentable sugars. Endoglucanases randomly hydrolyze internal β-1,4-glycosidic linkages in the amorphous regions of the cellulosic microfibrils (Taha et al. 2016; Lopes et al. 2018). Thus, oligosaccharides or smaller polysaccharides are released from long-chain cellulose molecules. Cellobiohydrolases are exo-acting enzymes and catalyze the hydrolysis of glycosidic bonds in the crystalline regions of cellulose. Cellobiohydrolases I and II release glucose or cellobiose molecules from reducing and non-reducing ends, respectively. The released cellobiose and oligosaccharide molecules are further cleaved into monomeric constituents by β-glucosidases (Lopes et al. 2018; Passos et al. 2018; Marques et al. 2018). β-Glucosidases are not directly acting on cellulose and thus are not considered as true cellulases. However, they are important components of the complete saccharification process.

Since cellulose is an insoluble substrate, therefore, endoglucanases have to diffuse, attach, and move a segment of the polymeric substrate to their active sites. Endoglucanases are modular proteins. They consist of catalytic domains connected with carbohydrate binding modules via flexible linker segment. The carbohydrate binding domain recognizes the substrate and adsorbs the enzyme onto the surface of cellulose microfibrils. In this way, they are increasing the effective concentration of the enzyme molecules at the substrate's surface. Additionally, it has also been postulated that carbohydrate binding modules have the ability to disrupt crystalline structure of cellulose (Shoseyov et al. 2006; Araki et al. 2010; Behera et al. 2017). The evidence supporting the above statement has accumulated from the assessment of adsorption and degradation ability of the carbohydrate binding module deleted enzymes. Such truncated enzymes showed a little change in their degradation activity at soluble substrates while their activity and adsorption declined significantly on crystalline cellulose (Behera et al. 2017). Chemical and site-directed mutagenesis studies demonstrated that aromatic amino acids play an important role in the adsorption process. Aromatic amino acids help in the attachment of the enzyme on the insoluble substrate surface by establishing accumulation forces between glucose rings and π-electrons of the aromatic rings (Chen 2014). Therefore, the overall impact of the presence of carbohydrate binding domain is to increase the enzyme's adsorption on crystalline cellulose substrate.

2.2 Hemicellulases

Hemicellulose is a complex and heterogenous polysaccharide mostly composed of xylan and mannan. The full depolymerization of hemicellulose requires synergistic action of an array of enzymes, e.g., xylanases, mannanases, and other side chain cleaving enzymes depend on the origin of the substrate. Xylanases group includes β-1,4-endo-xylanases (EC 3.2.1.8), β-xylosidases (EC 3.2.1.37), α-L-arabinofuranosidase (EC 3.2.2.55), α-glucuronidase (EC 3.2.1.131), acetyl xylan esterase (EC 3.1.1.6), ferulic acid esterase (EC 3.1.1.73), and *p*-coumaric acid esterase (EC 3.1.1.73). Among a variety of xylanases; β-1,4-endo-xylanases hydrolyze internal β-1,4-glycosidic linkage in the xylan backbone and release short xylooligosaccharides of varying length (Gong et al. 2013; Kim et al. 2014). The released xylooligosaccharides were hydrolyzed by β-xylosidases from the non-reducing end. The side chain acting enzymes; α-L-arabinofuranosidase releases terminal α-L-1, 2-, α-L-1, 3-, and α-L-1, 5-linked arabinofuranosyl residues, α-glucuronidase catalyzes the cleavage of α-1, 2 glycosidic bond between *O*-4-methyl glucuronic acid and main chain of glucuronoxylan, acetyl xylan esterase cleaves acetyl groups from acetyl xylan, and ferulic acid and coumaric acid esterases hydrolyze the ester linkage between arabinose and monomeric or dimeric ferulic acids, and arabinose and *p*-coumaric acid, respectively (Lopes et al. 2018; Marques et al. 2018). Xyloglucan endotransglycosylase/hydrolase (EC 2.4.1.270) enzyme catalyzes both transglycosylation and hydrolysis reactions. The hydrolytic activity degrades xyloglucans by breaking β-1,4-glucosyl linkages. Other enzymes assisting in the xyloglucan hydrolysis are xyloglucan-specific β-D-galactosidase (EC 3.2.1.23), xyloglucan-specific α-L-fucosidase (EC 3.2.1.51/63), and xyloglucan oligosaccharide-specific α-D-xylosidase (EC 3.2.1.37).

As with xylan, mannan deconstruction also requires a collection of enzymes including β-mannanase (EC 3.2.178), β-mannosidase (EC 3.2.1.25), β-glucosidase (EC 3.2.1.21), acetyl mannan esterase (EC 3.1.1.6), and α-galactosidase (EC 3.2.1.22). β-Mannanase is an endo-active enzyme which catalyzes hydrolysis of internal β-1,4-glycosidic bonds in the mannan backbone in a similar fashion as does endoxylanase. β-Mannosidase and β-glucosidase are exo-acting enzyme and release mannose and glucose, respectively, from the non-reducing ends of mannan and mannooligosaccharides (Inoue et al. 2015; Lopes et al. 2018). In addition to above, debranching enzyme α-galactosidase cleaves α-1,6-linked D-galactopyranoside residues from galactomannan and galactoglucomannan while acetyl mannan esterase releases acetyl groups from galactoglucomannan.

2.3 Ligninases

Ligninases are a group of broad substrate specificity enzymes which are capable of oxidizing randomly joined phenylpropanoids units in the lignin polymer. The four major lignin decomposing enzymes are manganese peroxidase (EC 1.11.1.13), heme-containing lignin peroxidase (EC 1.11.1.14), versatile peroxidase (EC 1.11.1.16), and Cu-containing laccase (EC 1.10.3.2) (Jaramillo et al. 2015; Andlar et al. 2018). Lignin peroxidases (LiP) mediate oxidation of phenolic compounds via utilizing hydrogen peroxide, and manganese peroxidases (MnP) oxidize Mn^{2+} to Mn^{3+} which oxidizes phenolic structures into unstable free radicals. Versatile peroxidases (VP) combine the activities of LiP and manganese-dependent peroxidases.

The combinatorial characteristics of versatile peroxidases render them to oxidize $Mn^{2+}$and phenolic/non-phenolic substrates including veratryl alcohol, methoxybenzenes, and lignin model compounds (Andlar et al. 2018; www.cazy.org). Laccases catalyze oxidation of phenolic and aromatic amines by using molecular oxygen as a terminal electron acceptor (Jaramillo et al. 2015; Lopes et al. 2018). The other enzymes showing synergism in the lignin degradation are aryl alcohol oxidase (EC 1.1.3.7), glyoxylate oxidase (EC 1.2.3.5,), pyranose 2-oxidase (EC 1.1.3.4), and cellobiose/quinone oxidoreductases (EC 1.1.5.1).

2.4 Auxiliary Activity Proteins/Oxidative Enzymes

In addition to lignin enzymes, the other proteins classified in the auxiliary activity families also exhibit oxidative characteristic and boost the glycosyl hydrolase activity either by improving the accessibility of lignocellulosic material or by removing inhibitory products. The other most studied oxidative enzymes involved in the lignocellulolytic processes are cellobiose dehydrogenases (EC 1.1.99.18) and lytic polysaccharide monooxygenases (LPMOs). Cellobiose dehydrogenases classified in the subfamily 1 of auxiliary activity 3 family (AA3_1) oxidize cellobiose, lactose, and in some cases glucose with the reduction of cofactor flavin adenine mononucleotide. Thus, relief of cellulase inhibition occurs with the removal of cellobiose, a product inhibitor (Lopes et al. 2018). Initially, LPMOs were believed to act on crystalline substrates chitin and cellulose, but now it has been established that they act on a variety of substrates including starch, xyloglucan, cellodextrins, and glucomannan (Johansen 2016). LPMOs have been placed in the auxiliary activity families 9, 10, 11, 13, and 14 (Levasseur et al. 2013; Couturier et al. 2018). These auxiliary activity family proteins are Cu-dependent LPMOs which cleave crystalline polymeric substrates with oxidation of C-1 and C-4 (www.cazy.org). Therefore, the major role of auxiliary activity proteins is to enhance the amorphogenesis and facilitate the action of lignocellulolytic enzymes (Andlar et al. 2018).

2.5 Accessory Proteins

The accessory proteins have a potential to support the feasibility of lignocellulolytic process by enhancing the degradation of lignocellulosic materials. However, there is no evidence favoring the direct involvement of accessory proteins in the cleavage of glycosidic linkage either by hydrolysis or oxidative roles. It is believed that accessory proteins break intra- and intermolecular hydrogen bonds, thereby reducing the crystallinity and recalcitrance of the substrates, enhancing the substrate access for hydrolytic and oxidative enzymes (Quiroz-Castañeda et al. 2011; Lopes et al. 2018; Andlar et al. 2018). In plants, the loosening of lignocellulosic matrix and the enlargement of plant cell wall have been assigned to the action of expansins which break down the hydrogen bonds between cellulose microfibrils or between cellulose molecules of neighboring cells (Lopes et al. 2018). Though initially reported in plants, recent studies emphasize the presence of expansins in bacteria and expansin-like proteins in fungi Quiroz-Castañeda et al. 2011. These expansin-like proteins in conjunction with conventional cellulases have greatly improved the hydrolysis of cellulose even though without any proven record of hydrolytic activity, pointing a clue for their synergistic action (Payne et al. 2015; Cosgrove 2017). Similarly, swollenin proteins produced by fungi show sequence similarity to expansins, and thus similar function can be anticipated for them. Congruently, swollenins promoted swelling of the cellulose fibers resulting in the reduction of particle size and crystallinity (Kubicek et al. 2014; Lopes et al. 2018). Likewise, positive synergism has also been observed between swollenins and cellulases (Santos et al. 2017; Andlar et al. 2018). Loosenins are other expansin-type proteins which enhance cellulase complex activity by disrupting intramolecular hydrogen bonds (Quiroz-Castañeda et al. 2011).

Hydrolytic and oxidative enzymes as well as accessory proteins have been separately discussed ascribing to the degradation of different components of lignocellulosic biomass. However, consideration of the interplay between the activities of these enzymes is noteworthy. The actions of one category of enzymes augment the action of another category of enzymes or vice-versa for efficient decomposition of lignocellulosic biomass.

3 Reaction Mechanisms of Enzymes Involved in Lignocellulose Degradation

Based on the type of reaction catalyzed, lignocellulose-degrading enzymes can be put into three major classes, i.e., glycosyl hydrolases which strictly carry out hydrolysis of glycosidic bond via utilizing water molecule, oxidative enzymes break glycosidic linkage by oxidizing one of the carbons of linked monomers, and accessory proteins disrupt intra- and intermolecular hydrogen bonds. Here in the following sections, reaction mechanisms of glycosyl hydrolases and oxidative enzymes are described briefly.

3.1 Reaction Mechanisms of Glycosyl Hydrolases

The reaction mechanism of glycoside hydrolases (cellulases, hemicellulases, and pectinases) is conserved among all the classes. They catalyze the reaction either by retention or inversion mechanism, i.e., either retaining or inverting the configuration of the anomeric carbon of the hydrolyzed bond (Biely et al. 1994; Davies and Henrissat 1995; Collins et al. 2005). Both types of reaction mechanisms utilize general acid-base catalysis with the help of carboxylate pair at the enzyme's active site. Either in retaining or inverting enzymes, one residue of the carboxylate pair localized at the distance of hydrogen bonding with glycosidic oxygen, acts as a general acid, and protonates oxygen of the O-glycosidic bond. However, another residue of carboxylate pair is located in the close vicinity of the sugar anomeric carbon in retaining enzymes which acts as nucleophile/base. This base is more distantly localized in inverting enzymes, and thus, a water molecule can be accommodated between the sugar anomeric carbon and the base. Briefly, the active site of retaining enzymes is constituted by two glutamic acid residues separated from each other with a distance of approximately 5.5 Å. However, carboxylate pair (glutamate and aspartate) of inverting enzymes' active site is approximately 7.5–10 Å apart from each other (Collins et al. 2005; Bhardwaj et al. 2012; Gupta et al. 2013). This distance creates micro-space to admit a water molecule between two catalytic residues in inverting enzymes. Mechanistically, retaining enzymes follow double displacement mechanism for the cleavage of glycosidic bond. In the process of double displacement, a covalent glycosyl enzyme intermediate is formed which is subsequently hydrolyzed by oxocarbenium-ion-like transition states (Rye and Withers 2000). The first step of the reaction is the protonation of the glycosidic oxygen via one of the carboxylate pair residues. Simultaneously, another residue attacks nucleophilically at the anomeric carbon which results in the removal of leaving group and formation of the α-glycosyl enzyme intermediate. In the second step, the first carboxylate residue (which protonated the glycosidic oxygen in the first step) acts as a general base and withdraws a proton from water molecule, thus activating water molecule for nucleophilic attack at the anomeric carbon. Overall step two events result in a second substitution in which the anomeric carbon forms oxocarbenium-ion-like transition state and is subsequently hydrolyzed to a β-configuration end product. Therefore, the ultimate effect is the retention of the anomeric carbon configuration as of the substrate. In contrary to the above, inverting enzymes perform hydrolysis via single displacement mechanism in which one of the carboxylate groups acting as general acid provides proton for the leaving group departure and another accepts a proton from the water molecule, thus activating the nucleophilic attack of the water molecule at anomeric carbon. It produces oxocarbenium-ion-like transition state, which is subsequently hydrolyzed into α-configuration end product (inversion β to α) (Collins et al. 2005).

3.2 Reaction Mechanism of Ligninases

As stated earlier, lignin is the most recalcitrant degraded by the different enzymes with varying specificities. The common feature of all enzymes is the utilization of oxidative reaction mechanism. Lignin decomposing peroxidase (LiP, MnP, and VP) enzymes follow general mechanism of horseradish peroxidases. The only difference is in their high oxidation potential. The large information has accumulated from several studies on the structure, reaction mechanisms, and kinetics of ligninases (Hammel et al. 1985; Glenn et al. 1986; Siddiqui et al. 2014; Lin et al. 2018; Fan et al. 2019) and reviewed by different authors which serve as the source of compiled information (Kirk and Farrel 1987; Martinez 2002; Kersten and Cullen 2007; Wong 2009; Bouacem et al. 2018). In the following section, we discuss reaction mechanisms of four types of lignin decomposing enzymes.

3.2.1 Reaction Mechanism of Lignin Peroxidases (LiPs)

LiPs are H_2O_2 requiring glycoproteins which contain one mole of iron protoporphyrin IX per protein molecule. The heme is sandwiched between proximal (C-terminal) and distal domains (N-terminal) and accessible from the solvent via two small channels. The heme iron is predominantly pentacoordinated at the distal side with histidine (Edwards et al. 1993; Claus 2004). The catalytic cycle starts with two electron oxidation of the ferric enzyme [Fe(III)] leading to the formation of intermediate compound I (LiP-I, ferry iron porphyrin radical cation). The electrons are accepted by $H_2O_2$and heterolytic cleavage of O-O of peroxide substrate is facilitated by coordination of heme Fe(III) and protonation by distal histidine 47. The leaving H_2O molecule promotes transient formation of Fe(III)-O^+, followed by electron transfer from Fe(III) and porphyrin π electron system to form ferry iron porphyrin radical cation. Thereafter, one electron reduction of LiP-I by different phenolic and aromatic non-phenolic substrates takes place in two successive steps (Renganathan and Gold 1986; Wong 2009). The first step of reduction of LiP-I by substrate forms compound II [Fe(IV)=O.porphyrin, LiP-II]. This compound II is converted into resting enzyme at the second successive step of reduction by another molecule of substrate. In some cases, LiP-I may also be directly renewed to the resting state via one equivalent of two electron substrate such as veratryl alcohol. In the reduction of LiP-I, susceptible aromatic nuclei are oxidized by one electron producing unstable cation radicals (Renganathan and Gold 1986; Lundell et al. 1993; Wong 2009). These cation radicals undergo a variety of non-enzymatic reactions leading to side chain cleavage, demethylation, intramolecular addition, rearrangement, and depolymerization of lignin molecules. Thus, the formation of radical cations is enzyme catalyzed, and subsequent steps are non-enzymatic which may take place spontaneously depending on the substitution of aromatic ring.

LiP-II may also be converted into inactive form III [LiP-III, Fe(III) O_2^-] in the presence of excess H_2O_2 and in the absence of reducing substrates. This inactive form of enzyme (LiP-III) is reactivated either spontaneously or by veratryl alcohol radical cation with the displacement of superoxide (Wong 2009). It has been observed that lignin peroxidase producing organism (*Phanerochaete chrysosporium*) also produces veratryl alcohol (Kirk and Farrel 1987). This veratryl alcohol is oxidized to veratryl alcohol radical cation during the reduction steps of LiP-I and LiP-II to the native form enzyme. Therefore, it may be an adaptive strategy of the lignin degrading organisms to overcome enzyme inactivation which has also been supported by the findings of increase in enzyme activity with the extracellular addition of veratryl alcohol(Harvey et al. 1986; Martinez 2002). Another suggested role of synthesized veratryl alcohol radical cation is redox mediator in the oxidative degradation of lignin (Kersten and Cullen 2007). It has been proposed that instead of direct lignin oxidation by LiP (large molecular weight enzyme, unable to diffuse into the vicinity of lignin molecules), veratryl alcohol radical cation (small molecule) diffuses into the interior of lignin and acts as direct oxidant for lignin degradation.

$$\text{LiP}\left[\text{Fe}(\text{III}).\text{porphyrin}\right] + \text{H}_2\text{O}_2 \rightarrow \text{LiP-I}\left[\text{Fe}(\text{IV}) = \text{O.porphyrin}^{\bullet+}\right] + \text{H}_2\text{O} \quad (6.1)$$

$$\text{LiP-I}\left[\text{Fe}(\text{IV}) = \text{O.porphyrin}^{\bullet+}\right] + \text{Subtrate} \rightarrow \text{LiP-II}\left[\text{Fe}(\text{IV}) = \text{O.porphyrin}\right] + \text{Free radical cation} \quad (6.2)$$

$$\text{LiP-II}\left[\text{Fe}(\text{IV}) = \text{O.porphyrin}\right] + \text{Substrate} \rightarrow \text{LiP}\left[\text{Fe}(\text{III}).\text{porphyrin}\right] + \text{H}_2\text{O} + \text{Free radical cation} \quad (6.3)$$

3.2.2 Reaction Mechanism of Manganese Peroxidases (MnP)

MnP contain heme and manganese [Mn(II)] ion as their integral structural components, and they are H_2O_2 requiring obligatory manganese oxidizing (Mn^{2+} to Mn^{3+}) enzymes (Glenn et al. 1986). MnP is similar to LiP, consisting of heme sandwiched between proximal and distal domains. Likewise, LiP heme iron is mainly pentacoordinated with distal histidine ligand. The Mn-binding site consists of three acidic amino acid residues (Glu35, Glu39, and Asp179) and Mn coordinates with carbonyl oxygen of these residues, heme propionate oxygen, and two other water oxygens (Sundaramoorthy et al. 1997). The catalytic cycle involves two electron oxidation of MnP into MnP-I in the presence of H_2O_2 leading to the heterolytic cleavage of O-O peroxide substrate. These two electrons are transferred from heme-porphyrin component of the enzyme in the same manner as with LiP. Transient-state kinetics has demonstrated that the reduction of MnP-I into MnP-II uses either Mn^{2+} or a variety of phenolic compounds as reducing substrate (Kersten and Cullen 2007; Wong 2009). However, the reduction of MnP-II into the native form enzyme is carried out only by Mn^{2+} demonstrating its unique characteristic of Mn^{2+} dependency (Martinez 2002). The overall process leads to the formation of Mn^{3+} ions which are potent oxidizers and can oxidize different types of phenolic compounds including lignin model compounds. However, Mn^{3+} is quite unstable in the aqueous solution, and its stabilization is achieved with chelating molecules (oxalic and other organic acids) secreted by the fungi. The chelated Mn^{3+} diffuses in the interior of lignocellulosic material and causes one electron oxidation of phenolic lignin constituents to produce unstable radicals (Kuan et al. 1993). These unstable radicals undergo rearrangement, bond cleavage, and non-enzymatic degradation to yield different products. The Mn^{3+} also produces thiol and lipid-derived free radicals in the presence of glutathione and lipids, respectively. The synthesized thiol and lipid-derived free radicals abstract hydrogen from non-phenolic substrates yielding benzylic radical (Kersten and Cullen 2007). The formed benzyl radicals yield different products by non-enzymatic reactions.

Similar to LiP, MnP-II can also be converted into an inactive Fe(III)-superoxo form (MnP-III) in the presence of excess H_2O_2. The reactivation uses Mn^{3+} to act either as oxidizing agent of iron coordinated superoxide or catalase type to remove H_2O_2. Therefore, the reactivation of MnP-III is different from LiP-III where it uses aromatic radical cation.

$$\text{MnP}\left[\text{Fe}(\text{III}).\text{porphyrin}\right] + H_2O_2 \rightarrow \text{MnP-I}\left[\text{Fe}(\text{IV}) = O^{\bullet+}\right] + H_2O \quad (6.4)$$

$$\text{MnP-I}\left[\text{Fe}(\text{IV}) = O^{\cdot+}\right] + Mn^{2+} \rightarrow \text{MnP-II}\left[\text{Fe}(\text{IV}) = \text{O.porphyrin}\right] + Mn^{3+} \quad (6.5)$$

$$\begin{aligned}&\text{MnP-II}\left[\text{Fe}(\text{IV}) = \text{O.porphyrin}\right] + Mn^{2+} \rightarrow \text{MnP}\left[\text{Fe}(\text{III}).\text{porphyrin}\right] \\ &\quad + Mn^{3+} + H_2O\end{aligned} \quad (6.6)$$

$$Mn^{3+} + \text{Phenolic substrate} \rightarrow Mn^{2+} + \text{Phenolic substrate radical cation} + H_2O \quad (6.7)$$

3.2.3 Reaction Mechanism of Versatile Peroxidases (VP)

The versatile peroxidases share the characteristics of LiP and MnP. They oxidize Mn^{2+} into Mn^{3+} as well as phenolic and non-phenolic substrates (veratryl alcohol, *p*-dimethoxybenzene, and lignin model dimer veratrylglycerol-β-guaiacyl ether). Owing to its catalytic properties, enzyme molecule contains Mn^{2+} binding site (formed by acidic residues Glu/Glu/Asp) similar to MnP and other amino residues which are involved in the interaction of phenolic and non-phenolic substrates (Kersten and Cullen 2007; Wong 2009; Siddiqui et al. 2014). The reaction of mechanism of versatile peroxidase is similar to that of LiP and MnP. The reaction is triggered in the presence of H_2O_2 involving the formation of two electron oxidized compound I and one electron oxidized compound II intermediates (Martinez 2002; Wong 2009). Compound I intermediate is reduced to compound II by one electron oxidation of either Mn^{2+} or phenolic/nonphenolic substrates depending on the availability; subsequently compound II is converted into native form enzyme in another round of one electron oxidation of the oxidizable substrates.

In the above sections, we had seen that all the lignin decomposing peroxidase enzymes (LiP, MnP, and VP) require H_2O_2. This necessity has prompted the research for the discovery of H_2O_2 producing enzymes. Soon, the observation of glucose oxidase (glucose-1-oxidase and pyranose 2-oxidase) activity in the ligninolytic mycelia was attributed to the primary source of H_2O_2 (Kelley and Reddy 1986; Daniel et al. 1994). Additionally, several putative aryl alcohol oxidases (structurally related to glucose oxidase) have been implicated in the extracellular production of H_2O_2 to start the redox reaction. The exploration of other possibilities revealed the involvement of fatty acyl-CoA oxidase in the washed and starved mycelia in the absence of substrates. Glyoxylate oxidase activity observed in the ligninolytic cultures may also be responsible for H_2O_2 production by oxidizing several simple aldehydes, α-hydroxy carbonyl and dicarbonyl compounds (Kersten and Cullen 2007). The most important implication in the start of a catalytic cycle has been evidenced with H_2O_2 production by MnP which oxidizes glutathione, dithiothreitol, and dihydroxymaleic acid with the coupled reduction of O_2 to H_2O_2 (Kirk and Farrel 1987; Wong 2009).

3.2.4 Reaction Mechanism of Laccases

Laccases are blue multicopper oxidases structurally characterized by the presence of three distinct domains, each of which displays Greek key β-barrel architecture known as cupredoxin-like domains (Mot and Silaghi-Dumitrescu 2012). The active site is composed of three types of four copper centers per molecule, i.e., one type 1, one type 2, and two type 3. Type 1 (T1) copper is characterized by paramagnetic behavior and strong electronic absorption at 614 nm due to covalent bond between copper and cysteine and thus imparts the blue color. Type 2 (T2) copper is also paramagnetic but shows no absorbance in the visible range. Type 3 (T3) copper center is a pair of two copper ions designated as T3α and T3β which gives a weak absorbance at 330 nm and shows no EPR signal. Thus, the copper pair of T3 center is linked by anti-ferromagnetic coupling and maintained by a hydroxyl bridge (Claus 2004; Sitarz et al. 2016). T2 is strategically positioned close to T3 and forms a T2/T3 trinuclear cluster consisting of three of the four copper ions in three reaction nuclei sites. T1 is located in domain 3 with copper positioned in a shallow depression on the surface of the enzyme, whereas T2/T3 trinuclear cluster is located at the interface between domains 1 and 3. T1 copper coordinates with two histidines and one cysteine as conserved equatorial ligands. Additionally, the axial position contains non-coordinated phenylalanine or leucine and isoleucine residues in the vicinity of T1 copper center. The presence of these non-coordination ligands in fungal laccases has been considered as an advantage in determining the higher redox potential of T1 copper site which is approximately +790 mV (Claus 2004; Wong 2009; Sitarz et al. 2016). Due to high redox potential, T1 copper is the primary electron acceptor site where successively four one electron oxidations of substrate take place. T2 copper coordinates with two histidines and one oxygen ligand (hydroxide or water). Each of the T3 copper ions symmetrically coordinated with three histidines and a bridging oxygen (either hydroxyl or O^{2-}). T2/T3 trinuclear cluster is the active center for the reduction of molecular oxygen into water. Therefore, four successive electrons abstracted from substrate at T1 site need to be transported to T2/T3 trinuclear site. However, T1 and T2/T3 trinuclear sites are 13Å distant from each other. Thus, a conserved tripeptide (His-Cys-His) is involved in the electron transfer pathway between these sites (Mot and Silaghi-Dumitrescu 2012). Additionally, two conserved amino acid residues are present in the close proximity of the catalytic pocket (deprotonated glutamic acid close to T3 and protonated aspartic acid close to T2).

The resting form of the enzyme is fully oxidized, and all the four copper centers are in their (++) oxidation state. The initial phase of the catalysis takes place at T1 center with the abstraction of one electron from the reducing substrate. Successively four electrons are abstracted and transferred to T2/T3 trinuclear cluster forming fully reduced form (active form) enzyme (Solomon et al. 2008; Wong 2009). The reduction of the T3 copper centers results in the weakening of the hydroxide bridge between T3α and T3β. Thus, the hydroxide is being protonated from the bulk solvent and released as water molecule (Claus 2004). Simultaneously, catalytic site is vacant to accept molecular oxygen; however, before the binding of the molecular oxygen to T2/T3 site, glutamic acid needs to be protonated from the bulk solution and aspartic acid deprotonated to lose its H^+ to the hydroxide bound with T2 (forming water molecule). Thereafter, dioxygen docks into the catalytic cavity and preferentially binds with T3βCu^+ due to its trigonal pyramidal geometry (Augustine et al. 2010; Sitarz et al. 2016). The bound oxygen is reduced by two electrons with one electron transfer from each T3βCu^+ and T2Cu^+, forming peroxide intermediate. Still the two electron reduced dioxygen is tightly bound to the trinuclear cluster by coordinating with T3βCu^{2+}, T3αCu^+, and T2Cu^{2+}. The next step is the rapid one electron transfer from T1Cu^+ to T2$Cu^{2+}$with the help of T1-Cys-His-T3β pathway (Piontek et al. 2002). Thus, T2Cu^{2+} is reduced to T2Cu^+ which loosens the bond with peroxide. Thereafter, two electrons, one from T2Cu^+ and another from T3αCu^+, are transferred to the antibonding orbitals of peroxide. It increases the energy and destabilizes the peroxide structure, leading to the breakage of peroxide (O-O) bond (Sitarz et al. 2016). This step is facilitated by concurrent one proton transfer to peroxide from glutamic acid and protonation of aspartate by abstracting proton from water molecule bound with T2Cu^+. The overall stoichiometry of this step is the production of fully oxidized native intermediate state of the active site in which two oxides bridge in the catalytic cavity. The native intermediate form is converted to either fully reduced active form or fully oxidized resting form containing one oxide bridge (Solomon et al. 2008; Sitarz et al. 2016). It depends on the availability of reducing substrate. In the presence of enough reducing substrate, the native intermediate is converted to fully reduced form of enzyme by accepting four electrons and four protons. Three of these four protons are directly transferred to oxide bridges releasing two water molecules and one protonated glutamate; at the same time, one proton from aspartic acid is transferred to OH bonded with T2 copper center. Thus, it produces the fully active form of the enzyme ready for another cycle of reaction (Wong 2009). However, in the lack of sufficient substrate, two protons are accepted by one of the oxide bridges of native intermediate, and one water molecule is released leading to the formation of fully oxidized resting form enzyme. The next cycle will start when substrate is available with the abstraction of four successive electrons, protonation of oxide bridge, and release of water.

3.3 Reaction Mechanism of Auxiliary Activity Family LPMO Enzymes

LPMOs are single copper containing enzymes which boost lignocellulose depolymerization by oxidative cleavage of polysaccharide chains. Highly diverse LPMOs acting on different substrates exist in the microbial world. These were classified into four auxiliary activity protein families of carbohydrate active enzyme (CAZy) database: (1) AA9 family contains cellulose and hemicellulose active fungal LPMOs, (2) AA10 contains cellulose and chitin-active bacterial LPMOs, (3) AA11 contains chitin-active fungal LPMOs, (4) AA13 contains starch-active fungal LPMOs, and (5) AA14 is a newly discovered family which contains xylan degrading LPMOs with C-1 oxidation from *Pycnoporus coccineus* (Levasseur et al. 2013; Couturier et al. 2018). Further, they have been categorized into three groups based on catalyzing the oxidation of specific carbons in the backbone. For example, type 1 and 2 LPMOs oxidize C-1 and C-4, respectively, whereas type 3 LPMOs are mixed nature and can oxidize both C-1 and C-4 (Beeson et al. 2015). Further, low sequence identity in the LPMO family members prompts diverse catalytic mechanisms. Even for the specific enzyme molecule, there may be more than one route of catalysis (Hedegard and Ryde 2018). Therefore, it is difficult to propose a unified mechanism of catalytic action, and mechanistic details of LPMOs are not fully known. However, a considerable research progressed in the understanding of fungal LPMOs reaction mechanisms in recent years. Here, we summarize achievements in the understanding of cellulose-active fungal LPMOs with a few structural considerations.

Initially LPMOs were supposed glycosyl hydrolases and wrongly placed in GH61 family of CAZy database. The first greatest advancement came from X-ray crystallographic structure of cellulose-active *Tr*Cel61B which revealed the presence of the active site on flat surface with lack of carboxylate residues (necessary for hydrolase action). This finding provoked scientific community that it may utilize different mechanism for glycosidic bond cleavage (Karkehabadi et al. 2008); thereafter considering other research findings also, these enzymes were placed in the separate family of auxiliary activity proteins (Levasseur et al. 2013). The flat surface exposed active site contains copper which is essential for the catalytic activity (Quinlan et al. 2011; Phillips et al. 2011). Usually this copper is in the Cu(II) oxidation state in the resting enzymes and coordinated with three ligands provided by two histidines (histidine brace moiety) as well as tyrosine and two water molecules. The N-terminal histidine residue coordinates with side chain $N^{\delta 1}$ and backbone amino N atoms while another histidine coordinates with $N^{\varepsilon 2}$ of the side chain only (Quinlan et al. 2011). Tyrosine and one water molecule are coordinated at the axial positions opposite to each other, whereas another water molecule is equatorial and trans to amino nitrogen of N-terminal histidine (Kjaergaard et al. 2014). The plane of the active site is made of loops L2, LC, and LS which exhibit plasticity and probably involved in the binding of substrate/reductant (Beeson et al. 2015). Specifically loop regions LC and LS have been predicted to form CDH binding groove (Li et al.

2012). The resting form of the enzyme can bind polysaccharide with the displacement of axial water; however, the catalytic cascade starts with the reduction of Cu(II) to Cu(I) via external reducing agent. The reduced and activated LPMO-Cu(I) follows either of the two routes; it may bind molecular oxygen or hydrogen peroxide (Bertini et al. 2018; Hedegard and Ryde 2018). Firstly consider the binding of oxygen, and the bound oxygen is protonated by His147 located in the vicinity of active site. The next step proceeds through one electron reduction and protonation resulting in the elimination of water molecule. Thereafter, the next cycle of reduction and proton transfer from His147 takes place into the formation of Cu(I)-oxyl and Cu(II)-hydroxyl species. These Cu-oxyl and hydroxyl species (another route) can also be generated by binding of H_2O_2 as co-substrate in which His147 proximates H_2O_2 to Cu(I) before cleavage of O-O bond. The Cu(II)-hydroxyl abstracts a hydrogen atom from C-H of glycosidic linkage and with His147 assistance forms Cu(I)-hydroxyl which is converted into product involving transition from triplet spin-state to a closed-shell singlet species (Hedegard and Ryde 2018).

3.4 *Reaction Mechanism of Cellobiose Dehydrogenase (CDH)*

The diverse roles of CDH have been postulated. It may either directly provide reducing power to LPMOs or reduce oxygen to generate hydroxyl-free radical and hydrogen peroxide (Beeson et al. 2015). The produced hydrogen peroxide may act as co-substrate for LPMOs or boost the action of other H_2O_2 requiring lignocellulose depolymerizing enzymes. On the other hand, hydroxyl-free radicals can diffuse in the internal matrix of recalcitrant lignocellulosic biomass and cause depolymerization (Kersten and Cullen 2007).

CDH is exactly flavocytochrome CDH which consists of a catalytically functional dehydrogenase domain connected to *b*-type cytochrome domain by linker peptide. Dehydrogenase domain is composed of FAD and substrate binding subdomains. The active center of dehydrogenase domain is structurally similar with the active site of cholesterol oxidase and contains conserved His and Asn residue in close proximity of flavin N5 atom (Hallberg et al. 2003). The overall reaction mechanism involves hydride transfer to flavin N5 and lactone substrate formation. It is accomplished with the help of His689 which will act as general base and abstract a proton from C1-OH; it follows a cascade of reactions involving nucleophilic attack at N5, rearrangement, and elimination. The reduced flavin moiety can either donate the reducing equivalence to LPMOs or ferric ions via heme prosthetic group of the CDH. Thus, it participates in the regeneration of ferrous ions which will produce hydroxyl radicals (·OH) by Fenton chemistry reaction denoted below (Kersten and Cullen 2007).

$$H_2O_2 + Fe(II) + H+ \rightarrow H_2O + Fe(III) + \cdot OH$$

4 Recent Advancements in the Production of Lignocellulose-Degrading Enzymes

The major obstacle in the commercialization of lignocellulose-derived biofuel and other products is the cost of enzymes. It necessitates 10–100 times reduction of enzyme costs (Binod et al. 2019). Therefore, researchers continuously take up the challenges to cut down the production costs of lignocellulose-degrading enzymes by searching novel producer organisms as well as manipulating organisms or enzymes employing modern approaches such as genetic engineering, metabolic engineering, and strain improvement. In this section, we will summarize some of these recent developments in the production of lignocellulose deconstructing enzymes.

4.1 Recent Developments in the Production of Cellulose-Degrading Enzymes

Cellulose being the major component of lignocellulosic waste biomass, much attention has been paid for the cost-effective production of cellulases by using different fungal organisms isolated from diverse habitats, cheap substrates, screening of inducers as well as employing genetic modification approaches. Here we discuss some examples of different strategies being used for the cost reduction of the cellulases production. Endophytic fungi are proven rich sources of secondary metabolites, but they are less explored for the production of industrial enzymes. In the recent years, trend has changed towards the utilization of endophytic fungi for the production of lignocellulose decomposing enzymes. In a recent screening of 14 endophytic fungal strains, endoglucanase producing potential of *Cladosporium cladosporioides* PAJ 03, *Phomopsis stipata* SC 04, *Trichoderma viride* PAJ 01, and *Botryosphaeria* sp. AM 01 was revealed on a mixture of sugarcane bagasse and wheat bran. Higher activities of endoglucanase and β-glucosidase were secreted by *Botryosphaeria* sp. AM01 and *Saccharicola* sp. EJC 04 on cottonseed meal and wheat bran mixture. A blend of the enzymatic extracts from these organisms efficiently saccharified pretreated sugarcane bagasse (Marques et al. 2018). Among the 41 endophytic fungal strains screened from mangrove tropical forests, *A. niger* and *A. awamori* exhibited higher production of endoglucanase and β-glucosidase with an evidence of twofold more hydrolysis of cellulosic materials than reference *A. niger* and *T. reesei* organisms (Maroldi et al. 2018). Screening of common indoor fungi demonstrated the potential of *Cladosporium sphaerospermum* and *Penicillium chrysogenum* to produce cellulase, β-glucosidase, mannase, β-galactanase, and arabinanase. Thus, these organisms can be used as over-producers to supplement industrial enzyme blends for boosting lignocellulose degradation (Andersen et al. 2016). Prajapati et al. (2018) isolated *Aspergillus tubingensis* from dung compost which emerged high producer of cellulase. The enzyme preparation effectively degraded sugarcane bagasse and rice straw releasing the polymer constituent units. Mohapatra et al. (2018) isolated high cellulase producing *Aspergillus fumigatus* from the soil containing decomposed lignocellulosic waste. A mixture of equal proportions of partially purified CMCase and FPase enzymes from crude extract of *A. fumigatus* efficiently hydrolyzed lignocellulose with a release of 396 mg/g reducing sugars from *Pennisetum* biomass. An isolate (*T. asperellum*) collected from the African jungle soil produced higher activities of cellulase, β-glucosidase, and CMCase. It has also emerged excellent degrader of natural waste biomass; however, the degradation efficiency was higher for pre-treated biomass (Wang et al. 2017). *Penicillium subrubescens* FBCC1632, a new potential producer of hemicellulase, has been isolated which is capable of degrading crude lignocellulose feedstock, as well as polysaccharides and metabolizing their monomeric components. The cellulolytic activities of *P. subrubescens* FBCC1632 on different plant biomass substrates were comparable to those of *A. niger* (Makela et al. 2016). Okereke et al. (2017) has purified and characterized acidophilic cellulase from *P. ostreatus* and suggested it would be a promising renewable source of cellulase for biomass degradation. Han et al. (2017) utilized repeated fed-batch strategy to increase cellulase production by *Penicillium oxalicum* under submerged fermentation condition. Among the different strategies, mutagenesis is also the easiest way of strain improvement. Hyper cellulase producing *T. reesei* D-7 mutant was obtained by treating *T. reesei* Rut-C30 strains with ethyl methyl sulfonate (EMS) and plasma irradiation. The obtained mutant strain responds well to corn starch hydrolysate soluble inducer and insoluble cellulose inducer (Zhang et al. 2017a). Textile industries discharge huge amount of waste water rich in cotton fibers and polyester. Utilization of this waste is an alternative of cheap cellulase production by *A. niger* (Hu et al. 2018). A large amount of domestic waste water is also generated continuously which could serve as a cheap source of medium and utilization of raw domestic waste water as base medium resulted in the 1.4-fold enhancement of filter paper cellulase activity in comparison with synthetic medium (Libardi et al. 2017). A bamboo shoot shell pretreated with *Pleurotus ostreatus* promoted cellulase synthesis in the cultures of *A. niger*. Interestingly, analysis of the pretreated shell revealed that the inducer is modified soluble lignin polymer and not a traditional disaccharide (Zhang et al. 2017b). The supplementation of duckweed 50 g L^{-1} in the shake flasks of *T. reesei* culture equally induces cellulase production as avicel; however, the hydrolysis rate of steam-exploded cornstalk by duckweed induced cellulase is 28% higher than avicel. The combined application of both inducers enhanced cellulase production and hydrolysis rates, 39% and 36%, respectively (Li et al. 2019). One of the major challenge in the decomposition of lignocellulosic biomass is the inactivation of enzymes by the phenolic components. The cellulolytic activities of enzyme extracts prepared from *Chrysoporthe cubensis* and *Penicillium pinophilum* have much more tolerance to phenolic inhibitors than the enzyme extracts of *T. reesei* and *A. niger*. *T. reesei and A. niger* cellulolytic activities were inactivated at 0.3 and 1.5 mg phenol mg^{-1} protein, whereas 1:1 mixture of extracts from *C. cubensis* and *P. pinophilum* resisted inactivation at 35 mg phenols mg^{-1} protein (Azar et al. 2017). Thus inclusion of enzyme preparations from these organisms will reduce the enzyme loadings for the hydrolysis of alkali pretreated lignocellulosic substrates. Genetic regulation tunes up the utilization of resources for the synthesis of each and every protein in the cell; cellulases are not exception to this concept. Replacement of transcription repressor gene *ace1* with the coding region of endoglucanase gene *egl1* in *T. reesei* Rut-C30 has enhanced 90 and 132% total cellulolytic and endoglucanase activities, respectively, under flask cultures; however, cellulase activity enhancement remains 74% than wild-type organism in the 5 L fermenter (Meng et al. 2018). A widely used cellulase producer *T. reesei* has been engineered with an overexpression of *Trivb-1* gene encoding putative transcription factor which has increased 200% cellulase production by the recombinant strain. The crude extract from the recombinant strain released 40% more glucose than wild-type counterpart from the pretreated corn stover (Zhang et al. 2017c). *P. oxalicum* is a promising alternative of *T. reesei*, and its cellulolytic potential has further been enhanced by overexpression of nine genes encoding endoglucanase. All the overexpressing strains displayed higher β-glucosidase activities, and this activity reached 65-fold specifically with the use of inducible promoter of *bgl2* gene. The overexpressing mutant has substantially reduced enzyme loads for the saccharification of delignified natural lignocellulose materials (Yao et al. 2016). A bi-functional enzyme produced by the fusion of endoglucanase gene from *Teleogryllus emma* and xylanase gene from *Thermomyces lanuginosus* has great potential for lignocellulose degradation. It displayed mild reaction pH (5.5), temperature (50 °C) and long half-life for cellulase and xylanase activities. The supplementation of this bi-functional enzyme with commercial preparation Ctec2 has increased 10–20% release of fermentable sugars from the pretreated rice straw (Chen et al. 2018). The cloning and expression of an endoglucanase gene from *Aspergillus terreus* in *Aspergillus nidulans* resulted in the high yield secretion of the enzyme. The purified enzyme showed optimum activity at pH 5.0 and temperature 55 °C, and the best thing is the non-specificity of the substrates. The enzyme preparation displayed high activity against β-glucan and xyloglucan while moderate activity against glucomannan and carboxymethyl cellulose (Segato et al. 2017). The expression and characterization of two endoglucanases from diverging subfamilies of GH45 demonstrated differences in their substrate specificities and thus they can be used as new tools in the enzyme cocktails for cellulose hydrolysis (Berto et al. 2019) (Table 6.1).

4.2 Recent Developments in the Production of Hemicellulolytic Enzymes

Hemicellulose is also a major constituent of plant biomass after cellulose, and efforts are underway to achieve the cost-effective production of enzymes for its degradation into fermentable sugars. Isolation and characterization of new organisms producing hemicellulose degradation enzymes is one of the strategy. A qualitative screening

Table 6.1 Some examples of recent developments in the production of cellulases

S. no.	Enzyme name	Producer organism/ expression host	Uniqueness of the produced enzyme/ developed system	Type of modification/ new isolate	References
1.	Endoglucanase	*Cladosporium cladosporioides* PAJ 03, *Phomopsis stipata* SC 04, *Trichoderma viride* PAJ 01, *Botryosphaeria* sp. AM 01	Higher titers of enzyme production on lignocellulosic substrates	Unexplored endophytic fungi	Marques et al. (2018)
2.	β-Glucosidase	*Saccharicola* sp. EJC 04, *Paecilomyces* sp. SF 021, *Ustilaginoidea* sp. CV 04, *Ustilaginoidea* sp. XYA 04	High activity of the enzymes produced in solid state fermentation	Unexplored endophytic fungi assessed	Marques et al. (2018)
3.	Endoglucanase and β-glucosidase	*Aspergillus niger* and *Aspergillus awamori*	Higher saccharification performance than reference organisms	Endophytic isolates from mangrove plants	Maroldi et al. (2018)
4.	Cellulase and β-glucosidase	*Cladosporium sphaerospermum* and *Penicillium chrysogenum*	–	Screening of novel producer organisms	Andersen et al. (2016)
5.	Cellulase and β-glucosidase	*Aspergillus tubingensis* NKBP-55	High titers of enzyme production in SSF	Screening of new isolate from dung compost sample	Prajapati et al. (2018)
6.	Cellulase	*Aspergillus fumigatus*	High producer organism isolated from decomposed lignocellulosic waste soil	New isolate	Mohapatra et al. (2018)
7.	Cellulase and β-glucosidase	*T. asperellum* T-1	High yield of enzyme activities and excellent waste biomass degrading abilities	The producer organism was isolated from African jungle soil	Wang et al. (2017)
8.	Endoglucanase	*Penicillium subrubescens* FBCC1632	–	A new potential producer of (hemi)-cellulases	Makela et al. (2016)
9.	Cellulase	*Pleurotus ostreatus*	Acidophilic cellulase	–	Okereke et al. (2017)

(continued)

Table 6.1 (continued)

S. no.	Enzyme name	Producer organism/ expression host	Uniqueness of the produced enzyme/ developed system	Type of modification/ new isolate	References
10.	Cellulase	*Penicillium oxalicum* RE-10	Repeated fed-batch fermentation strategy	–	Han et al. (2017)
11.	Cellulase	T. reesei D-7	High cellulose producing strain	Ethyl methyl sulfonate and plasma irradiation mutagenesis of *T. reesei* Rut-C30	Zhang et al. (2017a)
12.	Cellulase	*Aspergillus niger*	Utilization of textile waste for enzyme production	–	Hu et al. (2018)
13.	Cellulase	*Trichoderma harzianum* HBA03	Raw domestic waste water used as base culture medium	–	Libardi et al. (2017)
14.	Cellulase	*Aspergillus niger*	A new type of soluble inducer of cellulase activity discovered which is a modified lignin polymer	Bamboo shoot shell pretreated with *Pleurotus ostreatus*	Zhang et al. (2017b)
15.	Cellulase	*Trichoderma reesei* Rut C30	The cellulase production and hydrolysis of substrates increased	Duckweed combined with avicel as an inducer	Li et al. (2019)
16.	Cellulase	*Chrysoporthe cubensis* and *Penicillium pinophilum*	Enzyme mixture is tolerant to high levels of monocomponent phenolic inhibitors	The enzyme extracts prepared from *Chrysoporthe cubensis* and *Penicillium pinophilum* were mixed in equal proportions	Azar et al. (2017)

(continued)

Table 6.1 (continued)

S. no.	Enzyme name	Producer organism/ expression host	Uniqueness of the produced enzyme/ developed system	Type of modification/ new isolate	References
17.	Endoglucanase	*T. reesei* QS305	Endoglucanase and total cellulase activity enhanced 90% and 132%, respectively	Transcription repressor gene *ace1* of *T. reesei* Rut-C30 was replaced with *egl1* gene encoding endoglucanase	Meng et al. (2018)
18.	Cellulase	*T. reesei* Vib-1	Cellulase production enhanced 200% by the recombinant strain	The gene *Trvib-1* encoding putative transcription factor was overexpressed in *T. reesei* Rut-C30	Zhang et al. (2017c)
19.	β-Glucosidase	*Penicillium oxalicum*	Enhanced saccharification	Overexpression of β-glucosidase encoding genes	Yao et al. (2016)
20.	Bi-functional fusion enzyme EG-M-Xyn	*Yarrowia lipolytica*	High potential in improving the enzymatic hydrolysis of lignocellulosic materials	Endoglucanase gene from *Teleogryllus emma* fused with xylanase gene from *Thermomyces lanuginosus*	Chen et al. (2018)
21.	Endoglucanase	*Aspergillus terreus*/*Aspergillus nidulans*	High yield secretion of enzyme	Non-specific endoglucanase from *Aspergillus terreus*	Segato et al. (2017)
22.	Endoglucanase	*Gloeophyllum trabeum*/*Aspergillus nidulans*	Stable at 70 °C for 24 h and have useful properties to be used in enzyme cocktail preparation	Heterologous expression	Berto et al. (2019)
23.	Endoglucanase	*Myceliophthora thermophila*/*Aspergillus nidulans*	Stable at 70 °C for 24 h and have useful properties to be used in enzyme cocktail preparation	Heterologous expression	Berto et al. (2019)

showed the potential of edible fungus *Tuber maculatum* as a new source of xylanase production which may be comparable to xylanases from other sources (Bedade et al. 2017). Among the 14 endophytic fungal strains, *P. stipata* SC 04 and *Botryosphaeria* sp. AM 01 stood best producers of xylanase and β-xylosidase, respectively. These organisms produced 69433 and 487 U/g xylanase and β-xylosidase enzymes on 1:1 ratio of mixed sugarcane bagasse and wheat bran substrate (Marques et al. 2018). During the screening of 15 potential xylanase producers, *A. niger* being the highest producer secreted 456 U/g in SSF on sugarcane bagasse and palm kernel cake supplemented with salt solution. Authors claimed 1.5-fold higher production of xylanase in comparison to the reported literature (Lee et al. 2018). Mangrove endophytes *A. niger* and *A. awamori* have also produced higher activities of xylanase in addition to cellulase and released more glucose from the cellulosic materials than the reference organisms (Maroldi et al. 2018). Makela et al. (2016) isolated and screened endoxylanase producing *P. subrubescens* FBCC1632 as potential producer of hemicellulases. Similarly, *T. asperellum* T-1 isolated from African jungle soil produced high yield of xylanase which possesses excellent waste biomass degrading abilities (Wang et al. 2017). *A. tubingensis* NKBP-55 isolated from dung compost produced endoxylanase, endo-mannanase, and α-galactosidase. The enzyme extract from the organism efficiently degraded sugarcane bagasse and rice straw releasing xylose, glucose, and cellobiose (Prajapati et al. 2018). A new isolate of *Penicillium chrysogenum* F-15 has also produced high activities of xylanase which showed higher potential of agro-industrial waste biomass degradation than commercial xylan, thus indicating its usefulness in food, feed, pulp, and paper industries (Terrone et al. 2018). A comparison of the biochemical properties of *Penicillium roqueforti* ATCC xylanase with commercial xylanase (Sigma) produced by *Thermomyces lanuginosus* revealed superiority of the enzyme from former organism (Souza et al. 2018). Kumar and Shukla (2018) optimized xylanase production from *Thermomyces lanuginosus* at the pilot scale, and characterization of the purified enzyme demonstrated high thermostability in the presence of polyols. A novel endoxylanase produced *A. tamarii* Kita has useful characteristics such as maximum activity at 60 °C and pH 5.5 (activity over broad pH 4.0–9.0) for its application in the food, feed, pharmaceutical, and paper industries (Heinen et al. 2018). Likewise, Ye et al. (2017) produced and characterized novel xylose and alkali-tolerant β-xylosidase from *P. oxalicum* with a suggestion that enzyme extract from this organism may promote hemicellulose degradation. Thermophilic fungus isolate *Thermoascus aurantiacus* has produced novel type of acidophilic and thermotolerant xylanase. The purified xylanase exhibited maximum activity at 75 °C and pH 5.0; it was also stable over a broad pH and temperature range (Ping et al. 2018). Purification and characterization of novel endoxylanase from non-toxic *A. flavus* demonstrated unique enzymatic properties, such as excellent stability at 50 °C, pH 3.5–10.5, and tolerance to various proteases, which may be suitable for biotechnological potential (Chen et al. 2019). Hemicellulose contains different types of side chain groups which may hinder complete degradation. Andersen et al. (2016) isolated common indoor fungi *Cladosporium sphaerospermum* and *Penicillium chrysogenum* which produces β-galactanase, mannanase, and arabinase. The supplementation of these enzymes may boost the bioconversion of lignocellulosic biomass.

The release of phenolic inhibitors from lignocellulose degradation makes bioconversion process cumbersome, and it necessitates enzymes which are tolerant to these inhibitors. Xylanases produced by *Chrysoporthe cubensis* and *Penicillium pinophilum* were tolerant to phenolic inhibitors in the presence of laccase; however, their activity declined twofold in the absence of laccase (Azar et al. 2017). A novel extremophilic xylanase produced by *Aureobasidium pullulans* NRRL Y-2311-1 has improved dough rheology and bread quality. Remarkably, bread specific volume increased 30% more than the commercial counterpart being used (Yegin et al. 2018). It can also be a good source of xylanase supplementation in the bioconversion of lignocellulose. Genetic modification of the existing organisms is also foreseen as feasible approach for the higher and efficient production of lignocellulose-degrading enzymes. Genetic engineering of hyper cellulase producing *T. reesei*; replacement of cis-acting promoter region of *xyn3* with cis-acting region of *xyn1* exhibited improved inductivity against cellulosic carbon over wild-type promoter and acquired inductivity against xylanosic carbon. Further expression of β-glycosidase gene of *A. aculeatus* under the control of *xyn3* chimeric promoter enhanced saccharification ability through increased cellobiase activity. Thus, *xyn3* chimeric promoter is very efficient in the expression of carbohydrolase genes and may reduce the production cost of the industrial enzymes (Hirasawa et al. 2018). Heterologous expression and characterization of endoxylanase and β-xylosidase from *P. ostreatus* and *Irpex lacteus*, respectively, revealed novel enzymes with synergistic hydrolysis properties. Both the recombinant enzymes showed synergistic hydrolysis on oat spelts xylan and alkali pretreated cornstalk reaching a value of 2.26-fold degree of synergy (Zhuo et al. 2018). Recently, a novel GH30 family endoxylanase has been expressed in heterologous host, and characterization revealed that it is active on arabinoxylan and glucuronoxylan. Its unique characteristic is the release of xylobiose from xylooligosaccharide while other GH30 enzymes mainly release xylose (Espinoza and Eyzaguirre 2018). The production of xylanase and arabinofuranosidase from genetically engineered *A. nidulans* has been optimized on soybean fibers as substrate, and subsequently the enzyme extract applied for the hydrolysis of soybean fiber to obtain xylooligosaccharides (Pereira et al. 2018). The xylooligosaccharide yield reached 28% mass fraction of xylan (Table 6.2).

4.3 Recent Developments in the Production of Lignin-Degrading Enzymes

Lignin is the most recalcitrant portion of lignocellulosic biomass, and efficient decomposition of lignin increases the accessibility of enzymes for hydrolysis of other polysaccharides. Therefore, production of lignin-degrading and inhibitory phenolic compound modifying enzymes at low cost will further decrease the bioconversion cost of plant biomass by reduced loading of other carbohydrases. Lignin is chiefly degraded by LiP, MnP, VP, and laccases; hereunder we discuss briefly

Table 6.2 Some examples of recent developments in the production of hemicellulases

S. no.	Enzyme name	Producer organism/ expression host	Uniqueness of the produced enzyme/ developed system	Type of modification/ new isolate	References
1.	Xylanase	*Tuber maculatum*	–	New producer organism	Bedade et al. (2017)
2.	Endoxylanase	*P. stipata* SC 04	High production of enzyme in solid state fermentation	Unexplored endophytic fungi used	Marques et al. (2018)
3.	β-Xylosidase	*Botryosphaeria* sp. AM 01	High production of enzyme in solid state fermentation	Unexplored endophytic fungi used	Marques et al. (2018)
4.	Xylanase	*Aspergillus niger*	High yield of the enzyme in SSF	A new isolate	Lee et al. (2018)
5.	Xylanase	*Aspergillus niger* and *Aspergillus awamori*	Higher saccharification performance than reference organisms	Endophytic isolates from mangrove plants	Maroldi et al. (2018)
6.	Endoxylanase	*Penicillium subrubescens* FBCC1632	–	A new potential producer of (hemi)-cellulases	Makela et al. (2016)
7.	Xylanase	*T. asperellum* T-1	High yield of enzyme activities and excellent waste biomass degrading abilities	Isolate from African jungle soil	Wang et al. (2017)
8.	Endoxylanase and α-galactosidase	*Aspergillus tubingensis* NKBP-55	High titers of enzyme production in SSF	Screening of new isolate from dung compost sample	Prajapati et al. (2018)
9.	Xylanase	*Penicillium chrysogenum* F-15	High activity of the enzyme on agroindustry biomass	New soil isolate	Terrone et al. (2018)
10.	Xylanase	*Penicillium roqueforti* ATCC 10,110	Produced enzyme is superior than *Thermomyces lanuginosus* commercial preparation available from Sigma	–	Souza et al. (2018)

(continued)

Table 6.2 (continued)

S. no.	Enzyme name	Producer organism/ expression host	Uniqueness of the produced enzyme/ developed system	Type of modification/ new isolate	References
11.	Xylanase	*Thermomyces lanuginosus*	Enhanced thermostability with supplementation of polyols	A soil isolate	Kumar and Shukla (2018)
12.	Endoxylanase	*Aspergillus tamarii* Kita	A novel endoxylanase with useful physicochemical properties for application in food, feed, pharmaceutical, and paper industries	–	Heinen et al. (2018)
13.	β-Xylosidase	*P. oxalicum*	A novel xylose and alkali-tolerant enzyme with high catalytic activity	–	Ye et al. (2017)
14.	Xylanase	*Thermoascus aurantiacus* M-2	Novel acidophilic and thermostable enzyme	A new isolate	Ping et al. (2018)
15.	Endoxylanase	*Aspergillus flavus*	A new xylanase with excellent stability	–	Chen et al. (2019)
16.	β-Galactanase, mannanase, and arabinase	*Cladosporium sphaerospermum* and *Penicillium chrysogenum*	–	Screening of novel producer organisms	Andersen et al. (2016)
17.	Xylanase	*Chrysoporthe cubensis* and *Penicillium pinophilum*	Tolerance to phenolic inhibitors	Presence of laccase enhanced two fold tolerance	Azar et al. (2017)
18.	Xylanase	*Aureobasidium pullulans* NRRL Y-2311-1	A novel extremophilic xylanase	–	Yegin et al. (2018)
19.	Xylanase	*Trichoderma reesei*	The produced enzyme cocktail showed high saccharification efficiency	Chimeric promoter designed	Hirasawa et al. (2018)
20.	Endoxylanase	*Pleurotus ostreatus*	Novel enzyme	Heterologous expression	Zhuo et al. (2018)
21.	β-Xylosidase	*Irpex lacteus*	Novel enzyme	Heterologous expression	Zhuo et al. (2018)

(continued)

Table 6.2 (continued)

S. no.	Enzyme name	Producer organism/ expression host	Uniqueness of the produced enzyme/ developed system	Type of modification/ new isolate	References
22.	Xylanase	*Penicillium purpurogenum*	Novel xylanase of family GH30	Heterologous expression	Espinoza and Eyzaguirre (2018)
23.	Endoxylanase	*A. nidulans*	Increased xylolytic activity	Genetic modification of *A. nidulans* with *xynC* gene from *P. funiculosum*	Pereira et al. (2018)
24.	Arabinofuranosidase	*A. nidulans*	Increased xylolytic activity	Genetic modification of *A. nidulans* with *abfB* from *A. niger*	Pereira et al. (2018)

recent advancements in the production and characterization of these enzymes. A new selectively lignin degrading white rot fungus *Echinodontium taxodii* produced novel type MnP on natural lignocellulosic material of moso bamboo. The enzyme showed optimum activity at pH 3.5 and 55 °C. The produced MnP oxidizes both phenolic and non-phenolic lignin units, and degradation efficiency increased when combined with laccase from the same producer organism (Kong et al. 2016). Makela et al. (2013) produced lignin-modifying enzymes in the semi-solid cultures with milled grey alder, Norway spruce, and charcoal. Production of LiP and MnP significantly enhanced with the supplementation of wood (milled grey alder) under low nitrogen; however, optimum laccase production has only been obtained in the presence of 1.5 mM Cu^{2+} with different nitrogen sources. Production of lignin-modifying enzymes and removal of phenolic content assessment for seven white rot fungi revealed the high potential of *Cerrena unicolor* in submerged fermentation of olive mill by-products. *C. unicolor* produced laccase and MnP in the fermentation of olive pomace while only laccase was produced on olive tree sawdust. Interestingly, both laccase and MnP were produced from 50% diluted olive mill wastewater, and their production was further enhanced with the addition of Cu^{2+} and Mn^{2+}. The produced laccase has reduced 60–70% total phenols of olive pomace phenolics (Elisashvili et al. 2018). The assessment of different additives to enhance the production of ligninolytic enzymes by *Coriolus versicolor* on sweet sorghum bagasse demonstrated stimulatory effect of metallic salts and phenolic compounds. Specifically syringic acid increased LiP and laccase, gallic acid increased MnP, and Cu^{2+} increased laccase production, and the combinations of these supplements synergistically enhanced laccase, MnP, and LiP production. These enzymes had increased 2.43-fold release of fermentable sugars from pretreated sweet sorghum bagasse (Mishra et al. 2017). The optimum production of ligninolytic enzymes by *Schizophyllum commune* on rice straw was obtained at 35 °C, pH 5.0, C:N ratio

20:1, and in the presence of mediator $MnSO_4$. The crude extract has removed lignin 47–72% from different agro-industrial residues which enhanced 47% cellulose hydrolysis of untreated biomass by the crude cellulase extract from *T. harzianum* (Asgher et al. 2016). White rot fungus *Inonotus obliquus* degraded 72%, 39%, and 47% lignin from wheat straw, rice straw, and corn stover, respectively. MnP and LiP production was very high on wheat straw, and their respective production reached maximum after 2 and 4 days (Xu et al. 2017). A dye decolorizing fungus *Bjerkandera adusta* produced novel type LiP and MnP which have higher catalytic efficiencies (Bouacem et al. 2018). Consideration of ligninolytic enzyme production from cheap substrate utilized potato peel as sole nutrient source. *Trametes hirsuta* grown on potato peel produced high yield of laccase which has the ability of 87% anthraquinone dye decolorization (Schalchli et al. 2017). In another study, *P. ostreatus* produced much higher yield of laccase and MnP on dry potato peel pretreated with distilled water (Ergun and Urek 2017). Although fungi are heterotrophic, different wavelength light radiation can induce cascade of reactions which may influence the production of specific enzyme. White rot fungus *P. chrysosporium* exposed to various wavelength radiation produced LiP maximally under green light in the submerged fermentation (Ramírez et al. 2010). In addition to the conventional methods, genetic engineering is also an accurate and effective approach of attaining targeted yield. Therefore, a hyper lignin degrading *Phanerochaete sordida* has been genetically engineered for the expression of MnP under the promoter control of a highly expressed protein of the same organism (Sugiura et al. 2012). *Ceriporiopsis subvermispora* produces all known types of MnP, and their utilization is hampered due to lack of efficient preparation methods. The soluble expression of *C. subvermispora* MnP in *E. coli* may offset the problem. The efforts have resulted in the high yield of MnP under the control of cold shock promoter (Lin et al. 2018). As the name denotes, VPs are much more efficient than any other peroxidases, and their production seems fruitful for the efficient lignin degradation. VP from *Physisporinus vitreus* has been overexpressed in *E. coli* in inclusion bodies and refolded in vitro which has displayed efficient catalysis (Liu et al. 2019). Free hydroxyl-phenolic compounds are also known inhibitors of lignin peroxidases and hinder the efficient ligninolysis. However, these inhibitory phenolic compounds may be modified into non-inhibitory compounds by enzymes. One of the enzymes, *O*-methyltransferase, discovered and characterized from *P. chrysosporium*, methylated free hydroxyl-phenolic compounds at para and meta positions, rendering them non-toxic to the lignin-degrading enzymes (Mai and Kim 2016) (Table 6.3).

4.4 Recent Developments in the Production and Characterization of LPMOs

LPMOs are a recently discovered class of enzymes which plays an important role in the enhancement of lignocellulolytic processes. In spite of their less proportional contribution in the cleavage of glycosidic linkage, LPMO activity is significant due

Table 6.3 Some examples of recent developments in the production of lignin deconstruction enzymes

S. no.	Name of lignin-degrading enzyme	Producer organism/ expression host	Uniqueness of the produced enzyme/developed system	Type of modification/new isolate	References
1.	MnP	*Echinodontium taxodii* 2538	A novel enzyme oxidizes phenolic and non-phenolic lignin units	New selective lignin degrading white rot fungus	Kong et al. (2016)
	LiP, MnP, and laccase	*Phlebia radiata*	Production of the enzymes optimized with the supplementation of copper, nitrogen, and wood as nutrients	–	Makela et al. (2013)
	LiP and laccase	*Cerrena unicolor*	Fermentation of olive mill by-products for enzyme production	–	Elisashvili et al. (2018)
2.	LiP, MnP, and laccase	*Coriolus versicolor*	Synergistic action of the produced enzymes enhanced 2.43-fold lignin degradation	Addition of metallic salts and phenolic compounds	Mishra et al. (2017)
3.	MnP, LiP, and laccase	*Schizophyllum commune*	Delignification of agro-industrial residues	–	Asgher et al. (2016)
	MnP and LiP	*Inonotus obliquus*	High activities of LiP and MnP were observed in wheat straw culture	–	Xu et al. (2017)
	LiP and MnP	*Bjerkandera adusta*	Novel peroxidases from dye decolorizing fungus	–	Bouacem et al. (2018)
	Laccase	*Trametes hirsuta*	Potato peels used as sole nutrient source	–	Schalchli et al. (2017)
	MnP and laccase	*Pleurotus ostreatus*	Potato peel waste used as substrate support		Ergun and Urek (2017)
	LiP	*Phanerochaete chrysosporium*	Illumination with green light significantly enhanced enzyme production	–	Ramírez et al. (2010)
	MnP	*Phanerochaete sordida*	High yield of MnP activities	Promoter of a highly expressed protein used	Sugiura et al. (2012)
	MnP	*Ceriporiopsis subvermispora/E. coli*	Soluble expression of active enzyme in *E. coli*	Expression of MnP genes under control of a cold shock promoter	Lin et al. (2018)
	VP	*Physisporinus vitreus PF18 E. coli*	Efficient enzyme for biocatalysis and biodegradation	Expressed in *E. coli* as inclusion body and then refolded in vitro	Liu et al. (2019)
	O-methyltransferase	*Phanerochaete chrysosporium/E. coli*	Methylates para and meta positions of lignin peroxidase inhibitory free hydroxyl-phenolic compounds	Heterologous expression	Mai and Kim (2016)

to their action on the crystalline regions of the polysaccharides. Presently attention is being paid to these enzymes, and still their production and characterization need to be elucidated in detail. A very few studies have been undertaken for the enhanced production of LPMOs than other enzymes involved in the lignocellulose deconstruction. Most of the studies have either just evaluated the boost of lignocellulose hydrolysis with LPMO augmentation or overexpressed the enzyme in heterologous host for structural determination and understanding of catalytic mechanisms. However, one of the necessary components for feasible application in industry is the cheap availability of the enzyme. Attempts need to be made for optimization of their production on different substrates by various isolates because LPMOs are differentially regulated by multiple substrates. AA9 family LPMOs of *A. nidulans* are differentially regulated by xylan, xyloglucan, glucan, pectin, and cellulose (Jagadeeswaran et al. 2016). *A. nidulans* LPMO overexpressed in the same organism showed activity against hemicellulose xyloglucan. Thermophilic fungus *Myceliophthora thermophila* secretes over a dozen of different LPMOs. One of the enzymes from this organism expressed in *A. nidulans* and named *Mt*LPMO9J is a single domain and shares 63% sequence similarity with the catalytic domain of *Nc*LPMO9C. The enzyme is stable over pH 5.0 and temperature up to 50 °C and active against cellulose, soluble cello-oligosaccharides, and xyloglucan (Kadowaki et al. 2017). Two LPMOs of mangrove-associated fungus *Pestalotiopsis* sp. were expressed in yeast, and their heterologous production has reached up to 260 mg/L in the bioreactor. They had showed enhanced stability and activity in the presence of sea salt (Patel et al. 2016). White rot fungus *Heterobasidion irregulare* contains ten genes putatively encoding AA9 LPMOs. Two of these (*Pi*LPMO9H and *Pi*LPMO9I) were expressed in *P. pastoris* and characterized. *Pi*LPMO9H breaks glycosidic linkage of cellulose by C1 oxidation while *Pi*LPMO9I cleaves linkage with strict oxidation at C4 carbon of glucose unit and is also active against glucomannan (Liu et al. 2017). One of the limitations of heterologous production is the N-terminal processing and generation of catalytically active enzyme. Tanghe et al. (2015) tried several secretion signals and found that inclusion of native secretion signal is best. It has enabled correct processing of the N-terminal and high yield of the protein in the heterologous host *P. pastoris*. Recently Bulakhov et al. (2017) expressed β-glucosidase of *A. niger* and LPMO of *T. reesei* under the control of inducible glucoamylase promoter *gla1* in *Penicillium verruculosum*. The heterologous LPMO expression preparations boosted 10–43% lignocellulose hydrolysis (Table 6.4).

5 Concluding Remarks

The cost reductions in the production of bio-based valuable products are seen on the availability of low priced lignocellulolytic processes, which can only be sustained with the higher yields of efficient enzymes per capita following different strategies such as isolation and screening of high producer organisms, modification of producer organisms by classical and site-directed mutagenesis targeting the regulatory

Table 6.4 Some examples of recent developments in the production and characterization of LPMOs

S. no.	Producer organism/ expression host	Uniqueness of the produced enzyme/ developed system	Type of modification/new isolate	References
1.	*Myceliophthora thermophila/ Aspergillus nidulans*	Single domain LPMO with 63% similarity to the catalytic domain of *Nc*LPMO9C	Heterologous overexpression	Kadowaki et al. (2017)
2.	*Pestalotiopsis* sp. Nci6/*Pichia pastoris*	Salt tolerant	Heterologous expression	Patel et al. (2016)
3.	*Heterobasidion irregulare/P. pastoris*	–	Heterologous expression	Liu et al. (2017)
4.	*T. reesei/ P. pastoris*	Use of native secretion signal	Heterologous expression	Tanghe et al. (2015)
5.	*Penicillium verruculosum*	The enzyme preparation proved to be more effective in the hydrolysis of lignocellulosic substrates	*T. reesei* LMPO gene expressed under the control of inducible glucoamylase promoter (*gla1*)	Bulakhov et al. (2017)

components, screening of novel and efficient inducers, and metabolic and genetic engineering. In addition to traditional glycoside hydrolases and lignin-degrading enzymes, inculcation of inhibitory compound modifying enzymes, auxiliary activity oxidative enzymes, and crystallinity reducing accessory proteins in the enzyme cocktails can be helpful to make the process economical. Although recently the reaction mechanism of auxiliary activity family 9 LPMOs has been discovered, they follow more than one route of catalysis. It would be fruitful to decipher the efficient route of catalysis, and thereafter they can be engineered for unified mode of efficient catalysis. Simultaneously, much attention needs to be paid for the optimization of enhanced yields of the enzymes from the cheapest substrates. In this regard, production of LPMOs and accessory proteins can be prioritized.

References

Andersen B, Poulsen R, Hansen GH (2016) Cellulolytic and xylanolytic activities of common indoor fungi. Int Biodeterior Biodegrad 107:111–116

Andlar M, Rezić T, Mardetko N, Kracher D, Ludwig R, Šante KB (2018) Lignocellulose degradation: an overview of fungi and fungal enzymes involved in lignocelluloses degradation. Eng Life Sci 18:768–778

Araki Y, Karita S, Tsuchiya T, Kondo M, Goto M (2010) Family 17 and 28 carbohydrate binding modules discriminated different cell-wall sites in sweet potato roots. Biosci Biotechnol Biochem 74:802–805

Arevalo-Gallegos A, Ahmad Z, Asgher M, Parra-Saldivar R, Iqbal HMN (2017) Lignocellulose: a sustainable material to produce value-added products with a zero waste approach—a review. Int J Biol Macromol 99:308–318

Asgher M, Wahab A, Bilal M, Muhammad H, Iqbal N (2016) Lignocellulose degradation and production of lignin modifying enzymes by *Schizophyllum commune* IBL-06 in solid-state fermentation. Biocatal Agric Biotechnol 6:195–201

Augustine AJ, Kjaergaard C, Qayyum M, Ziegler L, Kosman DJ, Hodgson KO, Hedman B, Solomon EI (2010) Systematic perturbation of the trinuclear copper cluster in the multicopper oxidases: the role of active site asymmetry in its reduction of O_2 to H_2O. J Am Chem Soc 132:6057–6067

Azar RISL, Morgan T, dos Santos ACF, Ximenes EA, Ladisch MR, Guimaraes VM (2017) Deactivation and activation of lignocellulose degrading enzymes in the presence of laccase. Enzym Microb Technol 109:25–30

Bedade D, Berezina O, Singhal R, Deska J, Shamekh S (2017) Extracellular xylanase production from a new xylanase producer *Tuber maculatum* mycelium under submerged fermentation and its characterization. Biocatal Agric Biotechnol 11:288–293

Beeson WT, Vu VV, Span EA, Phillips CM, Marletta MA (2015) Cellulose degradation by polysaccharide monooxygenases. Annu Rev Biochem 84:923–946

Behera BC, Sethi BK, Mishra RR, Dutta SK, Thatoi HN (2017) Microbial cellulases—diversity & biotechnology with reference to mangrove environment: a review. J Genet Eng Biotechnol 15:197–210

Bertini L, Breglia R, Lambrughi M, Fantucci P, Gioia LD, Borsari M, Sola M, Bortolotti CA, Bruschi M (2018) Catalytic mechanism of fungal lytic polysaccharide monooxygenases investigated by first-principles calculations. Inorg Chem 57:86–97

Berto GL, Velasco J, Ribeiro CTC, Zanphorlin LM, Domingues MN, Murakami MT, Polikarpov I, de Oliveira LC, Ferraz A, Segato F (2019) Functional characterization and comparative analysis of two heterologous endoglucanases from diverging subfamilies of glycosyl hydrolase family 45. Enzym Microb Technol 120:23–35

Bhardwaj A, Mahanta P, Ramakumar S, Ghosh A, Leelavathi S, Reddy VS (2012) Emerging role of N- and C-terminal interactions in stabilizing (β/α)8 fold with special emphasis on Family 10 xylanases. Comput Struct Biotechnol J 2:e201209014

Biely P, Krernnicky L, Alföldi J, Tenkanen M (1994) Stereochemistry of the hydrolysis of glycosidic linkage by endo-β-1,4-xylanases of *Trichoderma reesei*. FEBS Lett 356:137–140

Bilal M, Asgher M, Iqbal HMN, Hu H, Zhang X (2017) Biotransformation of lignocellulosic materials into value-added products—a review. Int J Biol Macromol 98:447–458

Binod P, Gnansounou E, Sindhu R, Pandey A (2019) Enzymes for second generation biofuels: recent developments and future perspectives. Bioresour Technol Rep 5:317–325. https://doi.org/10.1016/j.biteb.2018.06.005

Bornscheuer U, Buchholz K, Seibel J (2014) Enzymatic degradation of (ligno)cellulose. Angew Chem Int Ed 53:2–20

Bouacem K, Rekik H, Jaouadi NZ, Zenati B, Kourdali S, El Hattab M, Badis A, Annane R, Bejar S, Hacene H, Bouanane-Darenfed A, Jaouadi B (2018) Purification and characterization of two novel peroxidases from the dye-decolorizing fungus *Bjerkandera adusta* strain CX-9. Int J Biol Macromol 106:636–646

Bulakhov AG, Volkov PV, Rozhkova AM, Gusakov AV, Nemashkalov VA, Satrutdinov AD, Sinitsyn AP (2017) Using an inducible promoter of a gene encoding *Penicillium verruculosum* glucoamylase for production of enzyme preparations with enhanced cellulose performance. PLoS One 12(1):e0170404. https://doi.org/10.1371/journal.pone.0170404

Chandel AK, Garlapati VK, Singh AK, Antunes FAF, da Silva SS (2018) The path forward for lignocellulose biorefineries: bottlenecks, solutions, and perspective on commercialization. Bioresour Technol 264:370–381

Chen H (2014) Biological fundamentals for the biotechnology of lignocellulose. In: Biotechnology of lignocellulose: theory and practice. Chemical Industry Press, Beijing, pp 73–141

Chen CC, Gao GJ, Kao AL, Tsai ZC (2018) Bi-functional fusion enzyme EG-M-Xyn displaying endoglucanase and xylanase activities and its utility in improving lignocellulose degradation. Int J Biol Macromol 111:722–729

Chen Z, Zaky AA, Liu Y, Chen Y, Liu L, Li S, Jia Y (2019) Purification and characterization of a new xylanase with excellent stability from *Aspergillus flavus* and its application in hydrolyzing pretreated corncobs. Protein Expr Purif 154:91–97

Claus H (2004) Laccases: structure, reactions, distribution. Micron 35:93–96

Collins T, Gerday C, Feller G (2005) Xylanases, xylanase families and extremophilic xylanases. FEMS Microbiol Rev 29:3–23

Cosgrove DJ (2017) Microbial expansins. Annu Rev Microbiol 71:479–497

Couturier M, Ladeveze S, Sulzenbacher G, Ciano L, Fanuel M, Moreau C, Villares A, Cathala B, Chaspoul F, Frandsen KE, Labourel A, Herpoel-Gimbert I, Grisel S, Haon M, Lenfant N, Rogniaux H, Ropartz D, Davies GJ, Rosso MN, Walton PH, Henrissat B, Berrin JG (2018) Lytic xylan oxidases from wood-decay fungi unlock biomass degradation. Nat Chem Biol 14:306–310

Cragg SM, Beckham GT, Bruce NC, Bugg TDH, Distel DL, Dupree P, Etxabe AG, Goodell BS, Jellison J, McGeehan JE, McQueen-Mason SJ, Schnorr K, Walton PH, Watts JEM, Zimmer M (2015) Lignocellulose degradation mechanisms across the tree of life. Curr Opin Chem Biol 29:108–119

Daniel G, Volc J, Kubatova E (1994) Pyranose oxidase, a major source of H_2O_2 during wood degradation by *Phanerochaete chrysosporium, Trametes versicolor*, and *Oudemansiella mucida.* Appl Environ Microbiol 60:2524–2532

Davies G, Henrissat B (1995) Structures and mechanisms of glycosyl hydrolases. Structure 3:853–859

de Jong E (2011) Bio-based chemicals: value added products from biorefineries. Rep Prep IEA Bioenergy Task 36. http://www.ieabioenergy.com/wp-content/uploads/2013/10/Task-42-Biobased-Chemicals-value-added-products-from-biorefineries.pdf

Edwards SL, Raag R, Wariishi H, Gold MH, Poulos TL (1993) Crystal structure of lignin peroxidase. Proc Natl Acad Sci U S A 90:750–754

Elisashvili V, Kachlishvili E, Asatiani MD (2018) Efficient production of lignin-modifying enzymes and phenolics removal in submerged fermentation of olive mill by-products by white-rot basidiomycetes. Int Biodeterior Biodegradation 134:39–47

Ergun SO, Urek RO (2017) Production of ligninolytic enzymes by solid state fermentation using *Pleurotus ostreatus*. Ann Agric Sci 15:273–277

Espinoza K, Eyzaguirre J (2018) Identification, heterologous expression and characterization of a novel glycoside hydrolase family 30 xylanase from the fungus *Penicillium purpurogenum.* Carbohydr Res 468:45–50

Fan Y, Zhang Z, Wang F, Li J, Hu K, Du Z (2019) Lignin degradation in corn stover catalyzed by lignin peroxidase from *Aspergillus oryzae* broth: effects of conditions on the kinetics. Renew Energy 130:32–40

Galbe M, Zacchi G (2002) A review of the production of ethanol from softwood. Appl Microbiol Biotechnol 59:618–628

Glenn JK, Akileswaran L, Gold MH (1986) Mn(II) oxidation is the principal function of the extracellular Mn-peroxidase from *Phanerochaete chrysosporium.* Arch Biochem Biophys 251:688–696

Gong X, Gruninger RJ, Forster RJ, Teather RM, McAllister TA (2013) Biochemical analysis of a highly specific, pH stable xylanase gene identified from a bovine rumen-derived metagenomic library. Appl Microbiol Biotechnol 97:2423–2431

Guerriero G, Hausman JF, Strauss J, Ertan H, Siddiqui KS (2016) Lignocellulosic biomass: biosynthesis, degradation, and industrial utilization. Eng Life Sci 16:1–16

Gupta R, Mehta G, Deswal D, Sharma S, Jain KK, Singh A, Kuhad RC (2013) Cellulases and their biotechnological applications. In: Kuhad RC, Singh A (eds) Biotechnology for environmental management and resource recovery. I.K. International Pvt. Ltd., New Delhi, pp 89–106

Hallberg BM, Henriksson G, Pettersson G, Vasella A, Dvine C (2003) Mechanism of the reductive half-reaction in cellobiose dehydrogenase. J Biol Chem 278:7160–7166

Hammel KE, Tien M, Kalyanaraman B, Kirk TK (1985) Mechanism of oxidative Cα-Cβ cleavage of a lignin model dimer by *Phanerochaete chrysosporium* ligninase. J Biol Chem 260:8348–8353

Han X, Song W, Liu G, Li Z, Yang P, Qu Y (2017) Improving cellulase productivity of *Penicillium oxalicum* RE-10 by repeated fed-batch fermentation strategy. Bioresour Technol 227:155–163

Harvey P, Schoemaker HE, Palmer JM (1986) Veratryl alcohol as a mediator and the role of radical cations in lignin biodegradation by *Phanerochaete chrysosporium*. FEBS Lett 95:242–246

Hedegard ED, Ryde U (2018) Molecular mechanism of lytic polysaccharide monooxygenases. Chem Sci 9:3866–3880

Heinen PR, Bauermeister A, Ribeiro LF, Messias JM, Almeida PZ, Moraes LAB, Vargas-Rechia CG, de Oliveira AHC, Ward RJ, Filho EXF, Kadowaki MK, Jorge JA, Polizeli MLTM (2018) GH11 xylanase from *Aspergillus tamarii* Kita: purification by one-step chromatography and xylooligosaccharides hydrolysis monitored in real-time by mass spectrometry. Int J Biol Macromol 108:291–299

Hirasawa H, Shioya K, Furukawa T, Tani S, Sumitani J, Kawaguchi T, Morikawa Y, Shida Y, Ogasawara W (2018) Engineering of the *Trichoderma reesei* xylanase 3 promoter for efficient enzyme expression. Appl Microbiol Biotechnol 102:2737–2752

Horn SJ, Vaaje-Kolstad G, Westereng B, Eijsink VGH (2012) Novel enzymes for the degradation of cellulose. Biotechnol Biofuels 5:45. https://doi.org/10.1186/1754-6834-5-45

Hu Y, Du C, Nattha P, Lin CSK (2018) Optimisation of fungal cellulase production from textile waste using experimental design. Process Saf Environ Prot 118:133–142

Inoue H, Yano S, Sawayama S (2015) Effect of β-mannanase and β-mannosidase supplementation on the total hydrolysis of softwood polysaccharides by the *Talaromyces cellulolyticus* cellulase system. Appl Biochem Biotechnol 176:1673–1686

Jagadeeswaran G, Gainey L, Prade R, Mort AJ (2016) A family of AA9 lytic polysaccharide monooxygenases in *Aspergillus nidulans* is differentially regulated by multiple substrates and at least one is active on cellulose and xyloglucan. Appl Microbiol Biotechnol 100:4535–4547

Jaramillo PMD, Gomes HAR, Monclaro AV, Silva COG, Filho EXF (2015) Lignocellulose degrading enzymes: an overview of the global market. In: Gupta VK, Mach RL, Sreenivasaprasad S (eds) Fungal biomolecules: sources, applications and recent developments, 1st edn. Wiley, Hoboken, pp 75–85

Johansen KS (2016) Lytic polysaccharide monooxygenases: the microbial power tool for lignocellulose degradation. Trends Plant Sci 21:926–236

Kadowaki MAS, VaÂrnai A, Jameson JK, Leite AET, Costa-Filho AJ, Kumagai PS, Prade RA, Polikarpov I, Eijsink VGH (2017) Functional characterization of a lytic polysaccharide monooxygenase from the thermophilic fungus *Myceliophthora thermophile*. PLoS One 13(8):e0202148. https://doi.org/10.1371/journal.pone.0202148

Karkehabadi S, Hansson H, Kim S, Piens K, Mitchinson C, Sandgren M (2008) The first structure of a glycoside hydrolase family 61 member, Cel61B from *Hypocrea jecorina*, at 1.6 Å resolution. J Mol Biol 383:144–154

Kelley RL, Reddy CA (1986) Identification of glucose oxidase activity as the primary source of hydrogen peroxide production in ligninolytic cultures of *Phanerochaete chrysosporium*. Arch Microbiol 144:248–253

Kersten P, Cullen D (2007) Extracellular oxidative systems of the lignin-degrading Basidiomycete *Phanerochaete chrysosporium*. Fungal Genet Biol 44:77–87

Kim DY, Shin DH, Jung S, Lee JS, Cho HY, Bae KS, Sung CK, Rhee YH, Son KH, Park HY (2014) Biocatalytic properties and substrate-binding ability of a modular GH10 β-1, 4-xylanase from an insect-symbiotic bacterium, *Streptomyces mexicanus* HY-14. J Microbiol 52:863–870

Kirk TK, Farrel RL (1987) Enzymatic combustion: the microbial degradation of lignin. Annu Rev Microbiol 41:465–501

Kjaergaard CH, Qayyum MF, Wong SD, Xu F, Hemsworth GR, Walton DJ, Young NA, Davies GJ, Walton PH, Johansen KS, Hodgson KO, Hedman B, Solomon EI (2014) Spectroscopic and computational insight into the activation of O_2 by the mononuclear Cu center in polysaccharide monooxygenases. Proc Natl Acad Sci U S A 111:8797–8802

Kong W, Chen H, Lyu S, Ma F, Yu H, Zhang X (2016) Characterization of a novel manganese peroxidase from white-rot fungus *Echinodontium taxodii* 2538, and its use for the degradation of lignin-related compounds. Process Biochem 51:1776–1783

Kuan IC, Johnson KA, Tien M (1993) Kinetic analysis of manganese peroxidase. J Biol Chem 268:20064–20070

Kubicek CP, Starr TL, Glass NL (2014) Plant cell wall-degrading enzymes and their secretion in plant-pathogenic fungi. Annu Rev Phytopathol 52:427–451

Kumar V, Shukla P (2018) Extracellular xylanase production from *T. lanuginosus* VAPS24 at pilot scale and thermostability enhancement by immobilization. Process Biochem 71:53–60

Lee SH, Limb V, Lee CK (2018) Newly isolate highly potential xylanase producer strain from various environmental sources. Biocatal Agric Biotechnol 16:669–676

Levasseur A, Drula E, Lombard V, Coutinho P, Henrissat B (2013) Expansion of the enzymatic repertoire of the CAZy database to integrate auxiliary redox enzymes. Biotechnol Biofuels 6:41–54

Li X, Beeson WT, Phillips CM, Marletta MA, Cate JHD (2012) Structural basis for substrate targeting and catalysis by fungal polysaccharide monooxygenases. Structure 20:1051–1061

Li Z, Liu G, Qu Y (2017) Improvement of cellulolytic enzyme production and performance by rational designing expression regulatory network and enzyme system composition. Bioresour Technol 245(Pt B):1718–1726

Li C, Li D, Feng J, Fan X, Chen S, Zhang D, He R (2019) Duckweed (Lemna minor) is a novel natural inducer of cellulase production in *Trichoderma reesei*. J Biosci Bioeng 127(4):486–491. https://doi.org/10.1016/j.jbiosc.2018.09.017

Libardi N, Soccol CR, Góes-Neto A, de Oliveira J, Vandenberghe LPS (2017) Domestic wastewater as substrate for cellulase production by *Trichoderma harzianum*. Process Biochem 57:190–199

Lin MI, Nagata T, Katahira M (2018) High yield production of fungal manganese peroxidases by *E. coli* through soluble expression, and examination of the activities. Protein Expr Purif 145:45–52

Liu B, Olson Å, Wu M, Broberg A, Sandgren M (2017) Biochemical studies of two lytic polysaccharide monooxygenases from the white-rot fungus *Heterobasidion irregulare* and their roles in lignocellulose degradation. PLoS One 12(1):e0170404. https://doi.org/10.1371/journal.pone.0170404

Liu J, Zhang S, Shi Q, Wang L, Kong W, Yu H, Ma F (2019) Highly efficient oxidation of synthetic and natural lignin-related compounds by *Physisporinus vitreus* versatile peroxidase. Int Biodeterior Biodegradation 136:41–48

Lopes AM, Filho EXF, Moreira LRS (2018) An update on enzymatic cocktails for lignocellulose breakdown. J Appl Microbiol 125:632–645

Lundell T, Wever R, Floris R, Harvey P, Hatakka A, Brunow G, Schoemaker H (1993) Lignin peroxidase L3 from *Phlebia radiata* pre-steady state and steady-state studies with veratryl alcohol and a non-phenolic lignin model compound l-(3,4-dimethoxyphenyl)-2-(2-methoxyphenoxy) propane-l,3-diol. Eur J Biochem 211:391–402

Mai PLT, Kim YH (2016) Discovery and characterization of new O-methyltransferase from the genome of the lignin-degrading fungus *Phanerochaete chrysosporium* for enhanced lignin degradation. Enzym Microb Technol 82:66–73

Makela MR, Lundell T, Hatakka A, Hilden K (2013) Effect of copper, nutrient nitrogen, and wood-supplement on the production of lignin-modifying enzymes by the white-rot fungus *Phlebia radiata*. Fungal Biol 117:62–70

Makela MR, Mansouri S, Wiebenga A, Rytioja J, de Vries RP, Hilden K (2016) *Penicillium subrubescens* is a promising alternative for *Aspergillus niger* in enzymatic plant biomass saccharification. New Biotechnol 33:834–841

Maroldi MMC, Vasconcellos VM, Lacava PT, Farinas CS (2018) Potential of mangrove-associated endophytic fungi for production of carbohydrolases with high saccharification efficiency. Appl Biochem Biotechnol 184:806–820

Marques NP, Pereira JC, Gomes E, da Silva R, Araújo AR, Ferreira H, Rodrigues A, Dussán KJ, Bocchini DA (2018) Cellulases and xylanases production by endophytic fungi by solid state fermentation using lignocellulosic substrates and enzymatic saccharification of pretreated sugarcane bagasse. Ind Crop Prod 122:66–75

Martinez AT (2002) Molecular biology and structure-function of lignin-degrading heme peroxidases. Enzym Microb Technol 30:425–444

Meng Q-S, Liu CG, Zhao XQ, Bai FW (2018) Engineering *Trichoderma reesei* Rut-C30 with the overexpression of *egl1* at the *ace1* locus to relieve repression on cellulase production and to adjust the ratio of cellulolytic enzymes for more efficient hydrolysis of lignocellulosic biomass. J Biotechnol 285:56–63

Mishra V, Jana AK, Jana MM, Gupta A (2017) Enhancement in multiple lignolytic enzymes production for optimized lignin degradation and selectivity in fungal pretreatment of sweet sorghum bagasse. Bioresour Technol 236:49–59

Mitrovic A, Glieder A (2015) Discovery of fungal enzymes and pathways. In: Gupta VK, Mach RL, Sreenivasaprasad S (eds) Fungal biomolecules: sources, applications and recent developments, 1st edn. Wiley, Hoboken, pp 51–62

Mohapatra S, Padhy S, Mohapatra PKD, Thatoi HN (2018) Enhanced reducing sugar production by saccharification of lignocellulosic biomass, *Pennisetum* species through cellulase from a newly isolated *Aspergillus fumigatus*. Bioresour Technol 253:262–272

Mot AC, Silaghi-Dumitrescu R (2012) Laccases: complex architectures for one-electron oxidations. Biochemistry 77:1395–1407

Okereke OE, Akanyab HO, Egwim EC (2017) Purification and characterization of an acidophilic cellulase from *Pleurotus ostreatus* and its potential for agrowastes valorization. Biocatal Agric Biotechnol 12:253–259

Passos DF, Pereira NJ, Castro AM (2018) A comparative review of recent advances in cellulases production by *Aspergillus, Penicillium* and *Trichoderma* strains and their use for lignocellulose deconstruction. Curr Opin Green Sustain Chem 14:60–66

Patel I, Kracher D, Ma S, Garajova S, Haon M, Faulds CB, Berrin JG, Ludwig R, Record E (2016) Salt-responsive lytic polysaccharide monooxygenases from the mangrove fungus *Pestalotiopsis* sp. NCi6. Biotechnol Biofuels 9:108. https://doi.org/10.1186/s13068-016-0520-3

Payne CM, Knottm BC, Mayes HB, Hansson H, Himmel ME, Sandgren M, Stahlberg J, Bekham GT (2015) Fungal cellulases. Chem Rev 115:1308–1448

Pereira GF, de Bastiani D, Gabardo S, Squina F, Ayub MAZ (2018) Solid-state cultivation of recombinant *Aspergillus nidulans* to co-produce xylanase, arabinofuranosidase, and xylooligosaccharides from soybean fibre. Biocatal Agric Biotechnol 15:78–85

Phillips CM, Beeson WT, Cate JHD, Marletta MA (2011) Cellobiose dehydrogenase and a copper dependent polysaccharide monooxygenase potentiate cellulose degradation by *Neurospora crassa*. ACS Chem Biol 6:1399–1406

Ping L, Wang M, Xi Y, Cui F, Huang D, Sun W, Zou B, Huo S, Wang H (2018) Production and characterization of a novel acidophilic and thermostable xylanase from *Thermoascus aurantiacus*. Int J Biol Macromol 109:1270–1279

Piontek K, Antorini M, Choinowski T (2002) Crystal structure of laccase from the fungus *Trametes versicolor* at 1.90 Å resolution containing a full complement of coppers. J Biol Chem 277:37663–37669

Prajapati BP, Suryawanshi RK, Agrawal S, Ghosh M, Kango N (2018) Characterization of cellulase from *Aspergillus tubingensis* NKBP-55 for generation of fermentable sugars from agricultural residues. Bioresour Technol 250:733–740

Quinlan RJ, Sweeney MD, Lo Leggio L, Otten H, Navarro Poulsen JC (2011) Insights into the oxidative degradation of cellulose by a copper metalloenzyme that exploits biomass components. Proc Natl Acad Sci U S A 108:15079–15084

Quiroz-Castañeda RE, Martínez-Anaya C, Cuervo-Soto LI, Segovia L, Folch-Mallol JL (2011) Loosenin, a novel protein with cellulose disrupting activity from *Bjerkandera adusta*. Microb Cell Factories 10:8

Ramírez DA, Muñoz SV, Atehortua L, Michel JFC (2010) Effects of different wavelengths of light on lignin peroxidase production by the white-rot fungi *Phanerochaete chrysosporium* grown in submerged cultures. Bioresour Technol 101:9213–9220

Renganathan V, Gold MH (1986) Spectral characterization of the oxidized states of lignin peroxidase, an extracellular heme enzyme from the white rot Basidiomycete *Phanerochaete chrysosporium*. Biochemistry 25:1626–1631

Rye CS, Withers SG (2000) Glycosidase mechanisms. Curr Opin Chem Biol 4:573–580

Santos CA, Ferreira-Filho JA, O'Donovan A, Gupta VK, Tuohy MG, Souza AP (2017) Production of a recombinant swollenin from *Trichoderma harzianum* in *Escherichia coli* and its potential synergistic role in biomass degradation. Microb Cell Factories 16:83

Schalchli H, Hormazábal E, Rubilar O, Briceño G, Mutis A, Zocolo GJ, Diez C (2017) Production of ligninolytic enzymes and some diffusible antifungal compounds by white -rot fungi using potato solid wastes as the sole nutrient source. J Appl Microbiol 123:886–895

Segato F, Dias B, Berto GL, de Oliveira DM, De Souza FHM, Citadini AP, Murakami MT, Damásio ARL, Squina FM, Polikarpov I (2017) Cloning, heterologous expression and biochemical characterization of anon-specific endoglucanase family 12 from *Aspergillus terreus* NIH2624. Biochim Biophys Acta 1865:395–403

Shoseyov O, Shani Z, Levy I (2006) Carbohydrate binding modules: biochemical properties and novel applications. Microbiol Mol Biol Rev 70:283–295

Siddiqui KS, Ertan H, Charlton T, Poljak A, Khaled AKD, Yang X, Marshall G, Cavicchioli R (2014) Versatile peroxidase degradation of humic substances: use of isothermal titration calorimetry to assess kinetics, and applications to industrial wastes. J Biotechnol 178:1–14

Sitarz AK, Mikkelsen JD, Meyer AS (2016) Structure, functionality and tuning up of laccases for lignocellulose and other industrial applications. Crit Rev Biotechnol 36:7086

Solomon EI, Augustine AJ, Yoon J (2008) O_2 reduction to H_2O by the multicopper oxidases. Dalton Trans 30:3921–3932

Souza LO, de Brito AR, Bonomo RCF, Santana NB, Ferraz JLAA, Aguiar-Oliveira E, Fernandes AGA, Ferreira MLO, de Oliveira JR, Franco M (2018) Comparison of the biochemical properties between the xylanases of *Thermomyces lanuginosus* (Sigma®) and excreted by *Penicillium roqueforti* ATCC 10110 during the solid state fermentation of sugarcane bagasse. Biocatal Agric Biotechnol 16:277–284

Sugiura T, Mori T, Kamei I, Hirai H, Kawagishi H, Kondo R (2012) Improvement of ligninolytic properties in the hyper lignin degrading fungus *Phanerochaete sordida* YK-624 using a novel gene promoter. FEMS Microbiol Lett 331:81–88

Sundaramoorthy M, Kishi K, Gold MH, Poulos TL (1997) Crystal structure of substrate binding site mutants of manganese peroxidase. J Biol Chem 272:17574–17580

Taha M, Foda M, Shahsavari E, Aburto-Medina A, Adetutu E, Ball A (2016) Commercial feasibility of lignocellulose biodegradation: possibilities and challenges. Curr Opin Biotechnol 38:190–197

Tanghe M, Danneels B, Camattari A, Glieder A, Vandenberghe I, Devreese B, Stals I, Desme T (2015) Recombinant expression of *Trichoderma reesei* Cel61A in *Pichia pastoris*: optimizing yield and N-terminal processing. Mol Biotechnol 57:1010–1017

Terrone CC, de Freitas C, Terrasan CRF, de Almeida AF, Carmona EC (2018) Agroindustrial biomass for xylanase production by *Penicillium chrysogenum*: purification, biochemical properties and hydrolysis of hemicelluloses. Electron J Biotechnol 33:39–45

Wang Q, Chen L, Yu D, Lin H, Shen Q, Zhao Y (2017) Excellent waste biomass-degrading performance of *Trichoderma asperellum* T-1 during submerged fermentation. Sci Total Environ 609:1329–1339

Wong DWS (2009) Structure and action mechanism of ligninolytic enzymes. Appl Biochem Biotechnol 157:174–209

Xu X, Xu Z, Shi S, Lin M (2017) Lignocellulose degradation patterns, structural changes, and enzyme secretion by *Inonotus obliquus* on straw biomass under submerged fermentation. Bioresour Technol 241:415–423

Yao G, Wu R, Kan Q, Gao L, Liu M, Yang P, Du J, Li Z, Qu Y (2016) Production of a high-efficiency cellulase complex via β-glucosidase engineering in *Penicillium oxalicum*. Biotechnol Biofuels 9:78. https://doi.org/10.1186/s13068-016-0491-4

Ye Y, Li X, Zhao J (2017) Production and characteristics of a novel xylose- and alkali-tolerant GH 43 β-xylosidase from *Penicillium oxalicum* for promoting hemicellulose degradation. Sci Rep 7:11600. https://doi.org/10.1038/s41598-017-11573-7

Yegin S, Altinel B, Tuluk K (2018) A novel extremophilic xylanase produced on wheat bran from *Aureobasidium pullulans* NRRL Y-2311-1: effects on dough rheology and bread quality. Food Hydrocoll 81:389–397

Zhang XY, Zi LH, Ge XM, Li YH, Liu CG, Bai FW (2017a) Development of *Trichoderma reesei* mutants by combined mutagenesis and induction of cellulase by low-cost corn starch hydrolysate. Process Biochem 57:96–101

Zhang JG, Li QM, Thakur K, Faisal S, Wei ZJ (2017b) A possible water-soluble inducer for synthesis of cellulase in *Aspergillus niger*. Bioresour Technol 226:262–266

Zhang F, Zhao X, Bai F (2017c) Improvement of cellulase production in *Trichoderma reesei* Rut-C30 by overexpression of a novel regulatory gene Trvib-1. Bioresour Technol 247:676–686

Zhuo R, Yu H, Qin X, Ni H, Jiang Z, Ma F, Zhang X (2018) Heterologous expression and characterization of a xylanase and xylosidase from white rot fungi and their application in synergistic hydrolysis of lignocellulose. Chemosphere 212:24–33

Chapter 7
Strategic Role of Fungal Laccases in Biodegradation of Lignin

Shiv Shankar, Shailja Singh, Shikha, Anuradha Mishra, and Siya Ram

1 Introduction

Lignocellulosic biomass is considered as a sustainable and renewable energy resource on earth. It is replenished at the rate of 60,000 million tons every year as organic carbon on earth (Singh 2006; Prasad 2018). Lignocellulosic biomass consists of carbohydrate polymers like cellulose, hemicellulose, and lignin (Singh 2006). The carbohydrate constituents of lignocellulosic biomass carry dissimilar monomers (six and five carbon sugars) which are firmly linked to the lignin. Lignin is highly inflexible polymer which is not easily degraded in nature because of its complex structure characterized by three-dimensional cross-linking (Ruiz-Dueñas and Martínez 2009).

The removal of lignin from biomass makes the hydrolysis of hemicellulose and cellulose easy thereby facilitating the use of its sugar proportion in generation of sustainable fuels like ethanol and other value-added products (Siqueira et al. 2012).

In pulp and paper industries, the removal of lignin from wood is required for the production of paper products. These industries remove lignin chemically from the wood pulp in delignification and bleaching stages and generate huge amount (50–60

S. Shankar
Department of Environmental Science, School of Vocational Studies and Applied Sciences, Gautam Budhha University (A State University), Greater Noida, Uttar Pradesh, India

S. Singh · Shikha (✉)
Department of Environmental Science, School for Environmental Sciences, Babasaheb Bhimrao Ambedkar University (A Central University), Lucknow, Uttar Pradesh, India

A. Mishra
Department of Applied Chemistry, School of Vocational Studies and Applied Sciences, Gautam Budhha University (A State University), Greater Noida, Uttar Pradesh, India

S. Ram
Department of Biotechnology, School of Biotechnology, Gautam Budhha University (A State University), Greater Noida, Uttar Pradesh, India

R. Naraian (ed.), *Mycodegradation of Lignocelluloses*, Fungal Biology,
https://doi.org/10.1007/978-3-030-23834-6_7

million tons) of lignin compounds in processed water (Zhu et al. 2016; Avanthi and Banerjee 2016; Gosz et al. 2018). Environmental problems caused by chemicals used in pulping and bleaching stages of the paper production have compelled paper industries to consider environmental friendly alternatives for removal of lignin from wood pulp.

It is therefore necessary to develop effective green pretreatment technologies for the removal of lignin from lignocellulosic biomass and waste water originating from pulp and paper industries (Avanthi and Banerjee 2016).

Enzymatic degradation of lignin is considered as green technology for the degradation of lignin compounds. In nature, biodegradation of lignin occurs by ligninolytic enzymes secreted by microbes. Microbes degrade lignin in two stages with the help of ligninolytic enzymes. In first step, de-polymerization of lignin takes place leading to formation of heterogeneous aromatic compounds. In second step, heterogeneous aromatic compounds are mineralized in water and carbon dioxide.

White rot fungi belonging to the class basideomycetous are major lignin degraders in nature. Fungi such as *Phanerochaete chrysosporium*, *Pleurotus ostreatus*, *Phlebia radiata*, *Trametes versicolor*, *Peniophora* sp., *Agaricus bisporus*, *Pycnoporus cinnabarinus*, *Trametes hirusuta*, and *Ceriporiopsis subvermispora* are some well-known lignin-degrading fungal species (Singh 2006; Shankar and Shikha 2012; Afreen et al. 2018). White rot fungi degrade lignin with the help of ligninolytic enzymes which include laccases, manganese peroxidase, lignin peroxidase, etc. These enzymes depolymerize lignin via formation of lignin phenoxy radicals (Brijwani et al. 2010).

Laccases are regarded as one of the most active enzymes in lignin degradation (Shankar and Shikha 2012; Avanthi and Banerjee 2016; Bilal et al. 2017; Agarwal et al. 2018). The enzyme was first identified in sap of Japanese lacquer tree *Rhus vercinifera* (Thurston 1994). Laccases are multinuclear enzymes with molecular weights ranging from 60 to 80 KDa (Kiiskinen et al. 2004; Khlifi-Slama et al. 2012; Afreen et al. 2018; Hilgers et al. 2018). Laccases effectively degrade phenolic moieties in lignin Cα oxidation, Cα - Cβ cleavage and aryl cleavage (Archibald et al. 1997; Shankar and Shikha 2015).

The oxidation of different types of substrates like polyphenols, lignins, aromatic amines, methoxy-substituted monophenols by laccase results in formation of oxygen-centered free radical, which is ultimately transformed into quinone. Quinone and free radical may be polymerized in later stages of the reaction (Thurston 1994).

In the presence of suitable mediators such as ABTS [2, 2′-azinobis (3-ethylbenz thiazoline-6-sylphonate)] or HBT (hydroxybenzotriazole), laccases hold the potential to oxidize non-phenolic proportion of lignin and other substrates (Shankar and Shikha 2015).

Owing to the aforesaid properties, laccases show potential for biological degradation of lignin. In the light of the aforesaid context, this chapter highlights production, characterization and role of laccases and laccase mediator system in degradation of lignin.

2 Lignin

Swiss botanist A. P. de Candolle mentioned lignin in 1813 for the first time as tasteless, fibrous material which was insoluble in water but soluble in weak alkaline solutions. He termed this substance as "lignine," which is derived from the Latin word *lignum*, meaning wood. Lignin in the most abundant organic polymer on earth after cellulose representing 30% non-fossil organic carbon and 20–35% dry biomass of wood on earth.

Lignin occurs in nature as most abundant aromatic polymer followed by cellulose. The elemental composition of lignin differs from plant species to species. In aspen sample, there is 5.9% hydrogen, 63.4% carbon, 0.7% ash (mineral components), and 30% oxygen (by difference) corresponding to the formula approximately $(C_{31}H_{34}O_{11})_n$. Depending on source of biomass, dry weight of a plant cell wall is constituted by 20–35% of lignin (Galbe and Zacchi 2007). Hardwood generally contains 20–25% lignin while softwoods contain 25–35% lignin (Faravelli et al. 2010).

Cellulose–hemicellulose matrix in plant cell is surrounded by lignin which imparts rigidity to the plant cell wall and connects plant cells together firmly. Lignin protects wood constituents from microbial decomposition. Because of hydrophobicity, lignin also acts as a barrier against the penetration of water (Balakshin et al. 2009).

Because of aforesaid reasons, lignin behaves as most recalcitrant biopolymer on earth (Ruiz-Dueñas and Martínez 2009). Different types of functional groups are present in lignin. Phenolic, methoxyl, hydroxyl carbonyl, and alcoholic hydroxyl are some prominent groups present in lignin. Lignin consists of phenylpropanoid building blocks/monolignols linked together by carbon–oxygen and carbon–carbon bonds.

Phenylpropanoid units present in lignin are coniferyl, sinapyl alcohols, and *p*-coumaryl. These units are precursors of guaiacyl (G), syringyl (S), and *p*-hydrophenyl (H) type units in biopolymer lignin (Zhu et al. 2016). In plants, laccases and peroxidases generates radicals which biosynthesize lignin via oxidative coupling reactions leading to synthesis of a growing polymer.

Aryl-glycerol, β-aryl ether structure with the β-O-4 linkages is the most abundant linkage in lignin constituting most (approximately 60%) of the total internal linkage in lignin (Mainka et al. 2015; Murciano Martínez et al. 2016; Agarwal et al. 2018).

3 Fungal Degradation of Lignin

For the past two decades, researchers have investigated role of ligninolytic enzyme system of fungi, bacteria, and actinomycetes in biological degradation of lignin.

Under natural environmental conditions, biological degradation of lignin results in generation of monomeric aromatic compounds which are employed as a source of carbon and energy by microorganisms. Oxidizing potential of enzymes of ligninolytic system regulates degradation of lignin in the presence of oxygen and hydrogen peroxide. However, laccases do not require hydrogen peroxide for its activity. This property of laccases makes them appropriate candidate for environmental bioremediation of toxic pollutants (Knezevic et al. 2013; Avanthi and Banerjee 2016; Munk et al. 2018).

Fungal degradation of lignin is catalyzed by highly efficient and unique oxidative extracellular ligninolytic enzymes like laccases, lignin peroxidases, manganese peroxidases, versatile peroxidases, glyoxal peroxidases, aryl alcohol oxidases, quinone reductases, and xylanases (dos Santos et al. 2019).

White rot fungi have drawn significant attention of the researchers in last decade as they are the dominant sources of lignin-degrading enzyme laccase (Saito et al. 2018; Rudakiya and Gupte 2017).

Under adverse environmental conditions, white rot fungi with the help of ligninolytic enzymes completely mineralize lignin into water and carbon dioxide. Brown rot fungi are not considered as efficient degraders of lignin (Wu et al. 1996; Cardoso et al. 2018). Lignin-degrading white rot fungi occupies diverse ecological niches and grows on different habitats, for instance, conifers, deciduous trees, forest litter, crops, grassland soils, and roots of plants. Prominent lignin-degrading white rot fungi include *Coriolus versicolor, Pycnoporus cinnabarinus, P. ostreatus, Peniophora* sp. *Ganoderma*, *Phlebia radiate*, *Lentinula edodes*, *Phanerochaete chrysosporium*, and *Armillaria* spp. (Hatakka 2005; Singh 2006; Sigoillot et al. 2012; Rytioja et al. 2014; Wang et al. 2018).

4 Ligninolytic Enzymes of Fungi

4.1 Laccases

Laccase (benzenediol oxygen oxidoreductase, EC 1.10.3.2) is a copper-containing phenol oxidase. Laccases catalyze the oxidation of a wide range of aliphatic and aromatic compounds with a consequent reduction of molecular oxygen to water with the help of oxygen-centered free radical. Laccases are widely distributed in higher plants, fungi, bacteria, and insects. It has been demonstrated that almost all fungi produce laccase except some physiological groups such as lower fungi, i.e., Chytridiomycetes and Zygomycetes (Morozova et al. 2007). Although brown rot fungi are not known to produce laccase, the oxidation of syringaldazine and ABTS have been detected in *Coniophora puteana* and *Laetiporus sulphureus* (Schlosser and Hofer 2002), respectively. Almost all species of wood rotting basidiomycetes causing white rot are the most widely known species that produce appreciable quantity of laccase to varying degree (Hatakka 1994; Baldrian 2006; Sharma et al. 2007; Kolomytseva et al. 2017; Young et al. 2018).

Laccases from white rot fungi are responsible for degradation of lignin, fructification, detoxification, sporulation, phyto-pathogenicity, etc. Fungal laccases generally have high redox potential as compared to bacterial laccases. Because of this reason, fungal laccases are widely preferred over bacterial laccases for degradation of lignin model compounds. For biotechnological and environmental applications, much preference is given to fungal laccases as compared to bacterial laccases since fungal laccases have high redox potential than bacterial laccases (Surwase et al. 2016).

4.2 Production of Fungal Laccases

For widespread industrial and biotechnological applications, laccases are required in large amounts. Heterologous expression of laccase gene, optimization of physical and chemical conditions (Kiiskinen et al. 2004), or a combination of both strategies is used to cater the need of the laccases for these applications. White rot fungi produce laccases during secondary metabolism under solid as well as submerged culture conditions (Robinson et al. 2001; Rivera-Hoyos et al. 2013). In solid state fermentation, the fungi grow on solid material in the absence or near absence of free flowing water, while in submerged culture conditions, the fungi grow on a continuous liquid phase (Musatti et al. 2017). Laccase production is greatly influenced by different cultivation and nutritional parameters such as concentration of nitrogen source, carbon source, and micronutrients as well.

4.3 Laccase Production Under Submerged Culture Conditions

Production of laccases under submerged culture conditions involves cultivation of fungi under aerobic conditions in aqueous medium containing nutrients required for growth (Shiv Shankar and Shikha 2011). In order to achieve high production, the earlier research attempts have been oriented towards optimization of culture conditions (Bertrand et al. 2013; Elisashvili et al. 2018). The production of laccase at industrial scale is undertaken by submerged culture conditions (Dong et al. 2005; Couto and Toca-Herrera 2007; Piscitelli et al. 2010).

Production of laccase under submerged culture conditions is disadvantageous as abundant growth of fungal mycelium in the culture medium affects enzyme yield due to metabolic rate limitation and mass transfer. In addition, extensively grown mycelia cover impellers, leading to obstruction in the system. This problem can be solved by immobilization of the cells on different solid supports (Couto 2018). The efficiency of laccase production requires a well-defined media composition optimized in terms of physico-chemical and nutritional parameters (carbon and nitrogen sources, inducers, pH and temperature, etc.).

5 Effect of Physico-chemical Factors on Laccase Production

The physico-chemical parameters affecting production of laccases under SmF are temperature, pH of the fermentation medium, inducers, and aeration level during the cultivation (Afreen et al. 2018). Majority of researchers have reported that the optimum pH for laccase production by fungi is 4.5–4.6. The optimum temperature for laccase production by fungi has been reported to be 25–30 °C (Pointing et al. 2000).

6 Influence of Carbon and Nitrogen Sources on Laccase Production

Easily available carbon sources, such as cellobiose, mannitol, and glucose, generally favor higher laccase production as compared to the carbohydrates like lactose and cellulose (Mikiashvili et al. 2006; Afreen et al. 2018).

Rodriguez Couto et al. (2006) reported that the incorporation of different sources of carbon, such as fructose followed by glycerol, resulted in increase of the rate of laccase production *T. hirsuta* as compared to glucose alone. Bettin et al. (2009) reported that the fructose as a source of carbon optimum production of laccase by *Pleurotus sajor-caju* (Bettin et al. 2009).

Low concentration of glucose in culture medium has been found to increase the rate of laccase production (Thiruchelvam and Ramsay 2007; Kannaiyan et al. 2012). The optimum concentration of nitrogen for fungal laccase production by fungi is more ambiguous than that of carbon (Kannaiyan et al. 2012; Afreen et al. 2018). Generally, nitrogen starved conditions trigger the production of ligninolytic enzymes like laccases. Nitrogen in the concentration of 2–3 mM in culture media significantly increases the production of the laccase.

7 Induction of Laccase

Laccase production under submerged culture conditions can be improved by the incorporation of some substances called inducers to the media. Use of an appropriate inducer has been shown to be a prerequisite to achieve effective production of the enzyme at an industrial scale. Compounds frequently used as inducer of laccase production by white rot fungi include 2, 5-xylidine (a xenobiotic compound), veratryl alcohol (a secondary metabolite), lignin, $CuSO_4$, etc. (Xavier et al. 2001; Shankar and Shikha 2015). Inducers, like ethanol, has been found to trigger laccase production indirectly.

8 Solid State Fermentation

Laccase production under solid state fermentation is carried out particularly using filamentous fungi. Selection of appropriate feed stock for SSF is essential since the selected substrate determines the process of laccase production by laccase (Reddy et al. 2003). Factors like cost, chemical composition of substrates, porosity and size of the substrate should be considered while going for laccase production under solid state fermentation. Substrates, which are used for laccase production under solid state fermentation, have been categorized into two classes: Non-inert support which not only provide site for attachment to fungi but also render nutrients to the fungi (minerals, proteins, fibers, and sugars). Inert supports are synthetic compounds without any nutritional value for fungi. The wide varieties of solid materials used in SSF are classified into two main classes: inert. Fluidized bed bioreactors, rotating horizontal drum bioreactors, rotating disc reactors and column tray bioreactors are frequently used for the production of laccases under solid state fermentation (Couto 2018).

9 Structure of Laccases

Fungal laccases exist in different forms viz. multimeric, heteromeric, homotetrameric, and monomeric. Depending upon microorganism, the molecular weight of laccases ranges from 50 to 130 kDa (Jaiswal et al. 2015). Fungal laccases generally contain 10–30% carbohydrate content of molecular weight while plant laccases contain 45% carbohydrate content (Baldrian 2006).

The stability of the protein component is ensured by the carbohydrate proportion of the laccases. In addition, carbohydrate moiety in laccases prevents inactivation of laccases (Morozova et al. 2007).

Greek key β barrel topology is the main structural feature of the laccases which is generally comprised of 500 amino acids arranged in three domains consecutively. In laccases, amino acid moieties are divided into three domains:

- Domain 1 (150 amino acids)
- Domain 2 (150–300 amino acids)
- Domain 3 (300–500 amino acids)

Structure of laccases is consolidated by disulfide linkage between area I and III and area I and II (Plácido and Capareda 2015). Single laccase unit comprised of four dissimilar copper catalytic forms (Fig. 7.1). These four copper ions are grouped into three classes:

9.1 Type 1: Paramagnetic "Blue" Copper Site

Multi-copper proteins in laccases appear blue due to type 1 copper and covalent copper–cysteine bond. In laccases, type 1 site has a rhombohedral co-ordination, with two histidines and one cysteine. Axial position ligand affects the oxidative nature of laccase, responsible for oxidation of different substrates (Kumar et al. 2003; Enguita 2011).

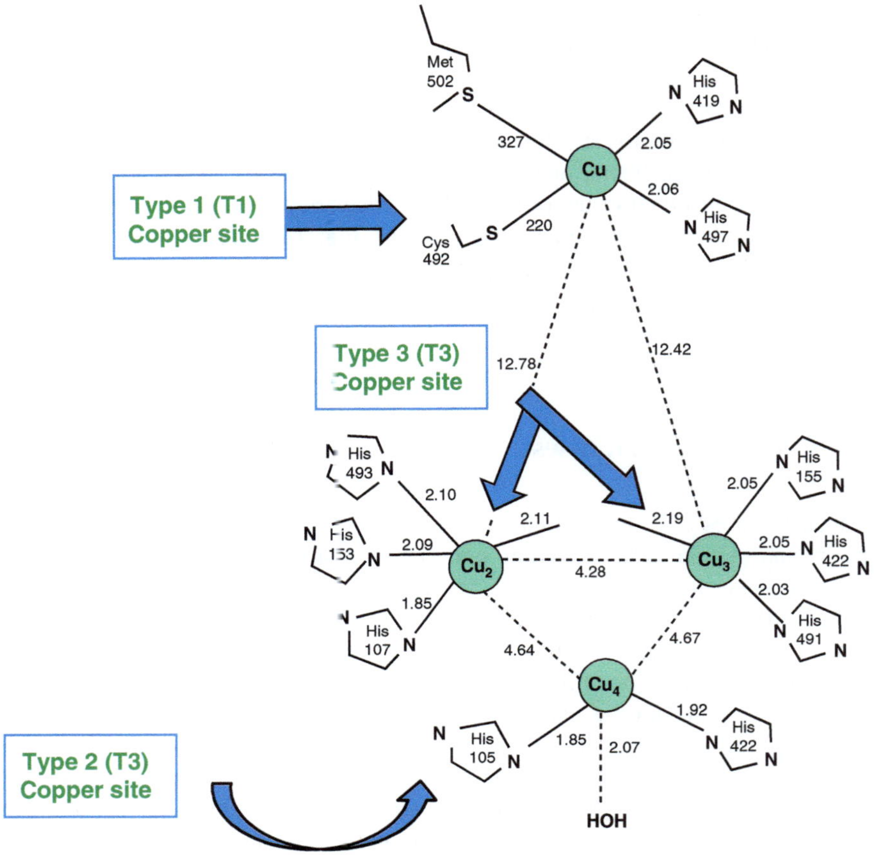

Fig. 7.1 Structure of laccase elaborating different active sites

9.2 Type 2: Paramagnetic "Non-blue" Copper

Non-blue copper site display normal electron paramagnetic resonance (EPR) spectra. Type 2 site dose not absorb intense radiation in visible region and unified by two histidine residues (Hilgers et al. 2018).

9.3 Type 3: Diamagnetic Spin-coupled Copper–Copper Pair

Type 3 site is a double nuclear center controlled by six histidines. This site is characterized spectroscopically as it absorbs electronic radiation at 330 nm. Type 3 site does not display electron paramagnetic resonance signal (Matera et al. 2008).

A try nuclear cluster is constituted by type 3 copper and type 2 copper. A try nuclear cluster reduction of oxygen takes place.

10 Categories of Laccases

10.1 White Laccases

White laccases show absorption peak at 400 nm, but they do not display peak at 330 nm and 605 nm. Abnormal metal content is the main property of white laccases. White laccases are optimally active at neutral pH (Palmieri et al. 1997). White laccase does not display T1 and T2 signals and electron paramagnetic resonance spectra. This happens because valency state of Cu^{2+} is changed (Zhao et al. 2012).

10.2 Yellow Laccase

These are artificially reduced blue laccases which do not absorb visible light and electron paramagnetic resonance spectra (Pozdnyakova et al. 2006a, b). Yellow laccases are formed due to reduction of type I copper site by aromatic products of lignin degradation (Mot and Silaghi-Dumitrescu 2012).

Several aspects of yellow laccases like their catalytic properties and characterization is still to be addressed by the researchers. Few yellow laccase producers include *Trametes* sp. F1635, *Stropharia aeruginosa*, *Leucoagaricus naucinus* LAC-04, *Aureobasidium pullulans* NAC8, *Sclerotinia sclerotiorum, Pleurotus ostreatus* D1 (YLPO), and *Sclerotinia sclerotiorum* (Pozdnyakova et al. 2004, 2006a, b; Mot and Silaghi-Dumitrescu 2012; Daroch et al. 2014; Ademakinwa and Agboola 2016; Ning et al. 2016; Wang et al. 2018).

11 Role of Laccases in Lignin Degradation and Their Mode of Action

11.1 Degradation of Phenolic Units in Lignin

Laccase-mediated degradation of phenolic units in lignin results in aryl-alkyl cleavage, Cα-Cβ cleavage, and C-α oxidation (Fig. 7.2). Laccases result in reduction of diatomic molecular oxygen to two water molecules. In this process, laccases facilitate one electron oxidation of different types of organic and inorganic substrates in general and aromatic amines, monophenols, and polyphenols in particular (Camarero et al. 1994; Bourbonnais et al. 1995; Moya et al. 2011). The oxidation of these substrates by laccases results in formation of an oxygen-centred free radical which is transformed in quinone ultimately. The quinone and the free radicals can then undergo polymerization (Archibald et al. 1997; Shankar and Shikha 2012).

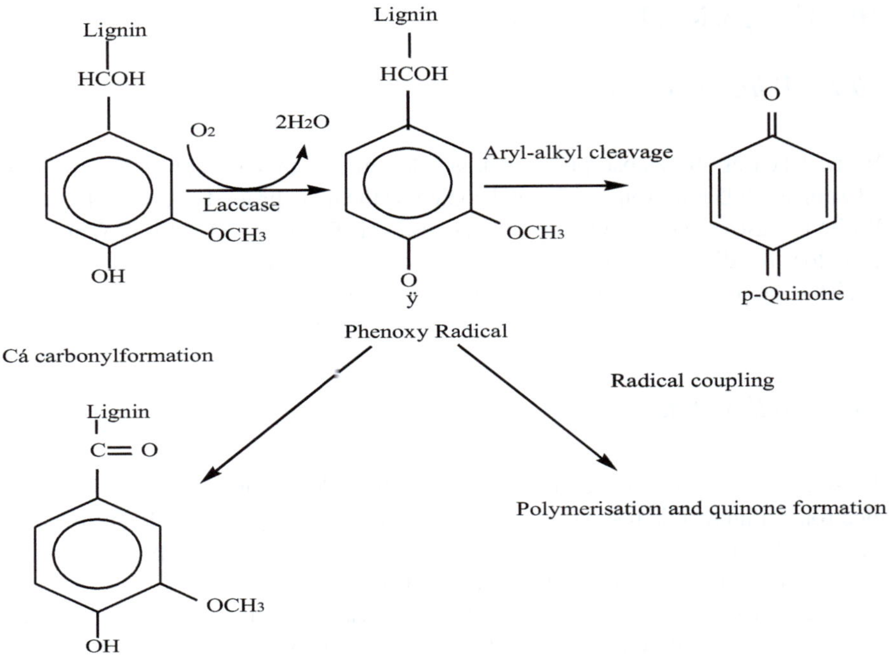

Fig. 7.2 Oxidation of phenolic subunits of lignin by laccase (Adopted from Archibald et al. 1997)

11.2 Degradation of Non-phenolic Subunits of Lignin in the Presence of Redox Mediators

The substrate range of laccase can be extended to non-phenolic subunits of lignin (Fig. 7.3) by the inclusion of a mediator such as 2, 2′-azinobis-(3-ethylbenzthiazoline-6-sulfonate) (ABTS), 1-hydroxybenzotriazole (HBT), and *N*-hydroxyphthalimide (NHPI) (Bourbonnais et al. 1997).

11.3 Biodegradation of Lignin with Laccase Mediator Systems

By using small molecular weight compounds called redox mediators, the oxidation potential of fungal laccases can be increased from 0.78 to 1.084 V (Bourbonnais et al. 1997; Zille et al. 2003). Redox mediators are low molecular weight compounds which are uninterruptedly oxidized by the laccase enzyme and eventually reduced by the substrate (Wu et al. 1996; Cañas and Camarero 2010).

High molecular weight substrates are difficult to enter into laccase active site. Redox mediators transfer electrons between substrate and laccase enzyme thereby surmounting steric interferences existing between enzyme and substrate (Li et al.

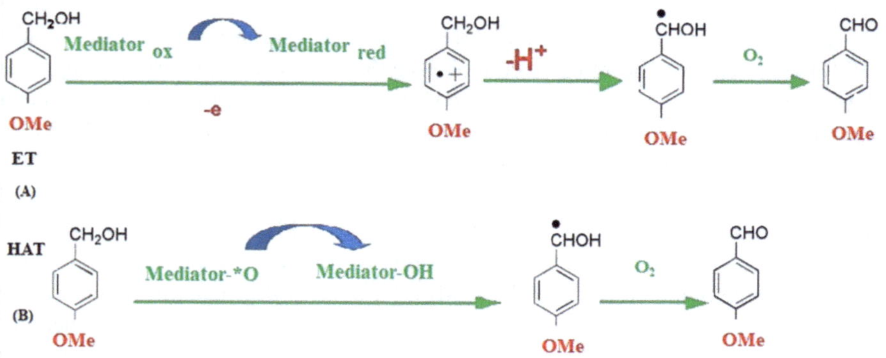

Fig. 7.3 Oxidation of 4-methoxy benzyl alcohol (non-phenolic lignin compound) to 4-methoxy benzaldehyde: (**a**) electron transfer (ET); (**b**) hydrogen atom transfer (HAT) (Barreca et al. 2004)

1999). Degradation of high molecular weight non-phenolic compounds is limited by laccases due to reduced enzyme activity and low accessibility.

The problem of limited substrate accessibility can be solved by employing appropriate redox mediators. Redox mediators with high redox potential are oxidized to stable intermediates in initial step by laccases. Thereafter, the diffusion of oxidized mediator from enzyme takes place. Mediators easily enter into plant cell wall because of their small size and act on target substrate lignin in general and non-phenolic content of the lignin in particular. An ideal redox mediator should be effective, atoxic, and cheap with consistent reduced and oxidized forms and should be capable of upholding the cyclic redox conversions (Morozova et al. 2007). The problem of instability and formation of toxic intermediates is the main problem in application of synthetic mediators in degradation of lignin model compounds. Instability of redox mediators sometimes results in unfinished redox cycles leading to inadequate oxidation of the substrates. It is difficult to re-utilize synthetic mediators as they may generate toxic intermediates. In addition they are toxic and expensive as well (Yaohua et al. 2019). At higher concentration of enzyme (more than 1 mM), synthetic redox mediators hamper the activity of laccases. Natural redox mediators can be the better option of synthetic mediators as they are cost-incentive and do not result in the formation of toxic intermediates. Few good examples of natural redox mediators of laccase include: vanillin, acetosyringone, syringaldehyde, *p*-coumaric acid, acetovanillone, and HBT (Andreu and Vidal 2011; Bibi et al. 2011).

Degradation of non-phenolic units using laccase mediator system in lignin takes place via three different types of mechanisms (a) Transfer of electrons, (b) transfer of radical hydrogen, (c) ionic mechanism. First two mechanisms have been shown in Fig. 7.3.

Bourbonnais and Paice (1990) reported for the first time that laccase oxidizes lignin model compounds such as 1-(3,4-dimethoxyphenyl)-2-(2-methoxyphenoxy)-propane 1,3-diol, veratryl alcohol into derivatives of α-carbonyl.

Bourbonnais et al. (1997) also reported that laccase in the presence of redox mediator ABTS results in demethylation and delignification of kraft pulp. Call and Mücke (1997) reported that the redox mediator 1-Hydroxybenzotriazole (HBT) improves degradation of lignin and bleaching of the kraft pulp (Call and Mücke 1997). Shankar and Shikha (2012) reported that the laccase from the *Peniophora* sp. (NFCCI-2131) results in delignification of kraft pulp effectively. Munk et al. (2018) reported that the laccase from *Myceliophthora thermophile and Trametes versicolor* in presence of four individual mediators, *N*-hydroxyphthalimide (HPI), 1-hydroxybenzotriazole (HBT), 2,2′-azino-bis(3-ethylbenzothiazoline-6-sulfonic acid) (ABTS), and 2,2,6,6-tetramethylpiperidin-1-oxyl (TEMPO) enhances degradation of lignin. The use of laccase mediator system has drawn significant attention of the researchers for diverse applications such as treatment of industrial waste water, decolorization of industrial and leather dyes, bioremediation water and soil contaminated with hydrocarbons and bio-deinking of waste paper (Shankar and Shikha 2012).

Redox mediator ABTS results in the oxidation of non-phenolic subunits of lignin model compounds via transfer of electrons while redox mediator 2,2′,6,6′-tetramethylpiperidine-1-oxyl (TEMPO) mediates the process of oxidation via ionic mechanism (Bourbonnais et al. 1997; Fabbrini et al. 2002). A coupling intermediate compound is formed by the redox mediator HBT which is subsequently degraded and release benzotriazole, a reduced form of HBT. Benzotriazole facilitates the connection of lignin to HBT (Potthast et al. 1997). The degradation of softwood lignin via HBT-laccase system has been found to result in reduction in G-units and aromatic component of the lignin while increase in *p*-hydroxyphenyl content.

Laccase mediator system results in the formation of degradation products like 2,6-dimethoxy-4-((E)prop-1-enyl)benzaldehyde, 4-ethyl-2,6-dimethoxybenzaldehyde, and 2,6-dimethoxy-4-methylbenzaldehyde during degradation of lignin (dos Santos et al. 2019; Couto 2018).

Laccase mediator system is widely used bio-bleaching step in pulp and paper industries. The system results in reduction in kappa number in the pulp and enhances its ISO brightness (Poppius-Levlin et al. 1999). Use of laccase mediator system during bio-bleaching of the pulp minimizes the use of chlorine compounds (elemental chlorine, sodium hypochlorite, chlorine dioxide, etc.) thereby reducing environmental risks in the process. The degree of degradation of lignin in wood pulp is determined by the type of mediator and pulp as well as dose of the laccase. Use of combination of more than one redox mediator has been found to increase the extent of the degradation of the lignin in soft wood and hard wood.

12 Conclusion

Lignin is an important constituent of lignocellulosic material. Lignin is highly resistant towards degradation because of its complex chemical structure and diverse chemical bonding. The removal of lignin from lignocellulosic material is required for the production of fermentable sugars and other value-added products.

In pulp and paper industries, the removal of lignin from wood is required for the production of paper products. Environmental problems caused by chemicals used in pulping and bleaching stages of the paper production have compelled paper industries to consider environmental friendly alternatives for removal of lignin from wood pulp. In addition, the degradation of lignin in waste water originating from pulp and paper industry is required by using green approaches as lignin and its derivative compounds cause adverse biochemical effects on aquatic flora and fauna.

Fungal laccases (benzenediol oxygen oxidoreductase, EC 1.10.3.2) in combination with natural redox mediator compounds degrade lignin in ecofriendly manner and offer sustainable way of pretreatment/disposal of lignocellulosic waste. Fungal laccases to be employed for degradation of lignin should be substrate unspecific, thermostable with high redox potential, cost-incentive, resistant towards toxic metal ions, halides, inhibitors, and free radicals.

Degradation of lignin using fungal laccases is an ecofriendly practice but availability of laccase for commercial applications is limited due to low production of enzyme by reported strains. In this respect, the exploration of novel laccase hypersecretary strains may be helpful. In addition, genetic improvement/mutagenesis in previously isolated strains and heterologous expression of laccase genes in appropriate host may be helpful in increasing the level of laccase production for different applications.

In this context, this chapter was an attempt to put forth the details of chemical structure of lignin, linkages consolidating the structure of lignin, production and characterization of fungal laccases, their mode of action, and role of laccase mediator system in degradation of phenolic and non-phenolic subunits of lignin.

References

Ademakinwa AN, Agboola FK (2016) Biochemical characterization and kinetic studies on a purified yellow laccase from newly isolated *Aureobasidium pullulans* NAC8 obtained from soil containing decayed plant matter. J Genet Eng Biotechnol 14(1):143–151

Afreen S, Anwer R, Singh RK et al (2018) Extracellular laccase production and its optimization from *Arthrospira maxima* catalyzed decolorization of synthetic dyes. Saudi J Biol Sci 25:1446–1453

Agarwal UP, Ralph SA, Reiner RS et al (2018) Production of high lignin-containing and lignin-free cellulose nanocrystals from wood. Cellulose 25:5791–5805

Andreu G, Vidal T (2011) Effects of laccase-natural mediator systems on kenaf pulp. Bioresour Technol 102(10):5932–5937

Archibald FS, Bourbonnais R, Jurasek L et al (1997) Kraft pulp bleaching and delignification by *Trametes versicolor*. J Biotechnol 53:215–236

Avanthi A, Banerjee R (2016) A strategic laccase mediated lignin degradation of lignocellulosic feedstocks for ethanol production. Ind Crop Prod 92:174–185

Balakshin MY, Capanema EA, Chang HM (2009) Recent advances in the isolation and analysis of lignins and lignin-carbohydrate complexes. In: Hu TQ (ed) Characterization of lignocellulosic materials. Blackwell Publishing Ltd, Oxford, pp 148–170

Baldrian P (2006) Fungal laccases-occurrence and properties. FEMS Microbiol Rev 30:215–242

Barreca AM, Sjögren B, Fabbrini M, Gentili P (2004) Catalytic efficiency of some mediators in laccase-catalyzed alcohol oxidation. Biocat Biotransform 22(2):105–112

Bertrand B, Martínez-Morales F, Trejo-Hernandez MR (2013) Fungal laccases: induction and production. Rev Mex Ing Quim 12:473–488

Bettin F, Montanari Q, Calloni R et al (2009) Production of laccases in submerged process by *Pleurotus sajor-caju* PS-2001 in relation to carbon and organic nitrogen sources, antifoams and Tween 80. J Ind Microbiol Biotechnol 36(1):1–9

Bibi I, Bhatti HN, Asgher M (2011) Comparative study of natural and synthetic phenolic compounds as efficient laccase mediators for the transformation of cationic dye. Biochem Eng J 56:225–231

Bilal M, Asgher M, Parra-Saldivar R et al (2017) Immobilized ligninolytic enzymes: an innovative and environmental responsive technology to tackle dye-based industrial pollutants—a review. Sci Total Environ 576:646–659

Bourbonnais R, Paice M (1990) Oxidation of non-phenolic substrates: an expanded role for laccase in lignin biodegradation. FEBS Lett 207:99–102

Bourbonnais R, Paice MG, Reid ID et al (1995) Lignin oxidation by laccase isozymes from *Trametes versicolor* and role of the mediator 2,2′-azinobis(3 ethylbenzthiazoline-6-sulfonate) in kraft lignin depolymerization. Appl Environ Microbiol 61:1876–1880

Bourbonnais R, Paice MG, Freiermuth B (1997) Reactivities of various mediators and laccases with kraft pulp and lignin model compounds. Appl Environ Microbiol 63(12):4627–4632

Brijwani K, Rigdon A, Vadlani PV (2010) Fungal laccase: production function and applications in food processing. Enzyme Res 10:149–748

Call HP, Mücke I (1997) History, overview and applications of mediated lignolytic systems, specially laccase-mediator-system (Lignozyme®-process). J Biotechnol 53:163–202

Camarero S, Galletti GC, Martínez AT (1994) Preferential degradation of phenolic lignin units by two white rot fungi. Appl Environ Microbiol 60:4509–4516

Cañas AI, Camarero S (2010) Laccases and their natural mediators: biotechnological tools for sustainable eco-friendly processes. Biotechnol Adv 28:694–705

Cardoso WS, Queiroz PV, Tavares GP et al (2018) Multi-enzyme complex of white rot fungi in saccharification of lignocellulosic material. Braz J Microbiol 49:879–884

Couto SR (2018) Solid-state fermentation for laccases production and their applications. In: Current developments in biotechnology and bioengineering. Elsevier, pp 211–234

Couto SR, Toca-Herrera JL (2007) Laccase production at reactor scale by filamentous fungi. Biotechnol Adv 25:558–569

Daroch M, Houghton CA, Moore JK et al (2014) Glycosylated yellow laccases of the basidiomycete *Stropharia aeruginosa*. Enzym Microb Technol 58(59):1–7

Dong JL, Zhang YW, Zhang RH et al (2005) Influence of culture conditions on laccase production and isozyme patterns in the white-rot fungus *Trametes gallica*. J Basic Microbiol 45:190–198

dos Santos AC, Ximenes E, Kim Y, Ladisch MR (2019) Lignin–enzyme interactions in the hydrolysis of lignocellulosic biomass. Trends Biotechnol 37(5):518–531. https://doi.org/10.1016/j.tibtech.2018.10.010

Elisashvili V, Kachlishvili E, Asatiani D (2018) Efficient production of lignin-modifying enzymes and phenolics removal in submerged fermentation of olive mill by-products by white-rot basidiomycetes. Int Biodeterior Biodegrad 134:39–47

Enguita FJ (2011) Structural biology of fungal multicopper oxidases. In: Leitao AL (ed) Mycofactories. Bentham Science Publishers, Emirate of Sharjah, pp 57–72

Fabbrini M, Galli C, Gentili P (2002) Comparing the catalytic efficiency of some mediators of laccase. J Mol Catal B Enzym 16:231–240

Faravelli T, Frassoldati A, Migliavacca G (2010) Detailed kinetic modeling of the thermal degradation of lignins. Biomass Bioenergy 34(3):290–301

Galbe M, Zacchi G (2007) Pretreatment of lignocellulosic materials for efficient bioethanol production. Adv Biochem Eng Biotechnol 108:41–65

Gosz K, Kosmela P, Hejna A et al (2018) Biopolyols obtained via microwave-assisted liquefaction of lignin: structure, rheological, physical and thermal properties. Wood Sci Technol 52:599–617
Hatakka A (1994) Lignin-modifying enzymes from selected white-rot fungi: production and role in lignin degradation. FEMS Microbiol Rev 13:125–135
Hatakka A (2005) Biodegradation of lignin. In: Alexander S (ed) Biopolymers online. Wiley-VCH Verlag GmbH & Co. KGaA, Weinheim
Hilgers R, Vincken JP, Gruppen H (2018) Laccase/mediator systems: their reactivity toward phenolic lignin structures. ACS Sustain Chem Eng 6:2037–2046
Jaiswal N, Pandey VP, Dwivedi UN (2015) Immobilization of papaya laccase in chitosan led to improved multipronged stability and dye discoloration. Int J Biol Macromol 8:288–295
Kannaiyan R, Mahinpey N, Mani T et al (2012) Enhancement of *Dichomitus squalens* tolerance to copper and copper-associated laccase activity by carbon and nitrogen sources. Biochem Eng J 67:140–147
Khlifi-Slama R, Mechichi T, Sayadi S et al (2012) Effect of natural mediators on the stability of *Trametes trogii* laccase during the decolourization of textile wastewaters. J Microbiol 50:226–234
Kiiskinen L-L, Palonen H, Linder M, Viikari L, Kruus K (2004) Laccase from *Melanocarpus albomyces* binds effectively to cellulose. FEBS Lett 576:251–255
Knezevic A, Milovanović I, Stajić M et al (2013) Lignin degradation by selected fungal species. Bioresour Technol 138:117–123
Kolomytseva M, Myasoedova N, Samoilova A et al (2017) Rapid identification of fungal laccases/oxidases with different pH-optimum. Process Biochem 62:174–183
Kumar SVS, Phale PS, Durani S (2003) Combined sequence and structure analysis of the fungal laccase family. Biotechnol Bioeng 83:386–394
Li K, Xu F, Eriksson KE (1999) Comparison of fungal laccases and redox mediators in oxidation of a nonphenolic lignin model compound. Appl Environ Microbiol 65:2654–2660
Mainka H, Täger O, Körner E et al (2015) Lignin—an alternative precursor for sustainable and cost-effective automotive carbon fiber. J Mater Res Technol 4:283–296
Matera I, Gullotto A, Tilli S et al (2008) Crystal structure of the blue multicopper oxidase from the white-rot fungus *Trametes trogii* complexed with p-toluate. Inorg Chim Acta 361:4129–4137
Mikiashvili N, Wasser SP, Nevo E (2006) Effects of carbon and nitrogen sources on *Pleurotus ostreatus* ligninolytic enzyme activity. World J Microbiol Biotechnol 22:999–1002
Morozova OV, Shumakovich GP, Gorbacheva MA et al (2007) "Blue" laccases. J Biochem 72(10):1136–1150
Mot AC, Silaghi-Dumitrescu R (2012) Laccases: complex architectures for one-electron oxidations. Biochemistry (Mosc) 77(12):1395–1407
Moya R, Saastamoinen P, Hernández M et al (2011) Reactivity of bacterial and fungal laccases with lignin under alkaline conditions. Bioresour Technol 102(21):10006–10012
Munk L, Punt AM, Kabel MA (2018) Laccase catalyzed grafting of-N-OH type mediators to lignin via radical-radical coupling. RSC Adv 7(6):3358–3368
Murciano Martínez P, Punt AM, Kabel MA et al (2016) Deconstruction of lignin linked p-coumarates, ferulates and xylan by NaOH enhances the enzymatic conversion of glucan. Bioresour Technol 216:44–51
Musatti A, Ficara E, Mapelli C (2017) Use of solid digestate for lignocellulolytic enzymes production through submerged fungal fermentation. J Environ Manag 199:1–6
Ning YJ, Wang SS, Chen QJ, Ling ZR, Wang SN, Wang WP, Zhang GQ, Zhu MJ (2016) An extracellular yellow laccase with potent dye decolorizing ability from the fungus *Leucoagaricus naucinus* LAC-04. Int J Biol Macromol 93:837–842
Palmieri G, Giardina P, Bianco C et al (1997) A novel white laccase from *Pleurotus ostreatus*. J Biol Chem 272:31301–31307
Piscitelli A, Pezzella C, Giardina P (2010) Heterologous laccase production and its role in industrial applications. Bioeng Bugs 1:252–262

Plácido J, Capareda S (2015) Ligninolytic enzymes: a biotechnological alternative for bioethanol production. Bioresour Bioprocess 2:23

Pointing SB, Jones EBG, Vrijmoed LLP (2000) Optimization of laccase production by *Pycnoporus sanguineus* in submerged liquid culture. Mycologia 92(1):139–144

Poppius-Levlin K, Whang W, Tamminen T et al (1999) Effects of laccase/HBT treatment on pulp and lignin structures. J Pulp Pap Sci 25:90–94

Potthast A, Koch H, Fischer K (1997) The laccase-mediator system—reaction with model compounds and pulp. In: Preprints 9th international symposium wood pulp (Montreal, Canada), S. F1-4–F1-8. Chem Tech Section F2-1–F2-4

Pozdnyakova N, Rodakiewicz-Nowak J, Turkovskaya O (2004) Catalytic properties of yellow laccase from *Pleurotus ostreatus* D1 J Mol Catal B Enzym 30:19–24

Pozdnyakova NN, Rodakiewicz-Nowak J, Turkovskaya OV, Haber J (2006a) Oxidative degradation of polyaromatic hydrocarbons and their derivatives catalyzed directly by the yellow laccase from *Pleurotus ostreatus* D1 J Mol Catal B Enzym 41:8–15

Pozdnyakova NN, Turkovskaya OV, Yudina EN (2006b) Yellow laccase from the fungus *Pleurotus ostreatus* D1: purification and characterization. Appl Biochem Microbiol 42(1):56–61

Prasad R (2018) Mycoremediation and environmental sustainability. Springer International Publishing, Cham

Reddy GV, Ravindra BP, Komaraiah P (2003) Utilization of banana waste for the production of lignolytic and cellulolytic enzymes by solid substrate fermentation using two Pleurotus species (*P. ostreatus and P. sajor-caju*). Process Biochem 38:1457–1462

Rivera-Hoyos CM, Morales-Alverez ED, Poutou-Pinales RA et al (2013) Fungal laccases. Fungal Biol Rev 27(3–4):67–82

Robinson T, Chandran B, Nigam P (2001) Studies on the production of enzymes by white-rot fungi for the decolourisation of textile dyes. Enzyme Microb Technol 29:575–579

Rodriguez Couto S, Rodríguez A, Paterson RRM, Lima N, Teixeira JA (2006) Laccase activity from the fungus *Trametes hirsuta* using an air-lift bioreactor. Lett Appl Microbiol 42(6):612–616

Rudakiya DM, Gupte A (2017) Degradation of hardwoods by treatment of white rot fungi and its pyrolysis kinetics studies. Int Biodeter Biodegr 120:21–35

Ruiz-Dueñas FJ, Martínez ÁT (2009) Microbial degradation of lignin: how a bulky recalcitrant polymer is efficiently recycled in nature and how we can take advantage of this. Microb Biotechnol 2:164–177

Rytioja J, Hildén K, Mäkinen S et al (2014) Saccharification of lignocelluloses by Carbohydrate Active Enzymes of the white rot fungus *Dichomitus squalens*. PLoS One 10(12):e0145166

Saito Y, Tsuchida H, Matsumoto T et al (2018) Screening of fungi for decomposition of lignin-derived products from Japanese cedar. J Biosci Bioeng 126:573–579

Schlosser D, Hofer C (2002) Laccase-catalyzed oxidation of Mn^{2+} in the presence of natural Mn^{3+} chelators as a novel source of extracellular H_2O_2 production and its impact on manganese peroxidase. Appl Environ Microbiol 68(7):3514–3521

Shankar S, Shikha (2011) Fungal laccases: production and biotechnological applications in pulp and paper industry. In: Jain PK, Gupta VK, Bajpai V (eds) Recent advances in environmental biotechnology. Dudweller Landstr. Saarbrücken, pp 73–111

Shankar S, Shikha (2012) Laccase production and enzymatic modification of lignin by a novel *Peniophora* sp. Appl Biochem Biotechnol 166:1082–1094

Shankar S, Shikha (2015) Effects of metal ions and redox mediator system on decolorization of synthetic dyes by crude laccase from a novel white rot fungus Peniophora sp. (NFCCI-2131). Appl Biochem Biotechnol 175(1):635–647

Sharma P, Goel R, Capalash N (2007) Bacterial laccases. World J Microbiol Biotechnol 23:823–832

Sigoillot JC, Berrin JG, Bey M et al (2012) Fungal strategies for lignin degradation. In: Jouanin L, Lapierre C (eds) Lignins: biosynthesis, biodegradation and bioengineering, vol 61. Academic, Elsevier, London, pp 263–308

Singh H (2006) Mycoremediation: fungal bioremediation, 1st edn. Wiley-Interscience. ISBN-13: 978-0471755012

Siqueira G, Várnai A, Ferraz A et al (2012) Enhancement of cellulose hydrolysis in sugarcane bagasse by the selective removal of lignin with sodium chlorite. Appl Energy 102:399–402

Surwase SV, Patil SA, Srinivas S et al (2016) Interaction of small molecules with fungal laccase: a surface plasmon resonance based study. Enzym Microb Technol 82:110–114

Thiruchelvam AT, Ramsay JA (2007) Growth and laccase production kinetics of *Trametes versicolor* in a stirred tank reactor. Appl Microbiol Biotechnol 74:547–554

Thurston CF (1994) The structure and function of fungal laccases. Microbiology 140:19–26

Wang SN, Chen QJ, Zhu MJ et al (2018) An extracellular yellow laccase from white rot fungus *Trametes* sp. F1635 and its mediator systems for dye decolorization. Biochimie 148:46–54

Wu F, Ozaki H, Terashima Y et al (1996) Activities of ligninolytic enzymes of the white rot fungus, and its recalcitrant substance degradability. Water Sci Technol 34:69–78

Xavier AMRB, Evtuguin DV, Ferreira RMP et al (2001) Laccase production for lignin oxidative activity. In: Proceedings of the 8th international conference on biotechnology in the pulp and paper industry, 4–8 June, Helsinki

Xu F (1997) Effects of redox potential and hydroxide inhibition on the pH activity profile of fungal laccases. J Biol Chem 272(2):924–928

Yaohua G, Ping X, Feng J et al (2019) Co-immobilization of laccase and ABTS onto novel dual-functionalized cellulose beads for highly improved biodegradation of indole. J Hazard Mater 365:118–124

Young D, Dollhofer V, Callaghan TM et al (2018) Isolation, identification and characterization of lignocellulolytic aerobic and anaerobic fungi in one- and two-phase biogas plants. Bioresour Technol 268:470–479

Zhao D, Zhang X, Cui DZ (2012) Characterisation of a novel white laccase from the deuteromycete fungus *Myrothecium verrucaria* NF-05 and its decolourisation of dyes. PLoS One 7:e38817

Zhu G, Ouyang X, Jiang L et al (2016) Effect of functional groups on hydrogenolysis of lignin model compounds. Fuel Process Technol 154:132–138

Zille A, Tzanov T, Gübitz GM (2003) Immobilized laccase for decolourization of reactive black 5 dyeing effluent. Biotechnol Lett 25:1473–1477

Chapter 8
Rapid Bioconversion of Lignocellulosic Biomass by Fungi

Adesh Kumar, Divya Srivastva, and Ramesh Chand

1 Introduction

A large amount of lignocellulosic residue (150 billion tons) is produced annually worldwide and the dominant biomass differs depending on the region (Asim et al. 2015). Lignocellulose is derived from unusable portions of plant biomass in the form of agricultural, industrial, domestic, and forest residues (Gupta et al. 2016). India accounts for 770 m tonnes of crop residue production annually. If these residues were not degraded, there would be an accumulation of carbon on Earth which would ultimately break carbon cycle and make life impossible on the biosphere. The recycling of the wastes is both ecological necessity and economic compulsion. The crop residue management in agriculture for sustaining high levels of crop productivity is the biggest challenge for researches. Composting is basically a process for decomposition of organic solid wastes. The decomposition process is accomplished by various microorganisms including fungi, bacteria, and actinomycetes. Primarily, nature has taught us that the breakdown of plant materials is mainly carried out by fungi, by means of potential cellulosic enzymes (Gupta et al. 2016; Solomon et al. 2016; Takasaki et al. 2013). Fungal cellulosic enzymes present efficient and gentle way of converting recalcitrant lignocellulosic materials into precursors for industrial purposes (Bischof and Seiboth 2016; Gupta et al. 2016; Solomon et al. 2016). Currently, several cellulolytic enzymes have been identified and are well characterized, but there may be many more cellulases, hydrolytic and non-hydrolytic enzymes available with a greater range of applications. Some previously uncharacterized

A. Kumar (✉) · D. Srivastva
Department of Plant Molecular Biology and Genetic Engineering, N.D. University of Agriculture and Technology, Faizabad, Uttar Pradesh, India

R. Chand
Department of Nematology, N.D. University of Agriculture and Technology, Faizabad, Uttar Pradesh, India

R. Naraian (ed.), *Mycodegradation of Lignocelluloses*, Fungal Biology,
https://doi.org/10.1007/978-3-030-23834-6_8

fungi may also possess unique enzymes or individual cocktails that could supersede commercial cellulase preparations, resulting in faster and more complete biomass hydrolysis at a reduced cost (Berrin et al. 2012; Gupta et al. 2016). Under optimal conditions, bioconversion of lignocelluloses converts at faster rate. Several fungi possess cellulolytic enzymes which can be utilized for bioconversion of agricultural residues. The main problem with bioconversion of agriculture residues is the presence of lignin. As compared to cellulose, lignin is recalcitrant and cellulose fibrils are always embedded in an amorphous matrix of lignin and hemicellulose. These three kinds of polymers bind strongly to each other by non-covalent forces as well as by covalent cross-links making a composite material, lignocelluloses, representing over 90% of dry weight of plant. For effective utilization of lignocelluloses residue, several physical and chemical pre-treatments are required to disrupt the bounds and hydrolysis of lignin. The delignification steps also increase the cost of process as well as generate toxic compounds due to hydrolysis (Batt 1991). Lignocellulolytic enzymes producing fungi are widespread and include species from the ascomycetes, e.g., *Trichoderma reesei*, basidiomycetes including white-rot fungi, e.g., *Phanerochaete. chrysosporium,* brown-rot fungi, e.g., *Fomitopsis palustris,* chytridiomycetes including *Caecomyces communis, Neocallimastix frontalis, Cyllamyces aberensis*, and mushroom fungi, e.g., *Pleurotus ostreatus* which can be used for solid state fermentation (SSF) of agricultural residues after delignification and finally a few anaerobic species e.g., *Orpinomyces* sp. which degrade cellulose in gastrointestinal tracts of ruminant animals (Kamra and Zadrazil 1986; Ozkose et al. 2001; Yoon et al. 2007; Doi 2008; Ljungdahl 2008; Griffith et al. 2009). Since none of the fungal strains including the best mutants are able to produce high levels of the enzymes at the same time necessary for bioconversion of lignocelluloses, there is a need to identify consortium of efficient cellulolytic and lignolytic microorganisms or fungal co-culturing which might act synergistically for rapid bioconversion of agricultural residues without any chemical pre-treatments. The mesophiles are not very successful for large-scale bioconversions through solid state fermentation process because temperature inside the residue during active fermentation reaches up to 50–70 °C. During this phase, thermophilic fungi play an important role in degradation of lignocellulose (Sharma and Johri 1992; Ahamed and Vermette 2008).

2 Composition of Lignocellulosic Biomass

Lignocellulose is the term used for biomass from woody or fibrous plant materials, being a combination of lignin, cellulose, and hemicellulose polymers interlinked in a heterogeneous but resilient matrix (Chen 2014; Guo et al. 2015). The composition of lignocellulosic biomass is highly complex and dynamic leading to its recalcitrance (Gandarias and Luis 2013; Himmel et al. 2007; Johansen 2016). A typical representation is given in Fig. 8.1. On a dry weight basis, cellulose, hemicellulose, and lignin are the primary constituents of cell walls, whereas proteins (which are highly glycosylated), lipids, soluble sugars, and minerals are minor components

(Chundawat et al. 2011; Guo et al. 2015). While cellulose and hemicellulose are polysaccharides that can be hydrolyzed to fermentable sugars, lignin has a three-dimensional phenolic structure that is highly resistant to microbial attack or to chemical hydrolysis (Gupta et al. 2016; Sims et al. 2008). These three compound fibrils and depolymerize cellulose and hemicellulose into simple sugars (Guo et al. 2015).

Cellulose is a complex macromolecule composed of linear β1-4glucan chains that tightly aggregate into microfibrils held together via strong intra- and intermolecular hydrogen bonds and van der Waals forces resulting from pyranose ring stacking (Kumar et al. 2009). It is an unbranched chain of glucose units linked head to tail (Fig. 8.2), and it is highly resistant to chemical and biological hydrolysis because of its structural arrangement (Gandarias and Luis 2013). The chain conformation of the glucose residues in cellulose forces the hydroxyl groups into radial (equatorial) orientation and the aliphatic hydrogen atoms into axial positions. As a result, there is strong inter-chain hydrogen bonding between adjacent chains in a cellulose sheet and weaker hydrophobic interactions between cellulose sheets. The hydrophobic face of cellulose sheets makes crystalline cellulose resistant to acid hydrolysis because it contributes to the formation of a dense layer of water near the hydrated cellulose surface while the strong inter-chain hydrogen-bonding network makes crystalline cellulose resistant to enzymatic hydrolysis (Beckham et al. 2011; Ding and Liu 2012).

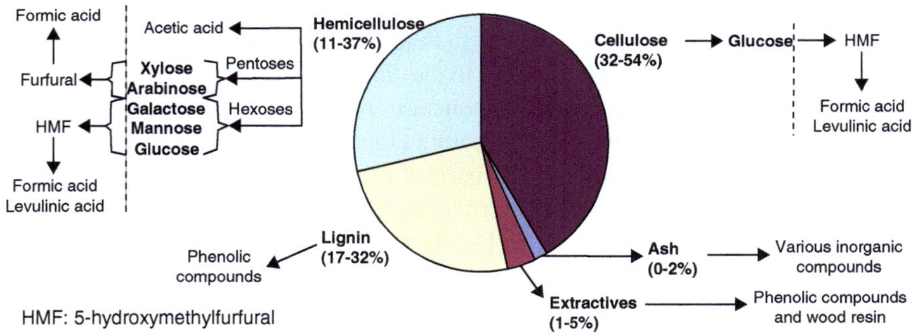

Fig. 8.1 Composition of lignocellulosic materials and their potential hydrolysis products and further degradation compounds. Source: (Guo et al. 2015)

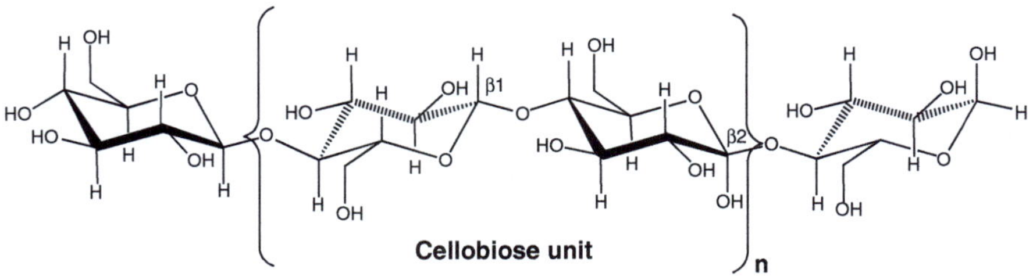

Fig. 8.2 Chemical structure of cellulose (Gandarias and Luis 2013)

Cellulose mainly contains carbon (44.44%), hydrogen (6.17%), and oxygen (49.39%). It is the most abundant material on Earth, being the main constituent of plants serving to maintain their structure and is also present in bacteria, fungi, algae, and even in animals (Chaplin 2016). The content of cellulose in these species except plant is negligible. The variability of cellulose is such that it forms linear crystals of about 3 nm thick in some plants, but which reach widths of more than 20 nm in certain algae. The degree of polymerization of cellulose varies, depending on its source, between 100 and 10,000 (Chundawat et al. 2011). Cellulose with high crystallinity tends to be of shorter polymeric units and is insoluble (Chaplin 2016). Pure cellulose exists in various polymorphic forms viz. cellulose Iα, Iβ, II, III1, III11, IV1, and IV11 (Zugenmaier 2001). The crystalline forms are Iα, Iβ, II, and III (Chaplin 2016). Cellulose I is the form found in nature while other crystal forms of artificial cellulose under artificial processing. Cellulose II, the second most extensively studied form, may be obtained from cellulose I by either of two processes: (a) regeneration, which is the solubilization of cellulose I in a solvent followed by reprecipitation by dilution in water to give cellulose II, or (b) mercerization, which is the process of swelling native cellulose in concentrated sodium hydroxide, to yield cellulose II on removal of the swelling agent. Celluloses III1 and III11 are formed, in a reversible process, from celluloses I and II, respectively, by treatment with liquid ammonia or some amines, and the subsequent evaporation of excess ammonia. Polymorphs IV1 and IV11 may be prepared by heating celluloses III1 and III11, respectively, to 206 °C, in glycerol (Chaplin 2016). These differences in glucan chain packing of these polymorphs have been shown to influence their hydrolysis rates (Chundawat et al. 2011). The crystalline structures of cellulose are aligned with each other, giving a structural bias to the cell wall (Cosgrove 2005).

Hemicellulose is another main component in plant fiber materials. It is the second most common polysaccharides in nature (Johansen 2016). Hemicelluloses are highly heterogeneous. These polysaccharides are very different from each other structurally and in physico-chemical properties (Chen 2014; Scheller and Ulvskov 2010). The necessity of the sugars β-(1→4)-linked backbones of an equatorial configuration was proposed in (Scheller and Ulvskov 2010). In this regard, polysaccharides such as β-(1→4)-galactan having axial configuration at C4 are not to be included with hemicelluloses.

On the other hand, some other researchers believe hemicelluloses may contain pentoses (β-D-xylose, α-L-arabinose), hexoses (β-D-mannose, β-D-glucose, α-D-galactose), and/or uronic acids (α-D-glucuronic, α-D-4-*O*-methylgalacturonic, and α-D-galacturonic acids). Other sugars such as α-L-rhamnose and α-L-fucose may also be present in small amounts and the hydroxyl groups of sugars can be partially substituted with acetyl groups (Girio et al. 2010). These confusions attest to the heterogeneity of hemicelluloses However, it is generally believed that hemicellulose is the glucan in the matrix of the cell, and the main components are xylan, xyloglucan, glucomannan, mannan, galactomannan, callose among others. The content and structure of hemicellulose in various plants are different, but on a broad classification, hardwood hemicelluloses contain mostly xylans that are partially acetylized, whereas softwood hemicelluloses contain mostly glucomannans with almost no acetyl groups (Scheller and Ulvskov 2010). The main chain may consist of one or more types of glycosyls, and the connections between glycosyl are also different. The frequency and composition of branches are dependent on the source of xylan (Saha 2003; Scheller and Ulvskov 2010). In actual fact, raw materials from different producing areas and different parts have different glycan compositions (Chen 2014). Hemicelluloses in the cell wall have the primary role of interacting with other polymers to ensure the proper physical properties of the wall. Other roles ascribed to hemicelluloses include functioning as seed storage carbohydrates—galactomannans, source of signal molecules—xyloglucans (Cosgrove 2005; Scheller and Ulvskov 2010). The schematic illustrations of the types of hemicelluloses found in plant cell walls indicating its complexity are shown in Fig. 8.3.

Lignin is a complex large molecular structure containing cross-linked polymers of phenolic monomers. It is present in the primary cell wall, imparting structural support, impermeability, and resistance against microbial attack. Thus making it one of the most significant contributors to biomass (Bugg and Rahmanpour 2015; Yuan et al. 2013). Lignin is the nature's dominant aromatic polymer, is found in most terrestrial plants in the approximate range of 15–40% dry weight to provide structural integrity (Wyman and Ragauskas 2015). Lignin is equally heterogeneous and their heterogeneity is manifested in different species of plants, length of growing season, and different parts of the plants. Even in the different morphologies of cells of the same xylem or different cell wall layers, the structures of lignin are not the same (Chen 2014). Generally, lignin is considered as being formed by the dehydrogenative polymerization of three hydroxycinnamyl alcohols: *p*-coumaryl, coniferyl, and sinapyl alcohols (Fig. 8.4). Alkyl-aryl, alkyl-alkyl, and aryl-aryl ether bonds link these phenolic monomers together (De Lima et al. 2016). The monolignols (monomeric units of lignin)—*p*-coumaryl alcohol, sinapyl alcohol, and coniferyl alcohol—give rise to the aromatic units *p*-hydroxyphenyl (H), syringyl (S), and guaiacyl (G), respectively, in the lignin polymer. In addition, it is now widely accepted that other monomers also participate in coupling reactions that give rise to the lignin macromolecule, such as coniferaldehyde, coniferyl alcohol, and acylated monolignols (De Lima et al. 2016; Yuan et al. 2013).

In general, herbaceous plants such as grasses have the lowest contents of lignin, whereas softwoods have the highest lignin contents (Pan et al. 2005). Hardwood lignins are predominantly G and S monolignols with trace amounts of H units. Softwood lignins are composed of mostly G units, whereas monocots incorporate equivalent amounts of G and S units along with significantly higher amounts of H monolignols (Chen 2014; Yuan et al. 2013). The composition of the functional groups of lignin and physical distribution, such as syringyl (S) unit to guaiacyl (G) unit ratio have also been demonstrated to affect the enzymatic hydrolysis of cellulose. In general, the yield of enzymatic hydrolysis is believed to be enhanced by low lignin content and a high S/G ratio of the substrate. Lignin appears to limit cellulose hydrolysis by two distinct mechanisms: first by forming a physical barrier that prevents enzyme access and second by non-productive binding cellulolytic enzymes (Dixon 2013; Yuan et al. 2013)

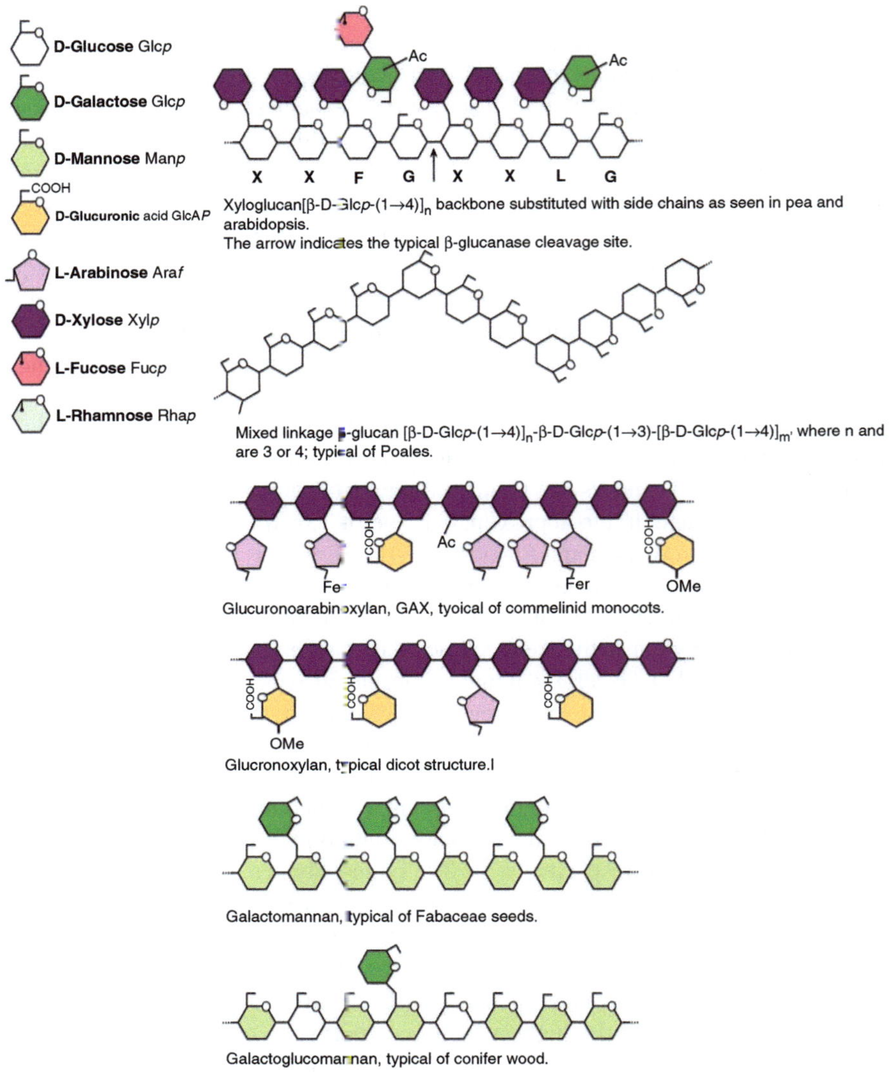

Fig. 8.3 Schematic illustrations of the types of hemicelluloses found in plant cell walls (Scheller and Ulvskov 2010)

3 Lignocellulose-Degrading Fungi

In nature, lignocellulose accounts for the major part of biomass and consequently its degradation is essential for operation of global carbon cycle. During decomposition, microorganisms transform organic matter into CO_2 biomass thermo-energy and humus-like end-product. The organic substrate, bulking agents, and the amend-

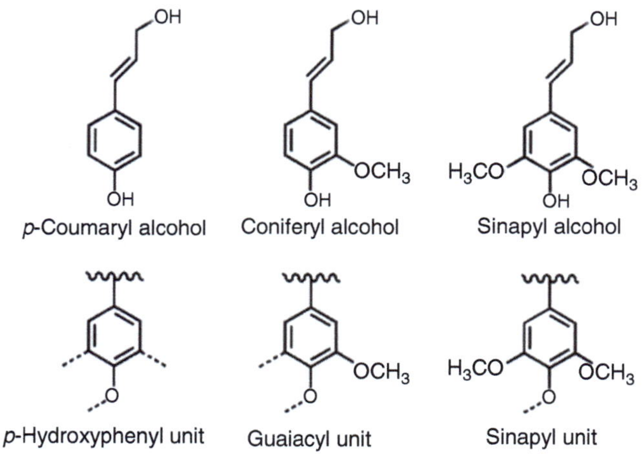

Fig. 8.4 Basic structural unit of lignin including three primary lignin monomers, *p*-coumaryl alcohol, coniferyl alcohol, and sinapyl alcohol and lignin structural *p*-hydroxyphenyl, guaiacyl, and syringyl units. Source: (Yuan et al. 2013)

ments used in composting are mostly from plant material. The capacity of microorganisms to assimilate organic matter depends on their ability to produce the enzymes needed for degradation of the substrate components, e.g., cellulose, hemicelluloses, and lignin. The more complex the substrate, the more extensive and comprehensive is the enzyme system required. Through the synergistic action of microorganisms, complex organic compounds are degraded to smaller molecules which can be utilized by microbial cells (Golueke 1992). Fungal strategies for lignocellulose depolymerization are substantially very complex, due to the high complexity of the raw materials. Recalcitrance to saccharification is a major limitation for the enzymatic conversion of lignocellulose-containing raw materials to get valuable end-products. The combination of hemicellulose and lignin forms a protective barrier around the cellulose, which must be modified before the hydrolysis of cellulose. But also crystalline structure of cellulose makes it insoluble and resistant for enzyme decomposition. Nevertheless, the removal of lignin is a key challenge to increase enzyme access to hemicellulose and cellulose (Jaramillo et al. 2015). Plant cell wall-degrading filamentous fungi have an important role in recycling nutrients in forest ecosystem. They are known to produce a broad variety of extracellular enzymes with diverse catalytic activities for the hydrolysis of renewable lignocelluloses containing raw materials. Predominantly responsible for lignocellulose degradation are wood-degrading fungi. They decompose and assimilate the most recalcitrant organic polymers, which is mainly attributed to their highly adaptive lifestyles, reflected by a large phylogenetic and phenotypic diversity (Kracher and Ludwig 2016). If they want to overcome the physical and chemical stability of lignocellulose, the fungi have to employ large sets of enzymes, which they release into the environment during their growth. Fungi have two types of degradation systems: intracellular, together with the outer cell envelope layer, and extracellular, important for polysaccharide degradation. Furthermore, the extracellular enzymatic system includes two types of enzymes: hydrolytic, responsible for polysaccharide degradation; and oxidative, which degrade lignin and open phenyl rings. Three groups of fungi, with different effects and degradation mechanisms onto the lignocellulose, have been described: soft-rot, brown-rot, and white-rot fungi (Sanches 2009). Soft-rot fungi are mostly ascomycete fungi that can degrade polysaccharides in the surface layers of plants. Degradation leads to darkening and softening of the wood by the produced laccases and peroxidases involved in lignin modifications. These enzymes are unspecific and more limited in function than those isolated from white-rot and brown-rot fungi. The soft-rot fungi belong to genera *Aspergillus* and *Neurospora.* Still, little is known about the degradation mechanisms of lignocellulose by soft-rot fungi (Woiciechowski et al. 2013). Brown-rot fungi are basidiomycetes that rapidly metabolize cellulose and hemicellulose while only slightly modifying lignin. They have no lignin-degrading enzymes except small molecule reactive species to depolymerize lignin. At an advanced stage of degradation, wood residue exhibits cube shape and has a brownish color due to the predominant presence of oxidized lignin. Disruption of the lignocellulose matrix by brown-rot fungi can be demonstrated using iron-dependent Fenton chemistry known as chelator-mediated Fenton system (CMF). The CMF system is a unique substrate deconstruction system based on oxygen radical chemistry that allows nonenzymatic deconstruction of the cellulose. Briefly, brown-rot fungal hyphae during the growth in the lumen area of plant cells produce oxalic acid, iron-reducing compounds (RC), and hydrogen peroxide (H_2O_2). The oxalic acid binds to a Fe^{3+} ion formatting the complex that diffuses into cell wall along with H_2O_2 and RC. With the pH change, RC sequesters Fe^{3+} from the Fe-oxalate complex and reduces it to Fe^{2+}. Fe^{2+} then reacts with H_2O_2 (Fenton reaction) and produces hydroxyl radicals (–OH). Upon attack of –OH radicals, lignocellulose matrix is disrupted. Models for the study of brown-rot fungi are *Gloeophyllum trabeum, Coniophora puteana,* and *Postia placenta* (Arantes and Goodell 2014). White-rot fungi are able to decompose all lignocellulose constituents: lignin, cellulose, and hemicellulose. Degradation of lignin is more efficient than in the case of brown-rot and soft-rot fungi because they possess a unique ability to its complete mineralization to CO_2 (Couturier and Berrin 2013). Therefore, white-rot basidiomycetes could be an interesting source of lignocellulose-active enzymes to supplement the commercial cocktails of hemicellulases and cellulases originated from ascomycetes such as *Aspergillus niger* or *Trichoderma reesei* (Sanches 2009). Some of the white-rot fungi capable of causing selective delignification of wood are *Phanerochaete chrysosporium, Phanerochaete carnosa, Pleurotus ostreatus, Pycnoporus cinnabarinus, Botrytis cinerea, Stropharia coronilla,* and *Trametes versicolor.* Several thermophilic fungi were isolated by researches which are able to degrade different agricultural residues. An alkalothemophillic *Thermomonospora* sp. producing high levels of xylanase was isolated from self-heating compost (George et al. 2001). *Malbranchea pulchella* and *Humicola grisea* were collected from wheat and broad bean straw composts 45 °C. Several thermophilic fungal genera represented by *Penicillium duponti,*

Myriococcum albomyces, Thermomyces lanuginosus and *Sporotrichum thermophile* appeared when temperature ranged between 51 and 54 (Moubasher et al. 1984). *Scytalidium thermophilum* is an important thermophilic fungus in the production of *mushroom compost* (Wiegant 1992). *Scytalidium thermophilum* was also isolated from fresh streak horse droppings and dominated the fungal biota of compost. *Myriococcum themophilum, Stilbella thermophila, Thielavia terrestris*, and other thermophilic fungi are considered for preparation of compost (Straatsma et al. 1994). Interaction of thermophilic fungi *Chaetomium thermophilum, Humicola insolens, Scytalidium thermophilum,* and *Talaromyces thermophilus* isolated from mushroom compost was studied for mutual influence (Bilay and Elliott 1995). Thermophilic and thermo-tolerant fungi *Thermomyces lanuginosus, Scytalidium thermophilum,* and *Paecilomyces variotii* were predominant in a well-defined material consisting of straw and an N source when composted in insulated containers (Klamer et al. 1998). Suyanto et al. (2003) isolated a novel thermophilic fungus *Chaetomium* sp. no. M.S.017 having palm-oil mill fiber-decomposing properties. Kalogeris et al. (2003) produced endoglucanase and β-glucosidase (extracellular cellulolytic enzymes) in solid state fermentation of wheat bran, wheat straw, rice straw, corn cobs, and oat bran by thermophilic fungus *Thermoascus aurantiacus.*

4 Enzymes Involved in Lignocellulosic Biomass Degradation

4.1 Cellulose Degrading

Once a substrate with high accessibility is generated, the enzymatic hydrolysis is required, involving a cellulolytic complex acting in synergy (De Lima et al. 2016; Johansen 2016; Kim et al. 2014; Payne et al. 2015). Cellulose being a very recalcitrant biopolymer needs consortia of enzymes to expedite it bioconversion after thermo-chemical or biological pre-treatment. In actual facts, that fungal cellulases—which were major representatives of free cellulases—acted approximately five times faster than cellulosomes against either untreated or delignified cell walls and were found to penetrate more into the pore structure of lignocellulose microfibril networks as against the larger cellulosome complexes, which could only penetrate the larger wall lamella gaps. These advantageous lignocellulose degradation properties exhibited by free cellusases on cell walls were compromised when digesting purified forms of crystalline cellulose—cellulosomes were much more active on purified cellulose (Dixon 2013). Interestingly, the two systems tend to complement each other. That is, these systems display dramatic synergistic lignocellulose degradation activities (Brunecky et al. 2013; Resch et al. 2013). In spite the markedly different paradigms, the chemistry of cellulose degradation between the two systems remains the same. All cellulases use one of two mechanisms to degrade cellulose: (1) hydrolysis with retention of the stereochemistry of the anomeric hydroxyl group, or (2) hydrolysis with inversion of the anomeric hydroxyl group. One important difference between these mechanisms is that most retaining enzymes can catalyze both transglycosylation and hydrolysis since a covalent intermediate is produced on the enzyme, while no known inverting enzyme catalyzes transglycosylation (Wilson 2008). Nevertheless, while multiple microorganisms across the various trees of life produce different categories of CAZymes to decrypt the underlying codes of the plant cell wall assembly, filamentous fungi seem to be better evolved for this ability—that is, they are able to attack and efficiently hydrolyze its constituent polymers to their oligomeric and monomeric building blocks through the collective action of multiple carbohydrate-active enzymes, typically acting together as a cocktail with complementary modes of action. This is their sacrosanct duty in nature as they are responsible for the vast majority the biomass degradation in nature. Fungi have colonized a vast range of terrestrial and marine environments and are thus of vital importance for the recycling of carbon on Earth (Couturier et al. 2016; Payne et al. 2015).

A two-component cellulolytic system was proposed designated as C1 occurs preliminary to the hydrolysis of the native cellulose straight chain by the second component (Cx). The C1 component helps in the conversion of the native cellulose molecule into linear anhydroglucose chains, as well as might be involved in splitting cross-linkages in native cellulose. On the other hand, the second component (Cx) corresponds to enzymes hydrolyzing the 1,4-β-glucosidic linkage in cellulose, is not produced in the absence of substances containing this particular linkage, and largely unable to hydrolyze starch, pectic acid, alginic acids, or bacterial dextran (Reese 1956). Follow up to the hypothesis, there were several propositions on the identity of the C1 and Cx components involved in the microbial conversion of native cellulose to simpler sugars. For example, the Cx component was designated as endoglucanase acting before C1, a cellobiohydrolase (Ladisch et al. 1983); the catalytic domain corresponds to the hydrolytic Cx components and the cellulose-binding domain corresponds to the non-hydrolytic C1 component (Din et al. 1994). Six decades after the proposition and confusions, the two-component system on the microbial strategy for native cellulose degradation is still widely acceptable. However, the general consensus on the elements of the component is that the C1 are largely accessory proteins—lytic polysaccharide monooxygenases, swollenins, expansins, hemicellulases—that helps to overcome the polysaccharide crystallinity and the Cx corresponds to the core cellulases—endoglucanases, exoglucanases, and betaglucosidases (Horn et al. 2012; Johansen 2016; Kang et al. 2013; Kim et al. 2014; Kracher et al. 2016; X. Liu et al. 2015; Martinez 2016; Qin et al. 2013). These groups of enzymes are generally for the core cellulases (Cx), endoglucanases (EC 3.2.1.4) randomly cleave β-1,4 bonds in amorphous areas of cellulose chains and generate soluble cellodextrins with new reducing and non-reducing ends. Endoglucanases display a two possible catalytic mechanisms with retention of configuration or with inversion of configuration and have a characteristic large cleft containing the catalytic amino acids. Cellobiohydrolases (EC 3.2.1.91) also known as exoglucanases are processive enzymes that release cellobiose (i.e., two units of glucose linked by a β-1,4 bond) from either reducing or non-reducing ends of cellodextrins produced from the actions of endoglucanases. The characteristic feature of cellobiohydrolases is the presence of tunnel-bearing structures, which allow the enzyme to slide along the cellulose chain to the next cleavage site as the product is released. β-glucosidases (EC 3.2.1.21) hydrolyze cellobiose or cello-oligosaccharides into glucose units. They are characterized by a pocket-containing topology for optimal detection of the non-reducing extremity and of a single sugar unit. β-Glucosidases are non-processive enzymes because the substrate must be released after each cleavage event to allow the new glucose molecule to exit the pocket. Both exoglucanases and β-glucosidases are strongly inhibited by their reaction products, cellobiose and glucose, respectively (Bischof and Seiboth 2016; Girard et al. 2013; Glass et al. 2013; Gupta et al. 2016; Himmel et al. 2010; Kim et al. 2014; Xu et al. 2011). In contrast to the core cellulases, the elements of the C1 components keep expanding with advancement in technologies and bioprospecting. However, major accessory proteins discovered till date are lytic polysaccharide monooxygenases (LPMOs), swollenins, expansins, and the various hemicellulases (Bischof and Seiboth 2016; Cragg et al. 2015; Espino-Rammer et al. 2013; Horn et al. 2012; Kang et al. 2013; Marjamaa et al. 2013; Martinez 2016; Payne et al. 2015; Qin et al. 2013). LPMOs are classified as fungal AA9 (previously GH61) enzymes and non-fungal members of family CBM33. They are implicated in the oxidative cleavage of cellulose and other plant polymers and have gained lot of tractions in recent times because of their huge synergistic effect on the activity of canonical cellulases in the saccharification of pre-treated lignocellulosic biomass. LPMOs are active on varieties of polysaccharides either crystalline or amorphous. Examples include chitin, cellulose, starch, xyloglucan, cellodextrins among others. LPMOs generally work in tandem with the enzyme cellobiose dehydrogenase (CDH), which serves as its redox partner to bring about its oxidative cleavages of polymeric substrates. Other potential extracellular electron source includes ascorbic acid, lignin, phenols procured from plant biomass or produced by fungi, and LPMO-reducing di- or triphenols among others. They have equally been reported to effect classical Fenton chemistry on varieties of substrates (Courtade et al. 2016; Cragg et al. 2015; Eibinger et al. 2014; Hemsworth et al. 2014; Johansen 2016; Kracher et al. 2016; Lo Leggio et al. 2015; Martinez 2016; Patel et al. 2016). Expansins, on the other hand, are non-catalytic, non-hydrolytic proteins that can play a role in cellulose degradation by the disruption of cellulose fibers and by breaking hydrogen bonds between cellulose microfibrils and cellulose–hemicellulose complexes, leading to polysaccharide slippage, also called creep activity. They are normally part of plant cell wall protein but have also been identified in varieties of bacteria and fungi actuating plant cell wall depolymerization activities. A representation of expansin in fungi is called swollenin (Cosgrove 2000; Cosgrove et al. 2002; Gourlay et al. 2013; Kang et al. 2013; Kim et al. 2014; Rocha et al. 2016; Saloheimo et al. 2002).

4.2 Hemicellulose-Degrading Enzymes

One other major accessory proteins fitting into the C1 modules are the group of proteins that acts on hemicellulose broadly categorized as hemicellulases. Hemicelluloses component of a biomass can be degraded by different kinds of enzymes viz. (1) endoezymes—which randomly cleave the bonds between the building blocks of a polymer (2), exoenzymes—which cleave either a single or monomer from the end of the polysaccharide chain, and (3) glycosidases—hydrolyze the oligomers of disaccharides of hemicellulose polymer (Alexander 1977). Hemicellulose being diverse requires a diverse set of enzymes acting on both terminal and/or internal glycosidic linkages. They include xylanases, β-xylosidases, xyloglucanases, acetyl xylan esterases, ferulic esterases, β-mannanases, β-galactosidases, arabinofuranosidases, and glucuronidases (Arantes and Saddler 2010, 2011; Gourlay et al. 2013; Gupta et al. 2016; Gusakov 2013; Linton and Greenaway 2004; Sun et al. 2015), which are produced by many microorganisms including fungi *Aspergillus niger* (Conrad 1981; Gokhale et al. 1986), *Aspergillus*, and *Trichoderma reesei* (Dekker 1983; Chahal 1985). Biswas et al. (1990) have characterized the xylanase produced by a hyperxylanolytic mutant of *Aspergillus ochraceus*. Bhat et al. (1987) screened several thermophilic fungi for extracellular xylanase. Of the thermophiles, *Thermomyces lanuginosus* RM-B performed best producing 154U xylanase. The enzyme was reported to have optimum temperature 60–70 °C at neutral pH (Bakalova et al. 2002).

4.3 Lignin-Degrading Enzyme

Lignin degradation is an oxidative process mainly attributed to the secondary metabolism, or to restricted availability of carbon, nitrogen, or sulfur, and it is normally not degraded as sole carbon and energy sources (Silva et al. 2010). In nature, it is generally versatile peroxidases (VPs; EC 1.11.1.16; CAZy AA2) that combine the substrate-specificity characteristics of the two other ligninolytic peroxidases (LiP and MnP). Unlike these two enzymes, it can oxidize phenolic and nonphenolic substrates including veratryl alcohol, methoxybenzenes, and lignin model compounds (Wong 2009). Peroxidases such as HRPs (EC 1.11.1.7, CAZy AA2), secreted by fungus *Phanerochaete chrysosporium*, are potential biocatalysts for bioremediation of environment polluted by harmful compounds (e.g., endocrine disrupting compounds: 17α-ethinylestradiol (EE2)) (Rathner et al. 2017). HRPs have also been implicated in the cell wall biosynthesis, indole-3-acetic acid (plant growth hormone) catabolism, and oxidation of toxic compounds (Falade et al. 2017). The lignin degradation can be further enhanced by the action of other enzymes such as aryl alcohol oxidases (AAO;EC1.1.3.7.; CAZy AA3) that oxidize many primary alcohols containing an aromatic ring and is described in *Pleurotus eryngii* (Guillén et al. 1992), glyoxylate oxidase (GOx, EC 1.2.3.5) (Kersten and

Cullen 2007), pyranose 2-oxidase (glucose 1-oxidase; EC1.1.3.4,CAZy AA3), and cellobiose dehydrogenase (CDH, EC 1.1.99.18; CAZy AA3). Fungal aryl-alcohol dehydrogenases (AAD; EC 1.1.1.90) and quinone reductases (QR, EC 1.6.99.2), tyrosinases (EC.1.14.18.1) and catechol oxidases (EC 1.10.3.1) are also involved in lignin degradation (Gutiérrez et al. 1994; Frommhagen et al. 2017). Significant amounts of peroxidases have been produced by fungi in the submerged or solid state fermentation. In the submerged culture, the fungus *Mucor racemosus* produced lignin peroxidases (75,376 U/L) and manganese peroxidases (4484 U/L) (Bonugli-Santos et al. 2010) while in a solid state fermentation of steam-exploded wheat straw by *Phanerochaete chrysosporium* manganese peroxidases (1375 U/L) were observed (Bonugli-Santos et al. 2010). Submerged fermentation of versatile peroxidases (7300 U/L) was realized by genetically modified *Pleurotus ostreatus* (Tsukihara et al. 2006). Finally to conclude, the lignin degradation is not catalyzed by any particular enzyme but concerted action of oxidative coupling of phenol oxidase, peroxidase, glucose oxidase, and ligninase. Ligninase catalyzes breakdown of the ether linkages with glucose oxidase providing necessary co-substrate peroxide, then peroxidase involved in radical formation, phenol oxidases polymerize and depolymerize phenols often, initial degradation depends on the level of glucose oxidases. Laccases are involved mainly in decarboxylation of side chains and to some extent in the oxidative cleavage.

5 Factors Affecting Lignocellulose Degradation

A wide range of fungi, bacteria, and actinomycetes are present on agricultural produce. The rate of their growth depends on the conditions of storage particularly the water activity of the substrate, aeration, ambient temperature, nutritional factors, pH, and type of substrate. If the energy released during the respiration of the plant and microbial cells cannot escape as quickly as it is produced, the substrate heats up and conditions which may favor the growth of thermophilic organisms (Sharma and Johri, 92).

5.1 Carbon to Nitrogen Ratio

The C/N ratio plays an important role in the decomposition of any organic matter. If a material has wider C/N ratio, then it is advisable to bring down C/N ration by adding nitrogen to enhance decomposition. The addition of nitrogen to materials increases their rate of decomposition. Addition of nitrogen to a substrate whose initial C/N ratio is above a critical ratio (20–30) causes a decrease in the substrates C/N ratio during its decomposition. If the initial C/N ratio is below the critical one, no change in the substrates C/N ratio will occur with time. Net mineralization of organic "N" occurs when the substrate being decomposed has an initial C/N ratio which is below the critical one (Henna 1975). Rajasekaran and Sampatkumar (1981) suggested that with proper C/N ratio adjustments, agricultural waste could be probably recycled with a greater degree of "N" enrichment. The lowering of C/N ratio of saw dust to 12: 1 with inorganic N enhanced decomposition of saw dust at 25 °C (Olayinka and Adebaya 1984). The addition of nitrogen in the form of fertilizer or cakes to nitrogen poor materials accelerated their decomposition. About 290 mg of CO_2 was evolved/100 g soil when wheat straw was supplemented with nitrogen as compared to wheat straw alone (120 mg of COp 00 g soil) (Bhardwaj and Gaur 1985). Nitrogen supplementation in the form of 0.08% NH_4CI was found to be optimum for *Pleurotus ostreatus* and *Sporotrichum pulverulentum* to degrade paddy straw (Kahlon and Dass 1987). Nigam et al. (1988) reported the effect of various nitrogenous substances on bioconversion of whole bagasse using four fungal cultures. Yeast extract was better source of nitrogen than NH_4CI for *Streptomyces badius* and *Streptomyces viridosporus* (Giroux et al. 1988). Urea nitrogen favored the degradation of lignin as well as cellulose and hemicellulose up to a certain level (1.5% sterile urea or 3% unsterile urea on dry weight basis) beyond which the degradation of lignin was relatively more adversely affected than cellulose (Tripathi and Yadav 1991). The application of paddy straw degraded with efficient fungi in combination with N_{80} produced a significant effect on wheat yield due to biodegradation of paddy resulting in nutrient-rich compost with a low C:N ratio and high humus content (Kanotra and Mathur 1994). Soybean meal increased xylanase production in mixed cultures of fungi (Gutierrez-Correa and Tengerdy 1998). The organic nitrogen source of soybean helped only growth but not enzyme production in mixed cultures of *Trichoderma reesei* and *Aspergillus niger* when grown on sugarcane bagasse (Gutierrez-Correa et al. 1999). George et al. (2001) reported that the use of yeast extract stimulated xylanase (15 IU/mL) by *Thermophilic fungi, Thermomonospora* sp. cellulolytic enzyme activities were higher with inorganic nitrogen, when *Thermoascus aurantiacus* was grown on organic residues. Ammonium sulfate, the least expensive nitrogen source, gave the best activities (Kalogeris et al. 2003). The C:N ratio in *Phanerochaete flavido-alba* inoculated lignocellulosic substrates decreased during the first month of composting and from then onwards increased because of C loss. The final C:N ratio reached with this fungus was close to 28 (Lopez et al. 2002). The main factor influencing enzymatic hydrolysis is lignin and hemicellulose content.

5.2 pH

For solid state fermentation of agricultural residues, pH adjustment is difficult and most of the agro-residues have pH near neutrality. The optimum pH for lignin decomposition by *Phanerochaete chrysosporium* was 4–4.5 with marked suppression above 5.5 and below 3.5 (Kirk et al. 1978). The optimum pH for *Coriolus versicolor* for lignin biodegradation was 4.5 (Ishihara 1983). Hegarty and Curran (1985) showed that different fungi used in their study produced greatest weight losses of wood in pH 5.0–8.0 range. Most xylanases known so far have their optimum pH around neutrality. Even xylanases produced by most alkaliphiles reported to date have their optimum pH around neutrality (Rajaram and Varma 1990; Park et al. 1992). Maximum xylanase activity by B. *circulans* AB 16 grown on modified medium containing wheat straw was between pH 6–7 retaining **91%** activity at pH 5 and 85 and 55% at pH 8.0 and 9.0, respectively (Dhillon and Khanna 2000). The xylanase from *Thermomonospora* sp. enzyme was active in a broad pH range (5–9) with maximum activity at pH 7.0, and retained 68% and 44% of its maximum activity at pH 8.0 and 9.0, respectively (George et al. 2001). The optimum pH for cellulase production was found to be resulting in high endoglucanase (1572 U/g) and *p*-glucosidase (106.6 U/g wheat straw) activities during solid state fermentation of wheat straw by *Thermoascus aurantiacus* (Kalogeris et al. 2003).

5.3 Aeration

Aeration greatly affects the composting process, during which organic matter undergoes biological and physiochemical reactions, mainly by oxidation, with the final production of carbon dioxide, water, and partially humified organic matter (He et al. 1992). The most important event during composting is the generation of heat. The requirement of oxygen increases in thermophilic range of temperature but normally remains constant at the temperature of 65 °C. The cause of heat generation above 65 °C is exothermic chemical reaction (Currie and Festenstein 1971). The thermophilic fungi can undergo growth as facultative aerobes with 0.7–1.05% aeration (Deploey and Fergus 1975). Solid state fermentation studies were undertaken to investigate the effect of O_2 supply, aeration rate, and carbon source, and level on the shelf life of *Lagenidium giganteum.* Growth rate as measured by CO_2 evolution rate was greater at higher aeration levels (Vander-Gheynst et al. 2000). Carrasco et al. (1999) evaluated a bioreactor for solid state fermentation of sugarcane bagasse and cellulolytic fungus strain *Aspergillus fumigatus* at the rate of 18 L/min aeration. Submerged fermentation was carried out in fermenter with controlled aeration (1.5 VVM) and dissolved oxygen level (30%), rifamycin production increased to about 1.5-fold greater than the flasks (Venkateswarlu et al. 2000). Lopez et al. (2002) reported that degradation was enhanced when low aeration (3 L of air per week) was supplied during SSF of cucumber and green bean waste inoculated with white-rot fungi.

5.4 Temperature

The rate of carbon mineralization of saw dust increased with temperature from 25 to 40 °C with the optimum being 30–35 °C (Olayinka and Adebaya 1984). Hegarty and Curran (1985) reported maximum wood decay by the majority of the test fungi was in the range of 15–25 °C. Heat pre-treatment of cereal straw at 60 and 90 °C was considered sufficient for commercial cultivation process with the added benefit of a fermentative pre-treatment reducing the risk of infection but best growth and highest speed of colonization were on the sterilized straw (Maziero and Zadrazil 1994). Horwath and Elliott (1996) composted rye grass for 45 days at a mesophilic temperature of 25 °C.

5.5 Water Activity/Moisture

Most species of fungi require at least 0.75_{aw} while no fungal growth occurs with less than 0.61_{aw}. With changes in temperature, vapor pressure over the substance did not change very much while with change in temperature, the extent of microbial activity of substrate was determined by the water activity (Christensen and Kaufman 1974). The number of moulds was reduced and *Pleurotus ostreatus* yield was increased (Stanek and Bisko 1982). Eighty per cent of water content seemed to be an important parameter for improvement of feed value for ruminants. Solid state fermentation of banana fruit stalk at 70% moisture content was found suitable for SSF with thermophilic *Bacillus* (Krishna and Chandrasekara 1996). The mixed culture of *Trichoderma reesei* with *Aspergillus niger* for solid state fermentation of sugarcane bagasse produced better results at 80% moisture content (Gutierrez-Correa et al. 1999). Initial moisture content (80%) of coffee pulp under solid state fermentation with *Aspergillus niger* had a significant effect on the total amino acid content.

5.6 Other Factors

Jeffries et al. (1981) reported that phosphorus starvation did not influence lignin degradation. These workers also suggested that the balance of trace metals Mg^{++} and Ca^{++} was important for lignin biodegradation. Shaikh et al. (1997) found that the divalents cations such as Co^{2+}, Mn^{2+}, Fe^{2+}, and Zn^{2+} had strong inhibitory effect on enzyme activity under solid state fermentation. The concentration of veratryl alcohol necessary for full protection of ligninase activity varied according to the concentration of H_2O_2 present in the medium, which in turn depends on carbon source (glucose or glycerol). It is proposed that the carbon source influences the overall ligninase activity, not only directly by affecting the rate and the type of synthesized ligninase but also by affecting the ratio of H_2O_2 production (Tandon and Odier 1988).

6 Fermentation

The limitations of physical and chemical methods of upgrading the nutritive value of straws and agro-residues have been overcome by the application of microbial process which can be used for bioconversion as well as production of lignocellulolytic enzyme by using thermophilic as well as mesophilic fungi. The processes are:

1. Composting
2. Submerged fermentation
3. Solid state fermentation.

7 Composting

The formation of compost from the degradation of organic matter depends on the abilities of microflora to produce and excrete specific degradative enzymes (Hankin et al. 1976). Crawford (1983) compiled the information regarding composting of agricultural residues. Several reports are available where inoculation with mesophilic fungi improved the quality of compost. Gaur et al. (1982) reported that inoculation of four mesophilic fungi, *Aspergillus niger, Aspergillus* sp., *Trichoderma viride*, and *Penicillium* sp., had increased total nitrogen, available phosphorus and humus content of composted sorghum stalk and wheat straw (5:3) and jamun leaves. Inoculation of cellulolytic fungi and actinomycetes increased the temperature faster and accelerated the process of rice straw composting (Hang Won et al. 1995). The mesophilic microorganisms were used for bioconversion of rice husks cane bagasse, wheat straw, maize cob and corn stalk, and many other agricultural residues listed in Table 8.1. Several workers have dealt with rice straw composting after

Table 8.1 Bioconversion of agricultural residues by different fungi

Substrate	Fungi	Reference
Sugarcane bagasse	*Trichoderma viride*	Sharma et al. (1991)
	Trichoderma reesei	Gutierrez-Correa et al. (1999)
	Pleurotus sp.	Zadrazil and Puniya (1995)
Rice straw	*Pleurotus florida, Sporotrichum* sp.	Mehta et al. (1990)
	Trichoderma viride	Kanotra and Mathur (1994)
Rice husk	*P. sajor-caju*	Beg et al. (1986)
Maize straw	*Phanerochaete chrysosporium*	Breccia et al. (1997)
	T. reesei	Ming et al. (1999)
Wheat straw	*Stropharia* sp.	Kamra and Zadrazil (1986)
	P. ostreatus, P. *flabellatus*	Rajarathnam et al. (1987)
Coir pith	*Polyporus* sp.	Mani and Marimuthu (1992)
	T. viride	Muniswaram and Charyulu (1994)
Saw dust	*Polyporus* sp.	Nandi et al. (1996)

supplementation with rock phosphate, glue waste, basic slag, and inorganic nitrogen sources (Jhorar et al. 1991; Wahyono and Sahwan 1998; Abdel-Azeem 2001). The use of cattle manure rather than relatively expensive microbial catalyst could be recommended as decreasing the composting degradation period of rice straw (Bhumibhamon et al. 1988). The inoculation of *Azotobacter* to 1-month-old decomposed rice straw increased the N-content and composting with rock phosphate increased both citrate and water-soluble phosphorus (Tiwari et al. 1988).

The two-step process of composting of rice straw was effective in accelerating humification, cation exchange capacity of the composted material (Kakezawa et al. 1992). Nandi et al. (1996) proposed a two-step process for composting of rice straw which results in better humus formation. The treatment of shredded rice straw supplemented with chemical accelerator, sheep manure, and EM solution (mixture of microbial culture, commercially known as EM solution) gave the most desirable characteristics of the final product particularly narrow C:N ratio and high WHC (Abdel-Azeem 2001). Two white-rot fungi, *Pleurotus ostreatus* and *Lentinus edodes* delignified and increased digestibility of corn straw *(Zea mays)* under solid state fermentation (Sermanni et al. 1994). Under fed-batch process of saccharification of short fiber waste material from the paper industry by *Penicillium* sp. and *Trichoderma reesei,* combined preparation of *T. reesei* and *Aspergillus* showed highest saccharification activity (Castellanos et al. 1995). Cellulolytic fungi such as *Trichoderma reesei* QM9414, *Trichoderma viride, Pleurotus sajor-caju,* and *Coprinus cinereus* lowered the C:N ratio of paddy straw during composting (Kanotra and Mathur 1994). Phospholipid fatty acid (PLFA) analysis study of compost samples collected from bench-scale grape pomace and rice straw composting showed that inoculation had little effect on the microbial community structure of the compost when temperature peaked (Lei-Fei et al. 1998). Thermophilic fungi are generally used for mushroom compost preparation either in tunnels or windrows (Gerrits et al. 1995; Straatsma et al. 1995; Sharma et al. 2000). Ross and Harris (1983) reported potential of *Torula thermophila* as a rapid composting organism. Natural thermophilic flora played a role during composting of wheat and broad bean straw (Moubasher et al. 1984). Biomass of these fungi was estimated by Wiegant (1992) in mushroom compost. Sharma and Johri (1992) have documented the role of thermophilic fungi in composting. *Scytalidium thermophilum* is an important thermophilic fungus in the production of mushroom compost (Wiegant 1992; Sharma et al. 2000).

8 Submerged Fermentation

It involves the growth of microorganisms as a suspension in a liquid medium in which various nutrients are either dissolved as in many cases of commercial media or suspended as particulate solid (Frost and Moss 1987). As the water content increases, the gas phase is reduced and gas exchange is thus increasingly impeded, finally conditions become anaerobic in the substrate suspension (Rajarathnam and

Bano 1989); hence, better aeration of the medium is a must, which can be done either by an air sprayer or an agitator. In submerged fermentation process, much care has to be taken in maintaining sterility throughout the period of process, and heat during the process has to be removed. Submerged fermentation process is being used for the production of many industrially important products such as antibiotics, alcoholic beverages, and single cell protein from agricultural residues. Garg and Neelakanthan (1982) fermented bagasse with *Aspergillus terreus* under submerged culture conditions. The fermented products contained 32.9% crude protein and protein recovery was 22.2%. Submerged fermentation of cellulose of solka floc with *Aspergillus terreus* for 48 h at 35 °C resulted in maximum reduction of cellulose and 37.8% increased crude protein content in the product (Miller and Srinivasan 1983). *Humicola aeruginosa* and *Sporotrichum thermophile* performed best in liquid fermentation; the higher productivity of xylanase (18.72 IU/mL) was obtained in liquid culture of *H. lanuginosa* (Grajek 1987). Dighe and co-workers (1988) utilized enzymatic hydrolysates of various crop wastes for the production of yeast biomass. Holder et al. (1989) cultivated *Geotrichum candidum* and *Candida utilis* ATCC 9256 in glucose containing bagasse hemicellulose hydrolysate. The cell yields of G. *candidum* and *C. utilis* obtained were 19.5 and 15 g per liter, respectively, in hemicellulose hydrolysates. Gulati (1989) studied various factors that affected the submerged fermentation of sugarcane bagasse by *Trichoderma reesei* OM 9414. The thermophilic fungus *Thermomyces lanuginosus* was investigated in semi-solid and liquid fermentation for production of cellulase-free endoxylanase from corn cobs. The best results were obtained with solid to liquid ratio 1:2 and inoculums size 20% (v/v) (Damaso et al. 2000).

9 Solid State Fermentation

Durand et al. (1988) defined solid state fermentation (SSF) as a process which refers to the culturing of microorganisms on and/or within the particles of solid matrix (solid substrate) where the liquid content bound with them is at the level corresponding to the water activity assuring growth and metabolism of cells, but not exceeding to the maximal water holding capacity of the solid matrix. The advantages and disadvantages of solid state fermentation are discussed by many workers (Moo-Young et al. 1983; Rosen and Schugerl 1983; Tengerdy 1985). Baca et al. (1993) studied the gas phase of compost during SSF of sugarcane bagasse. The evolution of CO_2 in the compost atmosphere (0–3.5 mmol L) paralleled temperature changes, whereas the highest O_2 concentrations (10–20%) coincided with the end of thermophilic phase. The accumulation of ethylene (0.01–0.4 ppm) was not markedly affected by changes in temperature. The maximum loss of organic matter of wheat straw was reported after 120 days of inoculation period during solid state fermentation process using two species of *Pleurotus* fungus (Zadrazil 1976). Solid state fermentation of sugarbeet pulp using *Thermoascus aurantiacus* and

Sporotrichum thermophile (Myceliophthora thermophila) increased two fold protein content in 48 h (Grajek 1988). Roche et al. (1994) reported that solid state fermentation of sugarbeet pulp with thermophilic fungus, *Thermoascus aurantiacus* ATCC 26904 increased the quantity of alpha-L-arabinofuranosidase. Maximum alpha-L-arabinofuranosidase production occurred when medium was inoculated with 5-day-old spores. Jain et al. (1999) produced xylanase by thermophilic fungus *Melanocarpus albomyces* IIS-68 during solid state fermentation on sugarcane bagasse and purified the enzyme in an aqueous two-phase system comprising polyethylene glycol (PEG 6000) and potassium salts.

10 Conclusion

The preponderance of inducible CAZymes as well as non-hydrolytic accessory proteins in the fungus secretome reiterates the concept of synergism between proteins for effective hydrolysis of lignocellulosic biomass. In addition, the preferential association of certain proteins and the attending differential biomass hydrolysis gives an insight into the concept of protein–protein interaction as well as clues about non-hydrolytic proteins from the fungus with possible roles in the synergistic deconstruction of lignocellulosic biomass.

Fungi involved in complex lignocelluloses containing raw materials degradation express a broad spectrum of enzymes. According to the enzyme composition and degradation mechanisms, three groups of fungi are described: soft-rot, brown-rot, and white-rot fungi. A huge effort has been done and is still ongoing to classify and group these fungi enzymes in a central database called the carbohydrate-active enzyme database or short CAZy. The CAZy classification groups the proteins in families according to amino acid sequence similarity. In the last few years, investigation attention was focused on the synergistic action of the fungal enzymes involved in the lignocelluloses degradation. One of the key players in those actions is LPMOs. These enzymes work in synergy with glycoside hydrolases (GHs) and stimulate their activity by increasing the accessibility to the substrate and enhancing overall polysaccharide conversion efficiency. In addition, LPMOs are able to degrade insoluble polysaccharides such as crystalline cellulose and soluble cello-oligomers. Importance of these enzymes is also confirmed by the phylogenetic investigations. Genomes of lignocelluloses-degrading fungi usually encode several LPMOs gens. Many cultivation strategies were studied to enhance efficiency of lignocellulose-containing raw materials degradation whereas delignification was indicated as the main barrier for the successful lignocellulose utilization. To remove lignin and enhance hemicellulose and cellulose degradation, one of the strategies was the solid state fermentation of brown-rot and white-rot fungi. This strategy approved the hypotheses of the synergistic action of the fungal enzymes and enhancement of the delignification efficiency.

References

Abdel-Azeem HHM (2001) Some factors affecting the composting process of rice straw. Ann Agricul Sci Cairo 46(2):525–542

Ahamed A, Vermette P (2008) Enhanced enzyme production from mixed cultures of *Trichoderma reesei* RUT-C30 and *Aspergillus niger* LMA grown as fed batch in a stirred tank bioreactor. Biochem Eng J 42:41–46

Alexander M (1977) Introduction to soil microbiology. Wiley-Eastern Ltd, New Delhi, pp 163–173

Arantes V, Saddler JN (2010) Access to cellulose limits the efficiency of enzymatic hydrolysis: the role of amorphogenesis. Biotechnol Biofuels 3(1):1–11

Arantes V, Saddler JN (2011) Cellulose accessibility limits the effectiveness of minimum cellulase loading on the efficient hydrolysis of pretreated lignocellulosic substrates. Biotechnol Biofuels 4:3

Arantes V, Goodell B (2014) Current understanding of brown-rot fungal biodegradation mechanisms: a review. In: Schultz TP, Goodell B, Nicholas DD (eds) Deterioration and protection of sustainable biomaterials. American Chemical Society, Mississippi, pp 4–21

Asim N, Emdadi Z, Mohammad M, Yarmo MA, Sopian K (2015) Agricultural solid waste for green desiccant applications: an overview of research achievement, opportunities and perspectives. J Clean Prod 91:26–35

Baca MT, Esteban E, Almendros G, Sanchez-Raya AJ (1993) Changes in the gas phase of compost during solid state fermentation of sugarcane bagasse. Bioresour Technol 44:5–8

Bakalova N, Petrova S, Atev A, Bhat M, Kolev D (2002) Biochemical and catalytic propoerties of endo-1, 4 xylanase from *Thermomyces lanuginosus* (wild and mutant strains). Biotechnol Lett 24:1167–1172

Batt CA (1991) Biomass. In: Mosses U, Cape RE (eds) Biotechnology, the science and the business. Harvard Academic Press, New York, pp S21–S36

Beckham GT, Bomble YJ, Bayer EA, Himmel ME, Crowley MF (2011) Applications of computational science for understanding enzymatic deconstruction of cellulose. Curr Opin Biotechnol 22(2):231–238

Beg S, Zafar SI, Saha FH (1986) Rice husk biodegradation by *Pleurotus ostreatus* to produce a ruminant feed. Agric Wastes 17:15–21

Berrin JG, Navarro D, Couturier M, Olive C, Grisel S, Haon M, LesageMeessen L (2012) Exploring the natural fungal biodiversity of tropical and temperate forests toward improvement of biomass conversion. Appl Environ Microbiol 78(18):6483–6490

Bhat KM, Smith DC, Wood TM, Grassi G. (ed), Delmon B (ed) Molle JF (ed), Zibetta H (1987) Screening of mesophilic and thermophilic fungi for extracellular xylanase and beta xylosidase activity. Biomass for energy and industry. In: 4th EC conference. Proceeding of the international conference, Orleans, 11–15 May 1987, pp 778–782

Bhardwaj KKR, Gaur AC (1985) Recycling of organic waste. All India coordinated research project on microbiological decomposition and reycling of farm and city wastes, pp 1–104

Bhumibhamon O, Chaiyapol K, Sirisansanee-Yakul S (1988) Studies on chemical and environmental changes during composting. Recent advances in biotechnology and applied biology. In: Proceedings of English international conference on global impacts of applied microbiology and international conference on applied biology and biotechnology, Hong Kong, 1–5 Aug 1988, pp 587–594

Bilay VT, Elliott TJ (1995) Interaction of thermophilic fungi from mushroom compost in different agar media and temperatures Mushroom Science xiv, volume 1. In: Proceedings of the 14th international congress on the science and cultivation of edible fungi, Oxford, 17–22 Sept 1995, pp 251–255

Bischof RJ, Seiboth B (2016) Cellulases and beyond: the first 70 years of the enzyme producer *Trichoderma reesei*. Microb Cell Fact 15(1):106

Biswas SR, Janar SC, Mishra AK, Nanda G (1990) Production, purification and characterization of xylanase from a hyperxylanolytic mutant of *Aspergillus ochraceus*. Biotechnol Bioeng 35:244–251

Bonugli-Santos RC, Durrant LR, Silva M, Sette LD (2010) Production of laccase, manganese peroxidase and lignin peroxidase by Brazilian marine-derived fungi. Enzyme Microb Technol 46:32–37

Breccia JS, Bettucci L, Piaggio M, Sinergi F (1997) Degradation of cane bagasse by several white rot fungi. Acta Biotechnol 17(2):177–184

Brunecky R, Alahuhta M, Xu Q, Donohoe BS, Crowley MF, Kataeva IA, Bomble YJ (2013) Revealing nature's cellulase diversity: the digestion mechanism of *Caldicellulosiruptor bescii* CelA. Science 342(6165):1513–1516

Bugg TD, Rahmanpour R (2015) Enzymatic conversion of lignin into renewable chemicals. Curr Opin Chem Biol 29:10–17

Carrasco T, Valino E, Medina I, Ravelo D (1999) Design and evaluation of a bioreactor for solid state fermentation. Cuba J Agric Sci 33(4):409–414

Castellanos OF, Sinitsyn AP, Vlasenko EY (1995) Evaluation of hydrolysis constitution of cellulosic materials by *Penicillium cellulase*. Bioresour Technol 52:109–117

Chahal DS (1985) Solid state fermentation with *Trichoderma reesei* for cellulase production. Appl Environ Microbiol 49:205–210

Chaplin M (2016) Water structure and science. http://www1.lsbu.ac.uk/water/cellulose.htm

Chen (2014) Chemical composition and structure of natural lignocellulose. In: Biotechnology of lignocellulose: theory and practice. Springer Netherlands, Dordrecht, pp 25–71

Christensen CM, Kaufman HH (1974) Microflora. In: Christensen M (ed) Storage of cereal grains and their products. American Association of Cereal Chemists, Eagan, pp 158–192

Chundawat SP, Beckham GT, Himmel ME, Dale BE (2011) Deconstruction of lignocellulosic biomass to fuels and chemicals. Annu Rev Chem Biomol Eng 2:121–145

Conrad D (1981) Enzymatic hydrolysis of xylans 1A high xylanase and xylosidase producing strain of *Aspergillus niger*. Biotechnol Lett 3:345–350

Cosgrove (2000) Loosening of plant cell walls by expansins. Nature 407(6802):321–326

Cosgrove LLC, Cho HT, Hoffmann-Benning S, Moore RC, Blecker D (2002) The growing world of expansins. Plant Cell Physiol 43(12):1436–1444

Cosgrove DJ (2005) Growth of the plant cell wall. Nat Rev Mol Cell Biol 6(11):850–861

Courtade G, Wimmer R, Rohr AK, Preims M, Felice AK, Dimarogona M, Aachmann FL (2016) Interactions of a fungal lytic polysaccharide monooxygenase with beta-glucan substrates and cellobiose dehydrogenase. Proc Natl Acad Sci U S A 113(21):5922–5927

Couturier M, Berrin JG (2013) The saccharification step: the main enzymatic components. In: Faraco V (ed) Lignocellulose conversion: enzymatic and microbial tools for bioethanol production. Springer, Berlin, Heidelberg

Couturier M, Bennati-Granier C, Urio MB, Ramos LP, Berrin JG (2016) Fungal enzymatic degradation of cellulose. In: Green fuels technology, pp 133–146

Cragg SM, Beckham GT, Bruce NC, Bugg TD, Distel DL, Dupree P, Zimmer M (2015) Lignocellulose degradation mechanisms across the tree of life. Curr Opin Chem Biol 29:108–119

Crawford JH (1983) Composting of agriculture waste—a review. Process Biochem 18:14–18

Currie JA, Festenstein GN (1971) Factors defining spontaneous heating and ignition of hay. J Sci Food Agric 22:223–320

Damaso MCT, Andrade CMMC, Pereira-Junior N, Finkelstein M, Davision BH (2000) Use of corncob for endoxylanase production by thermophilic fungus *Thermomyces lanuginosus* IOC-4145. Appl Biochem Biotechnol 84–86:821–834

De Lima DR, Silveira MHL, Del Rio L, Ramos LP (2016) Pre-treatment processes for cellulosic ethanol production: processes integration and modeling for the utilization of lignocellulosics such as sugarcane straw. In: Soccol RC, Brar KS, Faulds C, Ramos PL (eds) Green fuels technology: biofuels. Springer International Publishing, Cham, pp 107–131

Dekker RFH (1983) Bioconversion of hemicellulose: aspects of hemicellulose production by *Trichoderma reesei.* M. 9414 and enzymatic saccharification of hemicellulose. Biotechnol Bioeng 30:1127–1146

Deploey JJ, Fergus CL (1975) Growth and sporulation of thermophilic fungi and actinomycetes in 0 N atmospheres. Mycologia 67:780–797

Dhillon A, Khanna S (2000) Production of a thermostable alkali-tolerant xylanase from *Bacillus circulans* AV16 grown on wheat straw. World J Microbiol Biotechnol 16:325–327

Dighe AS, Khandeparkar VG, Berabet SM (1988) Production of single cell protein from enzymatic cellulosic hydrolysates. Indian J Microbiol 28:128–130

Din N, Damude HG, Gilkes NR, Miller RC, Warren RA, Kilburn DG (1994) C1-Cx revisited: intra molecular synergism in a cellulase. Proc Natl Acad Sci 91(24):11383–11387

Ding SY, Liu YS (2012) Imaging cellulose using atomic force microscopy. Methods Mol Biol 908:23–30

Dixon RA (2013) Microbiology: break down the walls. Nature 493(7430):36–37

Doi RH (2008) Cellulases of mesophilic microorganisms: cellulosome and noncellulosome producers. Ann N Y Acad Sci 1125:267–279

Durand A, Broise D, Blachere H, Broise D (1988) Laboratory scale bioreactor for solid state process. J Biotechnol 8(1):59–66

Eibinger M, Ganner T, Bubner P, Rosker S, Kracher D, Haltrich D, Nidetzky B (2014) Cellulose surface degradation by a lytic polysaccharide monooxygenase and its effect on cellulase hydrolytic efficiency. J Biol Chem 289(52):35929–35938

Espino-Rammer L, Ribitsch D, Przylucka A, Marold A, Greimel KJ, Herrero Acero E, Druzhinina IS (2013) Two novel class II hydrophobins from *Trichoderma* spp. stimulate enzymatic hydrolysis of poly(ethylene terephthalate) when expressed as fusion proteins. Appl Environ Microbiol 79(14):4230–4238

Falade AO, Nwodo UU, Iweriebor BC, Green E (2017) Lignin peroxidase functionalities and prospective applications. Microbiologyopen 6:e00394

Frommhagen M, Mutte SK, Westphal AH, Koetsier MJ (2017) Boosting LPMO driven lignocellulose degradation by polyphenol oxidase activated lignin building blocks. Biotechnol Biofuels 10:121

Frost GM, Moss DA (1987) Production of enzymes by fermentation. In: Rehn HJ, Reed G (eds) Biotechnology, vol 79. VCH. Pub., FRG, pp 65–221

Gandarias I, Luis P (2013) Hydrotreating catalytic processes for oxygen removal in the upgrading of bio-oils and bio-chemicals. https://doi.org/10.5772/52581

Garg SK, Neelakanthan S (1982) Effects of nutritional factors on cellulase enzyme and microbial protein production by *Aspergillus terreus* and its evaluation. Biotechnol Bioeng 24:109–125

Gaur AC, Sadasivam KV, Mathur RS, Magu SP (1982) Role of mesophilic fungi in composting. Agric Wastes 4:453–460

George SP, Ahmad A, Rao MB (2001) A novel thermostable xylanase from *Thermomonospora* sp. influence of conditions on thermostability. Bioresour Technol 78:221–224

Gerrits JPG, Amsing JGM, Straatsma G, Griensven LJLD, Elliott TJ (1995) Phase I process in tunnels for the production of *Agaricus bisporus* compost with special reference to the importance of water. Mushroom science and cultivation of edible fungi, Oxford, 17–22 Sept 1995, pp 203–211

Girard V, Dieryckx C, Job C, Job D (2013) Secretomes: the fungal strike force. Proteomics 13(3–4):597–608

Girio FM, Fonseca C, Carvalheiro F, Duarte LC, Marques S, Bogel-Lukasik R (2010) Hemicelluloses for fuel ethanol: a review. Bioresour Technol 101(13):4775–4800

Giroux H, Vialal P, Bouchard J, Lamy F (1988) Degradation of kraft indulin lignin by *Streptomyces viridosporus* and *Streptomyces badius*. Appl Environ Microbiol 54:3064–3070

Glass NL, Schmoll M, Cat JHD, Coradetti S (2013) Plant cell wall deconstruction by ascomycete fungi. Annu Rev Microbiol 67:477–498

Gokhale DV, Puntamabekar US, Deobagkar DN (1986) Xylanase and p-glucosidase production by *Aspergillus niger.* NCIM 1207. Biotechnol Lett 8:137–138

Golueke CG (1992) Bacteriology of composting. Biocycle 33:55–57

Gourlay K, Hu J, Arantes V, Andberg M, Saloheimo M, Penttila M, Saddler J (2013) Swollenin aids in the amorphogenesis step during the enzymatic hydrolysis of pretreated biomass. Bioresour Technol 142:498–503

Grajek W (1987) Production of D-xylanases by thermophilic fungi using different methods of culture. Biotechnol Lett 9:353–356

Grajek W (1988) Production of protein by thermophilic fungi from sugar beet pulp in solid state fermentation. Biotechnol Bioeng 32(1):255–260

Griffith GW, Ozkose E, Theodoroua MK, Davies DR (2009) Diversity of anaerobic fungal populations in cattle revealed by selective enrichment culture using different carbon sources. Fungal Ecol 2:87–97

Guillén F, Martinez AT, Martinez MJ (1992) Substrate specificity and properties of the aryl-alcohol oxidase from the ligninolytic fungus *Pleurotus eryngii*. Eur J Biochem 209:603–611

Gulati SL (1989) Physiological and enzymological problems of *Trichoderma reesei* a.M. 9414 associated with scale up of the process. In: Rameshwar S (ed). In: Proc. natl. seminar on biotechnol. lignin degradation. Indian Vet. Res. Inst., Izatnagar, 20–21 Dec, pp 64–71

Guo M, Song W, Buhain J (2015) Bioenergy and biofuels: history, status, and perspective. Renew Sust Energ Rev 42:712–725

Gupta VK, Kubicek CP, Berrin JG, Wilson DW, Couturier M, Berlin A, Ezeji T (2016) Fungal enzymes for bio-products from sustainable and waste biomass. Trends Biochem Sci 41(7):633–645

Gusakov AV (2013) Cellulases and hemicellulases in the 21st century race for cellulosic ethanol. Biofuels 4(6):567–569

Gutiérrez A, Caramelo L, Prieto A, Martínez MJ (1994) Anisaldehyde production and aryl-alcohol oxidase and dehydrogenase activities in ligninolytic fungi of the genus *Pleurotus*. Appl Environ Microbiol 60:1783–1788

Gutierrez-Correa M, Tengerdy RP (1998) Xylanase production of fungal mixed culture solid state substrate fermentation on sugarcane bagasse. Biotechnol Lett 20:45–47

Gutierrez-Correa M, Portal L, Moreno P, Tengerdy RP (1999) Mixed culture solid state fermentation of *Trichoderma reesei* with *Aspergillus niger* on sugarcane bagasse. Bioresour Technol 68:173–178

Hang Won K, Inkoo R, Minflee N, Hwang P, Fin Ho K, Kang HW, Rhee IK, Nam MH, Park HM, Kim JH (1995) J Agric Sci Soil Fertilizers 37(2):26 and 274

Hankin L, Anagniostakis SKL, Poincelot RP (1976) Compost by biodegradation of leaves. In: Sharply MJ, Kaplan AM (eds) Proc III international biodegradation symposium. Appl. Sci. Pub., London, pp 701–709

He X, Traina SJ, Logan TJ (1992) Chemical properties of municipal solid waste composts. J Environ Qual 54:1316–1323

Hegarty BM, Curran PMT (1985) The bio-deterioration of beach by marine and non-marine fungi in response to temperature, pH, light and dark. Int Biodeterior Bull 21:11–18

Hemsworth GR, Henrissat B, Davies GJ, Walton PH (2014) Discovery and characterization of a new family of lytic polysaccharide monooxygenases. Nat Chem Biol 10(2):122–126

Henna P (1975) Model for decomposition of organic material by microorganisms. Soil Biol Biochem 7:161–169

Himmel ME, Ding SY, Johnson DK, Adney WS, Nimlos MR, Brady JW, Foust TD (2007) Biomass recalcitrance: engineering plants and enzymes for bio-fuels production. Science 315:804–807

Himmel M, Xu Q, Luo Y, Ding S, Lamed R, Bayer E (2010) Microbial enzyme systems for biomass conversion: emerging paradigms. Biofuels 1(2):323–341

Holder NHM, Kilian SG, Preez JC (1989) Yeast biomass from bagasse hydrolysates. Biol Wastes 28:239–246

Horn SJ, Vaaje-Kolstad G, Westereng B, Eijsink VG (2012) Novel enzymes for the degradation of cellulose. Biotechnol Biofuels 5(1):45

Horwath WR, Elliott LF (1996) Rye grass straw component decomposition during mesophiles and thermophilic incubations. Biol Fertil Soils 21:227–232

Ishihara T (1983) Effect of pH in the oxidation of syringic acid by fungal laccase. Mokuzai Gakkaishi 29:801–805

Jain A, Johri BN, Jain A (1999) Partitioning of an extracellular xylanase produced by a thermophilic fungus *Melanocarpus albomyces* IIS-68 in an aqueous two phase system. Bioresour Technol 67(2):205–207

Jaramillo PMD, Gomes HAR, Monclaro AV, Silva COG (2015) Lignocellulose degrading enzymes: an overview of the global market. In: Gupta VK, Mach RL, Sreenivasaprasad S (eds) Fungal biomolecules: sources, applications and recent developments. Wiley, Chichester, pp 73–85

Jeffries TW, Choi S, Kirk TK (1981) Nutritional regulation of lignin degradation by *Phanerochaete chrysosporium*. Appl Environ Microbiol 42:290–296

Jhorar BS, Phogat V, Malik RS (1991) Kinetics of composting rice straw with glue waste at different carbon: nitrogen ratios in a semiarid environment. Arid Soil Res Rehabil 5(4):297–306

Johansen KS (2016) Lytic polysaccharide monooxygenases: the microbial power tool for lignocellulose degradation. Trends Plant Sci 21(11):926–936

Kahlon SS, Dass SK (1987) Biological conversion of paddy straw into feed. Biol Wastes 22:11–21

Kakezawa M, Mimura A, Takahara Y (1992) Application of two step composting process to rice straw compost. Soil Sci Plant Nutr 38(1):43–50

Kalogeris E, Christakopoulos P, Katapodis I, Alexiou A, Vlachou S, Kekos D, Macris BJ (2003) Production and characterization of cellulolytic enzymes from the thermophilic fungus *Thermoascus aurantiacus* under solid state cultivation of agricultural waste. Process Biochem 38:1099–1104

Kamra DN, Zadrazil F (1986) Influence of oxygen and carbon dioxide on lignin degradation in solid state fermentation of wheat straw with *Stropharia rugosoannulata*. Biotechnol Lett 7:345–340

Kang K, Wang S, Lai G, Liu G, Xing M (2013) Characterization of a novel swollenin from *Penicillium oxalicum* in facilitating enzymatic saccharification of cellulose. BMC Biotechnol 13:42

Kanotra S, Mathur RS (1994) Biodegradation of paddy straw with cellulolytic fungi and its application on wheat crop. Bioresour Technol 47:185–188

Kersten P, Cullen D (2007) Extracellular oxidative systems of the lignin-degrading Basidiomycete *Phanerochaete chrysosporium*. Fungal Genet Biol 44:77–87

Kim L, HJ CIG, Kim KH (2014) Synergistic proteins for the enhanced enzymatic hydrolysis of cellulose by cellulase. Appl Microbiol Biotechnol 98(20):8469–8480

Kirk TK, Schultz E, Connors WJ, Lorenz LF, Zeikus JG (1978) Influence of culture parameter on lignin metabolism by *Phanerochaete chrysosporium*. Arch Microbiol 117:277–285

Klamer M, Sochting U, Szmid, RAK (1998) Fungi in a controlled compost system with special emphasis on the thermophilic fungi. In: Proceedings of the international symposium on composting and use of composted materials for horticulture, Auchincruve, Ayr, V.K. 5–11 Apr Acta Horticulture, vol 469, pp 405–413

Kracher D, Scheiblbrandner S, Felice AKG, Breslmayr E, Preims M, Ludwicka K, Ludwig R (2016) Extracellular electron transfer systems fuel cellulose oxidative degradation. Science 352(6289):1098–1101

Kracher D, Ludwig R (2016) Cellobiose dehydrogenase: an essential enzyme for lignocellulose degradation in nature—a review. J Land Manag Food Environ 67:145–163

Krishna C, Chandrasekara M (1996) Banana waste as substrate for alpha-amylase production by *Bacillus subtilis* (B.T.K. 106) under solid state fermentation. Appl Microbiol Biotechnol 46(2):106–111

Kumar P, Barrett DM, Delwiche MJ, Stroeve P (2009) Methods for pre-treatment of lignocellulosic biomass for efficient hydrolysis and biofuel production. Ind Eng Chem Res 48(8):3713–3729

Ladisch MR, Lin KW, Voloch M, Tsao G (1983) Process considerations in the enzymatic hydrolysis of biomass. Enzyme Microbial Technol 5:8–16

Lei-Fei V, Gheynst JS, Fei L (1998) Community structure analysis of rice straw and grape pomace. Composting using phospholipid fatty acid analysis. In: ASAE annual international meeting, Orlando, 12–16 July 1998, p 13

Linton SM, Greenaway P (2004) Presence and properties of cellulase and hemicellulase enzymes of the gecarcinid land crabs *Gecarcoidea natalis* and *Discoplax hirtipes*. J Exp Biol 207:4095–4104

Liu X, Ma Y, Zhang M (2015) Research advances in expansins and expansion like proteins involved in lignocellulose degradation. Biotechnol Lett 37(8):1541–1551

Ljungdahl LG (2008) The cellulase/hemicellulase system of the anaerobic fungus Orpinomyces PC-2 and aspects of its applied use. Ann N Y Acad Sci 1125:308–321

Lopez MJ, Elorrieta MA, Vargas-Garcia MC, Suarez Estrella F, Moreno J (2002) The effect of aeration on the biotransformation of lignocellulosic wastes by white rot fungi. Bioresour Technol 81:123–129

Lo Leggio L, Simmons TJ, Poulsen JCN, Frandsen KEH, Hemsworth GR, Stringer MA, Walton PH (2015) Structure and boosting activity of a starch-degrading lytic polysaccharide monooxygenase. Nat Commun 22(6):5961

Mani MT, Marimuthu T (1992) Utilization of *Pleurotus* spp. for decomposing coconut coir pith. Mushroom Res 1(1):49–51

Marjamaa K, Toth K, Bromann PA, Szakacs G, Kruus K (2013) Novel Penicillium cellulases for total hydrolysis of lignocellulosics. Enzym Microb Technol 52(6–7):358–369

Martinez (2016) How to break down crystalline cellulose. Science 352(6289):1050–1051

Maziero R, Zadrazil F (1994) Effects of different heat pre-treatments of wheat straw on its microbial activity and colonization by different tropical and sub-tropical edible mushrooms. World J Microbiol Biotechnol 10(7):374–380

Mehta V, Bakshi A, Gupta JK, Kaushal SC (1990) Cellulolytic activities of *Pleurotus florida* on rice straw. Indian J Appl Pure Biol 5(1):15–18

Miller TF, Srinivasan VR (1983) Production of single cell protein from cellulose by *Aspergillus terreus*. Biotechnol Bioeng 25:1509–1519

Ming L, Cen X, Piel I, Cen PL (1999) Cellulose production by solid state fermentation on lignocellulosic waste from the xylose industry. Process Biochem 34:909–912

Moo-Young M, Moreira AR, Tengerdy RP (1983) Principles of solid state fermentation. In: Smith JE, Berry DR, Kristiansen B (eds) The filamentous fungi. Fungal technology, vol 4. Arnold Pub., London, pp 117–144

Moubasher AH, Abdel-Hafez S, Abdel-Fattah HM, Moharram AM (1984) Fungi of wheat and broad bean straw composts. Thermophilic fungi. Mycopathologia 84(2–3):65–71

Muniswaram PKA, Charyulu NCLN (1994) Solid state fermentation of coconut coir pith for cellulase production. Enzyme Microbial Technol 16:436–490

Nandi N, Hajra JM, Sinha NB (1996) Microbial synthesis of humus from rice straw following two step composting process. J Indian Soc Soil Sci 44(3):413–415

Nigam P, Pandey A, Prabhu KA (1988) Fermentation of bagasse by submerged fungal cultures—effect of nitrogen sources. Biol Wastes 23:313–317

Olayinka A, Adebaya A (1984) Effect of incubating temperatures and different sources of N and P on decomposition of saw dust in soil. Agric Waste 11:293–306

Ozkose E, Thomas BJ, Davies DR, Griffith GW, Theodorou MK (2001) *Cyllamyces aberensis* gen. nov. sp.nov, a new anaerobic gut fungus with branched sporangiophores isolated from cattle. Can J Bot 79:666–673

Pan X, Xie D, Gilkes N, Gregg DJ, Saddler JN (2005) Strategies to enhance the enzymatic hydrolysis of pretreated softwood with high residual lignin content. Appl Biochem Biotechnol 124(1):1069–1079

Park YS, Yum DY, Bai DH, Yu JH (1992) Xylanase from alkaliphilic *Bacillus* sp.YC-335. Biosci Biotechnol Biochem 56:1355–1356

Patel I, Kracher D, Ma S, Garajova S, Haon M, Faulds CB, Record E (2016) Salt-responsive lytic polysaccharide monooxygenases from the mangrove fungus *Pestalotiopsis*. Biotechnol Biofuels 20(9):108

Payne CM, Knott BC, Mayes HB, Hansson H, Himmel ME, Sandgren M, Beckham GT (2015) Fungal cellulases. Chem Rev 115(3):1308–1448

Qin YM, Tao H, Liu YY, Wang YD, Zhang JR, Tang AX (2013) A novel non-hydrolytic protein from *Pseudomonas oryzihabitans* enhances the enzymatic hydrolysis of cellulose. J Biotechnol 168(1):24–31

Rajaram S, Varma A (1990) Production and characterization of xylanase from *Bacillus thermoalkalophilus* grown on agricultural wastes. Appl Microbiol Biotechnol 34:141–144

Rajarathnam S, Bano Z (1989) *Pleurotus* mushrooms III Biotransformation of natural lignocellulosic wastes: chemical applications and implications. Crit Rev Food Sci Nutr 28:31–113

Rajarathnam S, Wankhede DB, Bano Z (1987) Degradation of rice straw by *Pleurotus flabellatus*. J Chem Technol Biotechnol 37:203–214

Rajasekaran P, Sampatkumar M (1981) Physico-chemical and microbiological properties of plant wastes treated with sewage sludge. Agric Waste 3:262–275

Rathner R, Petz S, Tasnadi G, Koller M (2017) Monitoring the kinetics of biocatalytic removal of the endocrine disrupting compound 17a-ethinylestradiol from differently polluted wastewater bodies. J Environ Chem Eng 5:1920–1926

Reese ET (1956) A microbiological process report: enzymatic hydrolysis of cellulose. Appl Microbiol 4:39–45

Resch MG, Donohoe BS, Baker JO, Decker SR, Bayer EA, Beckham GT, Himmel ME (2013) Fungal cellulases and complexed cellulosomal enzymes exhibit synergistic mechanisms in cellulose deconstruction. Energy Environ Sci 6(6):1858

Roche N, Desgranges C, Durand A (1994) Study on the solid state production of a thermostable alpha-L-arabinofuranosidase of *Thermoascus eurentiecus* on sugar beet pulp. J Biochem 38(1):43–50

Rocha VN, Maeda R, Pereira NF, Kern M, Elias L, Simister R, McQueenMason SJ (2016) Characterization of the cellulolytic secretome of *Trichoderma harzianum* during growth on sugarcane bagasse and analysis of the activity boosting effects of swollenin. Biotechnol Prog 32(2):327–336

Rosen, Schugerl (1983) In: Demain AL, Solomon NA (eds) Manual of industrial microbiology and biotechnology. Amer. Soc. Microbial, Washington, DC

Ross RC, Harris PJ (1983) The significance of thermophilic fungi in mushroom compost preparation. Sci Hortic 20(1):61–70

Saha BC (2003) Hemicellulose bioconversion. J Ind Microbiol Biotechnol 30(5):279–291

Saloheimo M, Paloheimo M, Hakola S, Pere J, Swanson B, Nyyssönen E, Penttilä M (2002) Swollen in, a *Trichoderma reesei* protein with sequence similarity to the plant expansins, exhibits disruption activity on cellulosic materials. Eur J Biochem 269(17):4202–4211

Sanches (2009) Lignocellulosic residues: biodegradation and bioconversion by fungi. Biotechnol Adv 27:185–194

Scheller HV, Ulvskov P (2010) Hemicelluloses. Annu Rev Plant Biol 61:263–289

Sermanni GG, Annibale AD, Lena GO, Vitale NS, Mattia ED, Minelli V (1994) The production of exo-enzymes by lentinus edodes and *Pleurotus ostreatus* and their use for upgrading corn straw. Bioresour Technol 48:173–178

Shaikh SA, Khire JM, Khan M (1997) Production of beta-galactosidase from thermophilic fungus *Rhizomucor* sp. J Ind Microbiol Biotechnol 19(4):239–245

Sharma HS, Johri BN (1992) The role of thermophilic fungi in Agriculture. In: Hand book of applied mycology, vol 4. pp 707–728

Sharma OK, Niwas S, Behera BK (1991) Solid state fermentation of bagasse for production of cellulase enzyme from cellulolytic fungi and extent of simultaneous production of reducing sugars in the fermenter. J Microbiol Biotechnol 6:7–14

Sharma HSS, Lyons G, Chambers J (2000) Comparison of the changes in mushroom *(Agaricus bisporus)* during window and bunker stages of phase I and lt. Ann Appl Biol 136(1):59–68
Silva IS, Menezes CR, Franciscon E, Santos EC (2010) Degradation of lignosulfonic and tannic acids by ligninolytic soil fungi cultivated under microaerobic conditions. Braz Arch Biol Technol 53:693–699
Sims TM, Saddler J, Mabee W (2008) From 1st to 2nd generation biofuel technologies: an overview of current industry and RD activities. https://www.iea.org/publications/freepublications/publication/2nd_Biofuel_Gen.pdf
Solomon KV, Haitjema CH, Henske JK, Gilmore SP, Borges-Rivera D, Lipzen A, O'Malley MA (2016) Early-branching gut fungi possess a large, comprehensive array of biomass-degrading enzymes. Science 351(6278):1192–1195. https://doi.org/10.1126/science.aad1431
Stanek M, Bisko NA (1982) Regulation of microbiological process in substrate for oyster mushroom *(Pleurotus ostreatus)* culture. Sbamik-UVT/Z-Zahradnictvi 9(3):221–233
Straatsma G, Samson RA, Olijnsma TW, Op-Den-Camp HJM, Gerrits JPG, Van-Griensven LJLD (1994) Ecology of thermophilic fungi in mushroom compost, with emphasis on *Scytalidium thermophilum* and growth stimulation of *Agaricus bisporus* mycelium. Appl Environ Microbiol 60(2):454–458
Straatsma G, Samson RA, Olijnsma TW, Gerrits JPG, Op-Den-Camp-HJM, Cariensven, LJLD (1995) Bioconversion of cereal straw into mushroom compost. Can J Bot 73:1019–1024
Sun FF, Hong J, Hu J, Saddle JN, Fang X, Zhang Z, Shen S (2015) Accessory enzymes influence cellulase hydrolysis of the model substrate and the realistic lignocellulosic biomass. Enzym Microb Technol 79-80:42–48
Suyanto T, Ohtsuki S, Yazaki S, Ui A, Mimura (2003) Isolation of a novel thermophilic fungus *Chaetomium* sp. novo MS.-017 and description of its palm-oil mill fibre-decomposing properties. Appl Microbiol Biotechnol 60:581–587
Takasaki K, Miura T, Kanno M, Tamaki H, Hanada S, Kamagata Y, Kimura N (2013) Discovery of glycoside hydrolase enzymes in an avicel-adapted forest soil fungal community by a metatranscriptomic approach. PLoS One 8(2):e55485. https://doi.org/10.1371/journal.pone.0055485
Tandon F, Odier E (1988) Influence of veratryl alcohol and hydrogen peroxide on ligninase activity and ligninase production of *Phanerochaete chrysosporium.* Appl Environ Microbiol 54:466–472
Tengerdy RP (1985) Solid state fermentation. Trends Biotechnol 3(4):96–99
Tiwari VN, Pathak AN, Lehri LK (1988) Manurial value of compost enriched with rock phosphate and microbial inoculants to green gram. J Indian Soc Soil Sci 36(2):280–283
Tripathi JP, Yadav JS (1991) Comparative ligninolytic and polysaccharolytic potential of an alkaliphilic basidiomycete on native lignocellulose. Int Biodeter Bull 27:49–59
Tsukihara T, Honda Y, Sakai R, Watanabe T (2006) Exclusive overproduction of recombinant versatile peroxidase MnP2 by genetically modified white-rot fungus, *Pleurotus ostreatus.* J Biotechnol 126:431–439
Vander-Gheynst JS, May BA, Karagosian M (2000) The effects of cultivation methods on the growth rate and shelf life of *Lagenidium giganteum.* In: ASAE annual international meeting Milwaukee, 9–12 July, p 5
Venkateswarlu G, Krishna PSM, Pandey A, Rao LV, Pandey A (2000) Evaluation of *Amycoatopsis mediterranei* VA18 for production of rifamycin B. Process Biochem 36(4):305–309
Wahyono S, Sahwan FL (1998) Solid waste composting trend and project. Biocycle 39(10):66–68
Wiegant WM (1992) Growth characteristics of the thermophilic fungi *Scytalidium thermophilum* in relation to production of mushroom compost. Appl Environ Microbiol 58(4):1301–1307
Wilson DB (2008) Three microbial strategies for plant cell wall degradation. Ann N Y Acad Sci 1125:289–297
Woiciechowski AL, Porto de Souza Vandenberghe L, Karp SG (2013) The pre-treatment step in lignocellulosic biomass conversion: current systems and new biological systems. In: Faraco V (ed) Lignocellulose conversion: enzymatic and microbial tools for bioethanol production. Springer, Berlin, Heidelberg

Wong DWS (2009) Structure and action mechanism of ligninolytic enzymes. Appl Biochem Biotechnol 157:174–209

Wyman CE, Ragauskas AJ (2015) Lignin bioproducts to enable biofuels. Biofuels Bioprod Biorefin 9(5):447–449

Xu Q, Luo Y, Ding SY, Himmel ME, Bu L, Lamed R, Bayer EA (2011) Multifunctional enzyme systems for plant cell wall degradation, pp 15–25

Yoon JJ, Cha CJ, Kim YS, Son DW, Kim YK (2007) The brown-rot basidiomycete *Fomitopsis palustris* has the endo-glucanases capable of degrading microcrystalline cellulose. J Microbiol Biotechnol 17:800–805

Yuan TQ, Xu F, Sun RC (2013) Role of lignin in a bio-refinery: separation characterization and valorization. J Chem Technol Biotechnol 88(3):346–352

Zadrazil F (1976) Release of water soluble compounds in the breakdown of straw by basidiomycetes as a basis for the utilization of straw. Z Acker Pflanzenbau 142:44–47

Zadrazil F, Puniya AK (1995) Studies on effect of particle size on solid state fermentation of sugarcane bagasse into animal feed using white rot fungi. Bioresour Technol 54:85–87

Zugenmaier P (2001) Conformation and packing of various crystalline cellulose fibers. Prog Polym Sci 26(9):1341–1417

Chapter 9
Recent Advancements in Mycodegradation of Lignocellulosic Biomass for Bioethanol Production

Divya Kumari and Barkha Singhal

Abbreviations

CBH	Cellobiohydrolases
CBP	Consolidated bioprocessing
CELF	Co-solvent enhanced lignocellulosic fermentation
COSLIF	Co-solvent-based lignocellulosic fractionation
CRISPR	Clustered regularly interspaced short palindromic repeats
DMC	Direct microbial conversion
EA	Extractive ammonia
GEMs	Genome-scale metabolic models
GVL	Gamma valerolactone
LPMOs	Lytic polysaccharide monooxygenases
MSW	Municipal solid waste
MTBE	Methyl tertiary butyl ether
NADH	Nicotinamide adenine dinucleotide (NAD) + hydrogen (H)
PGASO	Promoter-based gene assembly and simultaneous overexpression
SHF	Separate hydrolysis and fermentation
SPORL	Sulfite pretreatment to overcome recalcitrance of lignocellulose
SSCF	Simultaneous saccharification and co-fermentation
SSF	Simultaneous saccharification and fermentation
TEF	Translational elongation factor
USA	United States of America

D. Kumari · B. Singhal (✉)
School of Biotechnology, Gautam Buddha University, Greater Noida, Uttar Pradesh, India
e-mail: barkha@gbu.ac.in

R. Naraian (ed.), *Mycodegradation of Lignocelluloses*, Fungal Biology,
https://doi.org/10.1007/978-3-030-23834-6_9

1 Introduction

The phenomenal rise in the demand of environmental safety and energy security for attaining sustainability strongly favors the development of eco-friendly approaches and solutions for the energy resources at the global context (Wyman 2018). The clear anticipation regarding exhaustion of fossil fuels reserves in upcoming 50 years (Rastogi and Shrivastava 2018) due to their continuous demand and overwhelming use leads to the burgeoning quest for finding sustainable and economical substitutes. Apart from that, the burning of fossil fuels leads to the enhancement of greenhouse gases followed by global warming (Boykoff et al. 2019), loss of biodiversity, rise in sea as well as glacier level becoming a significant threat to the environmental safety and security (Opoku 2019). Therefore, the emergence of bio-based approaches (Dhyani and Bhaskar 2018) owing to their remarkable attributes with respect to eco-friendliness, non-toxicity, biodegradability, the production of biofuels has been considered promising and sustainable alternatives for energy sources. The development of biofuels plays a pivotal role for the building the roadmap of bioenergy-based economy (Demirbas 2008). The biofuels include bioethanol, biodiesel, and biogas (Xiong et al. 2019) and among these the development of bioethanol from various renewable natural sources leads to a paradigmatic shift in the field of bioenergy (Scarlat and Dallemand 2019). Currently, various western countries such as Brazil, China, the USA, Canada as well as EU member states (Kutas et al. 2007) have rigorously adopted the credentials of bioethanol and their commitments for the production of bioethanol has been increasingly realized at a global level. It was estimated that total ethanol production has been increased from 175 million gallons (Zabed et al. 2017) in 1980 to 14,810 million gallons in 2015 in the USA.

Till date, numerous renewable sources have been utilized for the production of bioethanol including sugars, starch, lignocellulosic biomass, and algae. The yield of bioethanol may vary with respect to types of sugars retrieved from the biomass due to diversed biochemical composition. The conversion of sugar and starch into bioethanol (Evcan 2012) has been considered to be first-generation bioethanol while utilization of lignocellulosic biomass and algae has been considered second- and third-generation bioethanol (Robak and Balcerek 2018), respectively. The sugar-based feedstock requires only extraction procedures to get fermentable sugars, while starchy feedstock requires hydrolysis to convert starch into glucose. Currently, the great potential has been envisaged in lignocellulosic biomass for bioethanol production that requires extensive pretreatment (Jönsson and Martín 2016) strategies prior to the hydrolysis in order to loosen cellulosic framework for enzyme accessibility for the bioconversion process. Moreover, the easy availability (Balat 2011) and low cost of lignocellulose biomass as well as proclaimed statement of Nova Institute of Germany regarding lignocellulosic resources for environmental sustainability lead to the enthusiastic drive for the bioethanol production. There are four major steps (Mosier et al. 2005) involved in the production of bioethanol from the lignocellulosic biomass that includes pretreatment, hydrolysis, and fermentation followed by distillation. Till date comprehensive methodologies have been developed for the pretreatment (Hsu 2018) as well as hydrolysis but for the effective bioconversion process (Lee 1997) especially fermentation of sugars greatly requires the involvement of microorganisms. A wide variety of bacterial as well as fungal microbial resources have been utilized (Kumar et al. 2008) for fermentation purpose but more prolific potential has been envisioned by exploring the world of fungi. The fungi possess the promising ability of hyphal penetration power as well as consist of extracellular enzymatic machinery that proves them interesting and excellent choice for better degradation of complex cell wall molecules into simplest form. Even though possessing the remarkable attributes, the field of mycodegradation of lignocellulosic biomass is still impeded by low ethanol yields and depicting lower tolerance towards higher titer of ethanol. Besides that, the recalcitrant structure framework of lignocellulose is costly and labor-intensive process. Therefore, in last decades the intervention of genetic and metabolic engineering methods (Mood et al. 2013) provided significant improvements in all of the four steps of bioethanol production and developed the concept of consolidated bioprocessing (CBP) that makes these resources as an amenable system for the fermentation of different sugars concurrently. Though the intervention of synthetic biology approaches led to a paradigmatic shift in engineering, the microbes for enhancing the process technology for bioethanol production still lagged behind the industrial utilization. Still, various technologies are in different degrees of development and industrial scaling has been reported for the production of bioethanol from lignocellulosic biomass. Therefore, this chapter examines recent advancements in mycodegradation of lignocellulosic biomass for bioethanol production. The chapter also delineates the current bottlenecks and future research priorities that paved the way for economic and sustainable production of bioethanol production at the commercial level.

2 Remarkable Attributes of Bioethanol

The enormous potential has been visualized in the production of bioethanol due to their excellent credentials that makes it the most suitable alternative source of bio-based energy. The conventional fuel that is commercialized globally is gasoline or petrol which has been produced through distillation process of petroleum (Bakshi et al. 2007) followed by mixing with different kind of additives to make suitable fuel source. Before the commitments for industrial scale production, scientists have comprehensively compared the properties of gasoline with bioethanol and found superior fuel properties. Bioethanol (Nagdeote and Deshmukh 2012) is biodegradable and oxygenated fuel comprising 34.7% oxygen (Malça and Freire 2006) that leads to higher combustion efficiency with respect to gasoline (C_7H_{17}) which is devoid of oxygen. Bioethanol devoid of sulfur serves as a safer alternative of methyl tertiary butyl ether (MTBE) has been used for the enhancement of octane number of gasoline. The only difficulty has been seen that limits its suitability for complete utilization as fuel is low cetane number that reduces the burning efficiency of bioethanol from compression ignition. The detailed comparative characteristics of

Table 9.1 Comparative account of various physicochemical properties of bioethanol

Properties	Bioethanol	Petrol/gasoline
Octane number	106–110	91–96
Higher evaporation enthalpy	1177 kJ/kg at 60 °C	348 kJ/kg at 60 °C
Laminar flame speed @100 kPa, 325 K	Near about 33	Near about 39
Heat of vaporization	840 kJ/kg	305 kJ/kg
Oxygenated fuel	34.7%	Absent
Combustion efficiency	Higher	Lower
Sulfur content	Negligible	Present
Density @15 °C	0.7904	0.7387
Vapor pressure @20 °C (kPa)	5.9	55

bioethanol versus gasoline that gave the idea of blending of bioethanol versus gasoline are summarized in Table 9.1. Therefore, effective measures have been implemented by mixing of bioethanol with ethylhexyl nitrate (Anderson et al. 2012) or diterbutyl peroxide to enhance the cetane number. The reduced pollution through the emission of the sulfur oxide, particulate matter, and nitrogen oxides has been seen when the blending of gasoline was done with bioethanol. The various countries based on their environmental and climatic condition adopted percentage-based blending of bioethanol with gasoline (Al-Hasan 2003) and the internationally recognized codes have been allocated that depicted the clear indication of blending percentage. The flow chart depicting various codes and their percentage has been depicted in Fig. 9.1.

3 Composition of Lignocelluloses

The potentiality of lignocellulose biomass has been realized as a promising source (Singh et al. 2010) of bioethanol production due to their easy availability, no interference in food and feed crop production as well as abrogating the need of land resources in the past decade. This biomass can be categorized into various groups that includes agricultural residues (wheat rice and other cereals straws, bagasse, and stovers), forest biomass (sawdust, hard wood, softwood, wood chips, pruning, and bark residues), energy crops (C_3 or C_4 plants like perennial grasses and other energy crops such as *Lantana camara*, *Ricinus communis*), aquatic plants (water hyacinth), and organic materials of municipal solid wastes (MSW). The major components of lignocellulosic biomass are cellulose (30–35%) (Sánchez 2009), hemicellulose (25–30%), and lignin (10–20%) (Wu et al. 2019) but smaller proportions of protein, pectins, lipids, minerals, and water are also present in these resources. The recalcitrant as well as robust structure (Zhao et al. 2012) of this biomass is due to different covalent and hydrogen bonding pattern among cellulose, hemicellulose, and lignin that confer crystallinity and a high degree of polymerization that leads to high fiber strength of lignocellulose biomass.

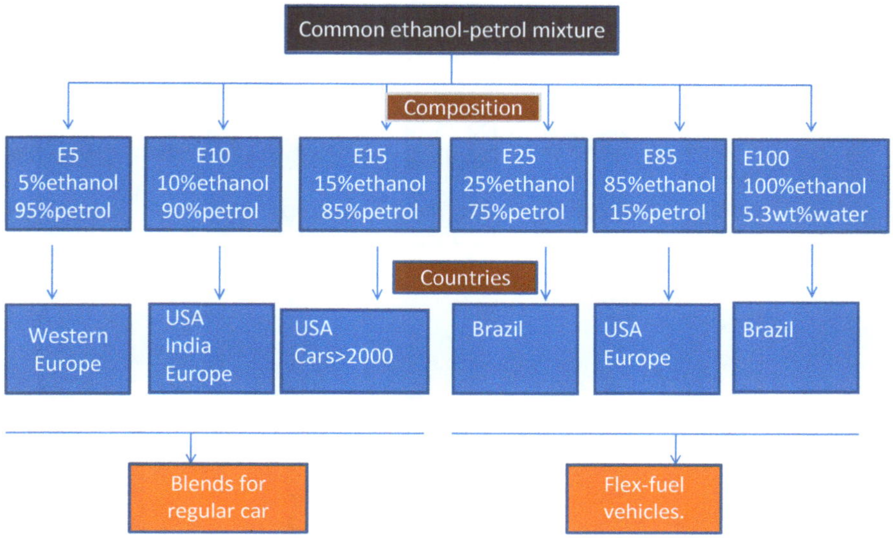

Fig. 9.1 The percentage of petrol-ethanol mixture and the international codes for their description and identification

Cellulose is one of the most rigid linear and crystalline homopolymers consists of D-glucospyranose subunits linked through β-1,4-glycosidic linkage (Bauer et al. 1999; Xu et al. 2019). The compactness of cellulose is due to the formation of micro-fibrils (Festucci-Buselli et al. 2007) that is stabilized by van der Waals forces and intra- and intermolecular hydrogen bonds. Hemicellulose is a heteropolymer composed of a variety of hexoses (glucose, mannose, galactose), pentoses (arabinose, xylose) (Kang et al. 2014) as well as sugar/uronic acids (D-galacturonic, D-glucuronic, methylgalacturonic) linked with each other through β-1,4- and β-1,3-glycosidic bonds. The major component of hemicellulose is xylan (Hendriks and Zeeman 2009) and its composition may vary within different feedstocks. Due to its varied composition, a wide variety of enzymatic resources has been required for the hydrolysis. Lignin is a complex heteropolymer consists of three structural monomers coumaryl alcohol, sinapyl alcohol, and coniferyl alcohol (Buxton and Russell 1988). It is a major constituent of wood (30–60% for softwoods and 30–55% for hardwoods) and linked with both cellulose and hemicellulose. It has been envisioned that a wide variety of lignocellulose feedstocks are available around the world but the production of bioethanol depends on the efficiency of glucan (for hexose sugars) xylan (for pentose sugars) hydrolysis (García-Aparicio et al. 2007) into soluble sugars, types of pretreatment strategies, and finally microorganisms involved in fermentation process. The various lignocellulose resources and their major advantages have been shown in Fig. 9.2 (Nasir and Ghazi 2015; Kumar et al. 2009; Dodo et al. 2017; Amin et al. 2017; Menegol et al. 2017; Kumar and Sharma 2017).

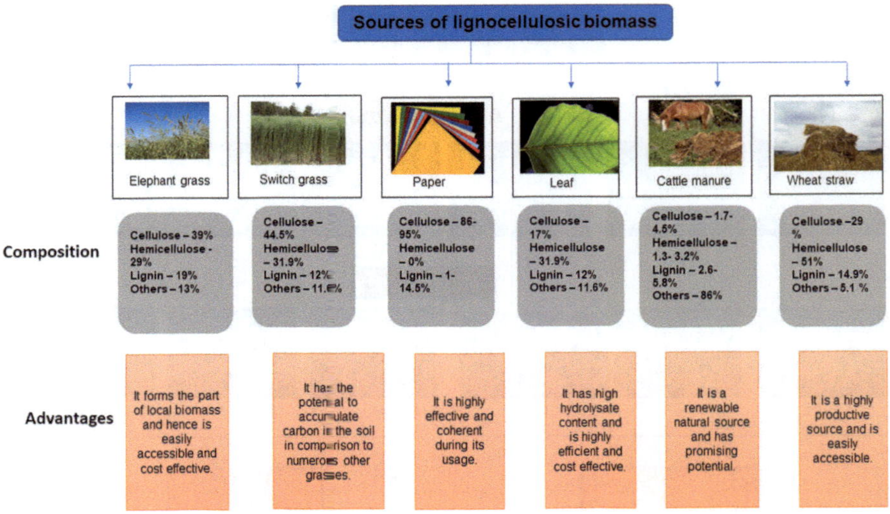

Fig. 9.2 Types of different lignocellulosic biomass, their composition, and advantages for bioethanol production

4 Basic Mechanism for the Conversion of Lignocellulosic Biomass to Bioethanol

The abundant availability (Gamage et al. 2010) of lignocellulosic biomass represents a long-term and sustainable solution for the rising energy demands and global climatic changes due to exhaustion of fossil resources. The versatile composition and physicochemical properties of lignocellulosic biomass, significantly affects the conversion process that considerably affects the yield of bioethanol; therefore, no universalized methods of bioconversion have been developed till date. In general, the conversion process has been initiated from grinding or pre-processing of the biomass for reduction in particle size and increment in the surface area of the particles. Thus, the conversion of lignocellulosic biomass to bioethanol centers around three broad interrelated steps including: (1) pretreatment of lignocellulose biomass (Moreno et al. 2017) for enhancing the accessibility biomass polysaccharides for hydrolytic breakdown, (2) enzymatic hydrolysis (Puttaswamy et al. 2016) of pretreated biomass to fermentable sugars, (3) microbes mediated fermentation of the sugars to produce ethanol. The detailed description is as follows:

4.1 Pretreatment of Lignocellulosic Biomass

Pretreatment step for the conversion of biomass into ethanol is considered as the most complex (Nasir Iqbal and Kamal 2012) and costly step that has the ability to alter macroscopic and microscopic structures of the biomass. This process has been significantly utilized for the removal of hemicellulose and lignin from the lignocellulosic feedstock; enhance the porosity and surface area of the biomass to reduce the crystallinity of cellulose. An ideal pretreatment strategy should possess the capability of generating reactive solid biomass with minimal reduction in sugars and fewer requirements of enzymes. There are broadly four methods for the pretreatment of lignocellulose biomass such as physical, chemical, physicochemical, and biological (Rezania et al. 2017) and overall success of the strategies are dependent on mode of reaction conditions and their mechanism of action. The extensive literature has been available for the pros and consequences on the pretreatment technologies that have been used previously but currently novel pretreatment technologies have developed for enhancing the efficiency of bioconversion process. These promising pretreatment methodologies include co-solvent-based lignocellulosic fractionation (COSLIF) (Sathitsuksanoh et al. 2010), co-solvent enhanced lignocellulosic fractionation (CELF) (Nguyen et al. 2016), extractive ammonia (EA) pretreatment (Chundawat et al. 2007), pretreatment applying ionic liquid(s) (Swatloski et al. 2002; Dadi et al. 2006; Singh et al. 2009; Li 2010; Cheng and Timilsina 2011; Perez-Pimienta et al. 2016; Singh and Simmons 2015; Konda et al. 2014), γ-valerolactone (GVL) pretreatment (Shuai et al. 2016; Wu et al. 2019), switchable butadiene sulfone pretreatment (De Frias and Feng 2013), and sulfite pretreatment to overcome recalcitrance of lignocellulose (SPORL) (Zhu et al. 2010). Though the above described methodologies have worked efficiently but require high capital cost, input energy as well as cause pollution. Therefore, an efficient, cost-effective, and eco-friendly pretreatment of lignocellulosic biomass has been developed by employing microorganisms (Sindhu et al. 2016) that do not require harsh conditions and specialized infrastructure. Various bacteria (Wilhelm et al. 2019) and fungi (Liu et al. 2019) have been explored for this purpose but great potential has been envisaged in fungi particularly white-rot, brown-rot, and soft-rot fungi. Brown-rot fungi cleave cellulose, while white- and soft-rot fungi cleave both lignin and cellulose. Various white-rot fungi (Baba et al. 2011), such as *Phanerochaete chrysosporium*, *Cyathus stercolerus*, *Ceriporia lacerata*, *Ceriporiopsis subvermispora*, *Pleurotus ostreatus*, *Pycnoporus cinnabarinus*, *Pleurotus chrysosporium*, *Trametes versicolor*, *Bjerkandera adusta*, and *Ganoderma resinaceum*, possess ligninolytic enzymatic machinery including lignin peroxidases, laccases, and manganese-dependent peroxidases that showed high delignification efficiency (Singh et al. 2014). Table 9.2 depicting various fungal species involved in pretreatment as well as hydrolysis of lignocellulosic biomass (Hahn-Hägerdal et al. 2006; Gray et al. 2006; Rouches et al. 2016; Banerjee et al. 2010). However, this pretreatment methodology suffers from serious limitation of low hydrolysis rate and requirement of longer incubation time; therefore, more comprehensive investigations must be required for the assessment of more white-rot fungi for enhanced ability for the delignification at shorter time interval. Table 9.3 depicting different types of pretreatment methodologies with their advantages and disadvantages (Harmsen et al. 2010; Mora-Pale et al. 2011; Brandt et al. 2011; Nitsos et al. 2013). Therefore, in a nutshell, till date the 100% conversion efficiency (Rocha-Meneses et al. 2017) has not been found in all pretreatment methodologies; therefore, promising potential have been seen by the

Table 9.2 Examples of various fungal species utilized for pretreatment and hydrolysis of lignocellulosic biomass

Phylum	Class	Family	Species
Ascomycota	Saccharomycetes	Saccharomycetaceae	*Saccharomyces cerevisiae*
	Saccharomycetes	Dipodascaceae	*Yarrowia lipolytica*
	Saccharomycetes	Saccharomycetaceae	*Hansenula polymorpha*
	Saccharomycetes	Saccharomycetaceae	*Pichia pastoris*
Neocallimastigomycota	Neocallimastigomycetes	Neocallimastigaceae	*Piromyces* sp. *Anaeromyces* sp.
	Neocallimastigomycetes	Neocallimastigaceae	
Basidiomycota	Agaricomycetes	Polyporaceae	*Trametes versicolor*
	Agaricomycetes	Nidulariaceae	*Cyathus stercoreus*
	Agaricomycetes	Phanerochaetaceae	*Phanerochaete chrysosporium*
	Agaricomycetes	Ganodermataceae	*Ganoderma resinaceum*
	Agaricomycetes	Polyporaceae	*Pycnoporus cinnabarinus*

Table 9.3 An overview of different pretreatment modalities for lignocellulosic biomass with their merits and demerits

Pretreatment types	Mode	Features	Merits	Demerits
Chemical	Acid hydrolysis	Acidic breakdown of lignocellulosic constituents via continuous or batch process	Proper eradication of hemicellulose and lignin Less time-taking process	Generation of toxic compounds Might result in corrosion of reaction vessel
	Alkaline hydrolysis	Lysis of lignin and hemicelluloses by water-soluble bases	Weakens the crystallinity of cellulose Enhances the available surface area	Formation of residual substances Not so effective for softwood
	Organosolvent	At 170–190 °C	Proper hydrolysis of lignocellulosic components	Expensive
Physical	Crushing/grinding/milling	Done at room temperature conditions	Low capital investment Easy and convenient method	Not so effective Very high energy demand in some samples

(continued)

Table 9.3 (continued)

Pretreatment types	Mode	Features	Merits	Demerits
Physicochemical	Wet oxidation	Temp: 150–200 °C for 30 min	Less energy input Promotes lignolysis	Expensive
	Ozonolysis	Done at room temperature for lignolysis	Lignin content is decreased Formation of toxic residues	Not cost-effective
	Carbon dioxide explosion	Takes place 50 °C at for 1 h 30 min	Surface area enhancement Almost negligible formation of inhibitors	Not applicable for those biomass which have lignin as their major constituent
	Ammonia recycle percolation method	160 °C for 15 min	Eradicates hemicellulose and lignin	Imperfect hydrolysis of hemicellulose
	Ammonia fiber explosion method	90 °C for 28 min	Efficient Surface area is exponentially increased	Hydrolysis is not at point
	Steam pretreatment	160–260 °C for 10 min	Cost-effective Automatic hemicellulose hydrolysis	Production of inhibitors Xylan damage
Biological	Some bacteria and fungi (major)	Brown-rot fungi cleave cellulose while white-rot cleaves both lignin (via lignolytic enzymatic machinery) and cellulose	Inexpensive Natural methodology	Low hydrolysis rate Takes energy

(continued)

Table 9.3 (continued)

Pretreatment types	Mode	Features	Merits	Demerits
Modern	Co-solvent-based lignocellulosic fractionation	Includes water, cellulosic solvent, and an organic solvent (involves reactions at moderate conditions)	High efficiency	–
	Co-solvent enhanced lignocellulosic fractionation	Sulfuric acid (dilute) and water act on cellulose	Cost-effective	–
	Extractive ammonia Pretreatment	Includes transformation of crystalline cellulose into soluble form	Effective method	–
	Pretreatment applying ionic liquids	1-Butyl-3-methyl imidazolium chloride is used	Inexpensive	High pollution
	Gamma-valerolactone pretreatment	A minimal harsh treatment of lignocellulosic biomass	High delignification rate Highly efficient methodology	Expensive
	Switchable butadiene sulfone pretreatment	Done in aqueous conditions to break cellulose	Efficient cellulose degradation	High water requirement Regular energy inputs needed
	Sulfite pretreatment to overcome recalcitrance of lignocellulose	Done to enhance enzymatic activity	Presence of reducing agents	Mostly done for woody biomass

synergistic action of two or more pretreatment methods. In spite of that, there is a pressing need for finding a novel efficient methods or modifying existing methodologies for harnessing best results. The recent interventions of artificial intelligence or predictive modeling can open a new avenue for the designing, optimization, and process configuration for pretreatment technologies (Zheng et al. 2009).

4.2 Enzymatic Hydrolysis and Saccharification

The pretreatment of lignocellulose biomass is followed by enzymatic hydrolysis for further de-polymerization of cellulose and hemicellulose (Humbird et al. 2011) to monomeric sugars that will be subsequently used for ethanol fermentation. This process requires either in situ production or utilization of commercial enzymes (cellulases and other enzymes) that enhance the capital cost for the bioethanol production. Besides that, the enzymes need to possess high activity, stability under reaction conditions including high temperatures, resistant to inhibitory molecules generated during the pretreatment process and finally possess a longer half-life. Extensive research has been done for increasing the stability, activity, and reducing the cost of enzymes in the last decade but availability at the cost of \$1.5–\$2.0 per kg has not reached economic viability. Till date, there is dearth of commercial enzymes that can withstand high-temperature applications for the hydrolysis as well as low accessibility to hemicellulose and lignin and susceptible to the inhibition by toxic compounds produced during pretreatment also stymied the complete and efficient hydrolytic conversion and demands high loading of enzymes. Therefore, various strategies like in situ production of hydrolytic enzymes from various microbial consortium described above section have been implemented as well as production of thermotolerant fungal enzymes has been tried. More recently, the discovery of novel non-hydrolytic enzymes (LPMOs) (Guerriero et al. 2016) polysaccharide monooxygenases add further dynamism in the hydrolysis process by alleviating the need of high loading of cellulase enzymes and reduction in the processing cost. The actual mechanism of action is still under investigation; however, the proposed mechanism is believed to be the oxidation of highly recalcitrant (Dien et al. 2003) crystalline regions of cellulose to aldonic acids, therefore provide more reducing/non-reducing ends for enzymatic hydrolysis. Though it was found that these enzymes are responsible for making less stable cellulase; therefore, more sophisticated studies are needed for their long-term commercial utility. Another major reason for slow hydrolysis rate is due to the mixture of hexoses and pentose sugar released during hydrolysis and most commonly ethalnogenic fungi *Saccharomyces cerevisiae* or bacteria *Zymomonas mobilis* cannot ferment pentose sugars (Seo et al. 2005) and the higher amount of sugars released cause substrate inhibition. Though methodologies have been developed for the conversion of pentose sugars into other molecules like furfural through chemical catalysis but that doesn't prove to be economically and technically sustainable process. Therefore, rigorous research is still centered on the co-fermentation of hexose and pentose sugars by utilizing the genetically engineered strains for bioethanol production.

4.3 Fermentation

The major impediment for the production of bioethanol from lignocellulosic biomass is fermentation. The process of conversion of hexose and pentose sugars into ethanol through various microorganisms is known as fermentation. Research studies speculated that the theoretical maxium yield of hexoses and pentoses is 0.51 L/kg ethanol (Chovau et al. 2013) and 0.489 kg CO_2/kg sugar and overall yield of ethanol becomes 0.719 and 0.736 L/kg of glucan (for hexoses) and xylan (for pentoses), respectively. A most popular and thoroughly investigated fungus, *S.*

cerevisiae has been explicitly utilized for the production of first-generation ethanol but suffers from the limitation of not having the pathway for pentose conversion. Therefore, the problem has been circumventing though more rigorous efforts for identifying other fungi and microbes that can co-ferment both C_5 and C_6 sugars at elevated operating temperatures or make the strains recombinant through genetic engineering that can metabolize pentose as well as hexose. However, the recent studies also suggested the possibility of the direct fermentation of cellobiose and cellodextrins (Kim et al. 2014) into ethanol also leads to progressive drive for the development of recombinant microbe for introducing cellodextrin transporter system from *Neurospora crassa* into *S. cerevisiae*. More recently, research was published for engineered yeast strain that can co-ferment cellobiose, xylose, and glucose for the bioethanol production (Zheng et al. 2018). Though engineered strains hold potential for metabolizing hexose as well as pentose sugars, low ethanol titers, and slow metabolism are the challenges that are still prevailing and need to be circumvented. Therefore, during enzymatic hydrolysis, some integrated methods for hydrolysis and fermentation are proposed.

5 Integrated Processes for Bioethanol Production Using Lignocellulosic Biomass

The basic steps described above for the bioconversion of lignocellulose to ethanol necessitates certain integrative strategies for combating the certain limitations of each process. The major four processes have been developed after pretreatment for the conversion of complex biomass into ethanol that includes (a) separate hydrolysis and fermentation (SHF), (b) simultaneous saccharification and fermentation (SSF) (Wyman 1994), (c) simultaneous saccharification and co-fermentation (SSCF) (Lynd et al. 2005), and (d) consolidated bioprocessing (CBP) (Zhao et al. 2018). The separate hydrolysis and fermentation (SHF) strategy is the most conventional two-stage process in which the pretreated biomass is cleaved into sugars in one reactor and fermentation of these sugars into ethanol has been performed in separate reactor. The method is advantageous in terms of carrying out the reactions at their optimum conditions but suffers from the limitation of high investment cost for maintaining two separate reactors as well as inhibition of fermentative microflora due to higher concentration of glucose. Therefore, to alleviate the inhibition of sugars and reduction in number of steps, two other approaches have been adopted in which the same reactor (Ballesteros et al. 2002) can be used for enzymatic saccharification and fermentation. Therefore, the sugars released from the biomass can be rapidly converted into ethanol that prevents the inhibition of sugars with faster hydrolysis rate and requires low enzyme loadings. Apart from that, the process requires less capital cost, easy operation conditions, and rapid conversion of ethanol leads to reduction in contamination. However, the major bottleneck of this process is the non-compatibility of the optimal conditions for enzymatic hydrolysis as well as fermentation. The higher temperatures usually 50–55 °C has been required for enzymatic hydrolysis while the optimum temperature for the growth of mesophilic fungi and bacteria is 28 °C and 37 °C, respectively, that makes this process less efficient as well as confers low ethanol yields (Sun and Cheng 2002). Though protein engineering approaches has been very well tried to reduce the optimum temperature for the enzymatic hydrolysis, the process has not been found to be economically viable; therefore, there is pressing demand for identifying thermotolerant strains for ethanol production at elevated temperature. The modification (Viñals-Verde et al. 2012) in the simultaneous saccharification and fermentation (SSF) has been more efficiently developed by the co-fermentation of both pentoses and hexoses formed during the pretreatment and hydrolytic processes of lignocellulosic biomass. This integrated technology is known as simultaneous saccharification and co-fermentation (SSCF). This process is accomplished either through addition of mixed cultures of yeast or addition of pure culture of fungi having the capability of utilizing both pentoses and hexoses.

However, the utilization of mixed culture leads to the generation of heterogeneous conditions in which the speedy growth was observed for hexose utilizing microorganisms as compared with the slow growth of pentose-utilizing microorganisms. Therefore, the latter approach of utilizing single microorganism is more viable option possessing the qualities of reduced cost and risk of contamination, less operation time as well as reduced substrate and product inhibition. Furthermore, the rigorous efforts have been pursued for consolidation of all the steps for bioethanol manufacturing in a one bioreactor utilizing the single microorganism that can produce the cellulolytic and ligninolytic enzymes, perform enzymatic saccharification and fermentation with the high yield of bioethanol. This strategy is known as direct microbial conversion (DMC) or consolidated bioprocessing (CBP) (Xu et al. 2009). A flowchart depicting CBP is shown in Fig. 9.3 (Alzate and Toro 2006; Stöcker 2008; Hamelinck et al. 2005; Aden et al. 2002). The success of this process lies on stringent selection of the microorganisms as no microorganisms in its native state possess all the desired properties till date. Therefore, the genetic engineering approaches have been well applied for enhancing the potential of the employed microorganisms. These efforts have been categorized the CBP into two subdivisions CBP I and CBP II. Presently, bacterial as well as fungi especially variety of yeast has been explored for this process. Therefore, CBP I involves the engineering of cellulase producing microbes to become ethanologenic and CBP II involves the engineering of ethanologenic microbes to be cellulolytic. In CBP I, the successful bioethanol productivity has been assessed through anaerobic cellulolytic thermophiles from the genus Clostridium (Roberts et al. 2010); however, the inability to metabolize xylose, low ethanol tolerance as well as production of by-products like lactate and acetate limits its applicability at commercial level. It is worth mentioning that metabolic engineering and directed evolutionary approaches have been rigorously improved ethanol titers by blocking the production of major by-products. At present, the inherent potential of various species of cellulolytic fungi like *Aspergillus*, (Bhardwaj et al. 2019), *Fusarium*, *Rhizobium*, and *Trichoderma* have been looked at as CBP I candidates. As for the second strategy, the primary concerns are expression and secretion of functional cellulases (Sukumaran et al. 2009) in ethanologenic

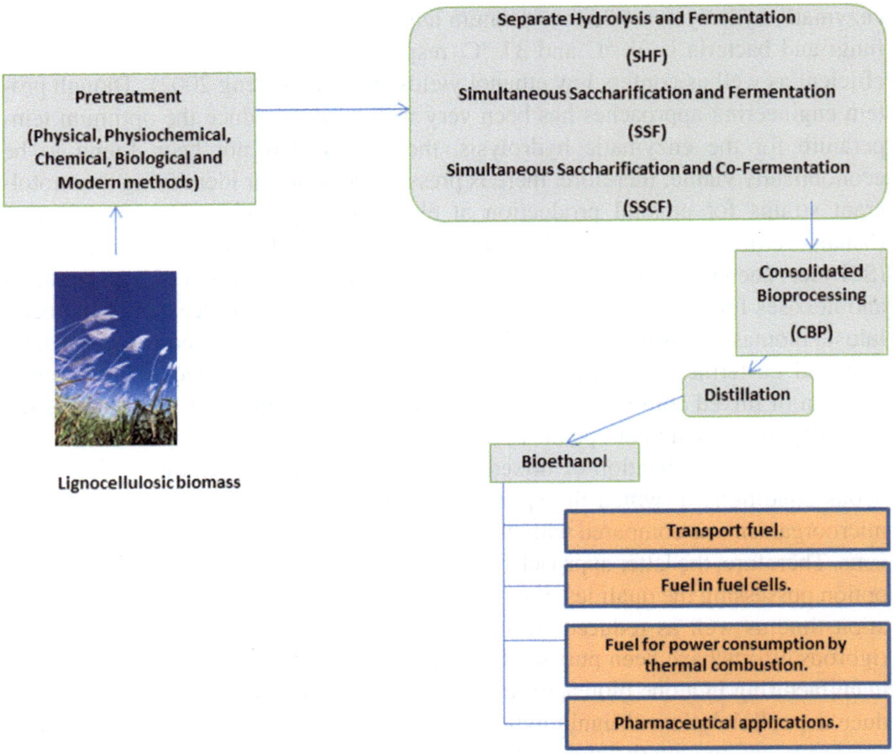

Fig. 9.3 The comprehensive strategy for the processing of bioethanol from lignocellulosic biomass

species, particularly *S. cerevisiae*, which has been engineered with genes encoding glycoside hydrolases including cellulases and hemicellulases through cell surface display techniques. Unfortunately, expression of the cellobiohydrolases (Van Tilbeurgh et al. 1986) (CBH I and CBH II) from *Trichoderma reesei* is generally poor (Meng et al. 2018); therefore, it's worth considering that engineering the species with more other enzymes or pathways required by the efficient production of bioethanol.

6 Recent Molecular Interventions for Improvement of Mycodegradation for the Bioethanol Production

The efficiency and abundant availability of lignocellulosic biomass is enhancing their potential utilization of bioethanol production. The scarcity and rising cost of non-renewable fuels and increasing carbon footprint lead to the dramatic increments in utilizing microbial factories for the bioconversion of lignocellulosic

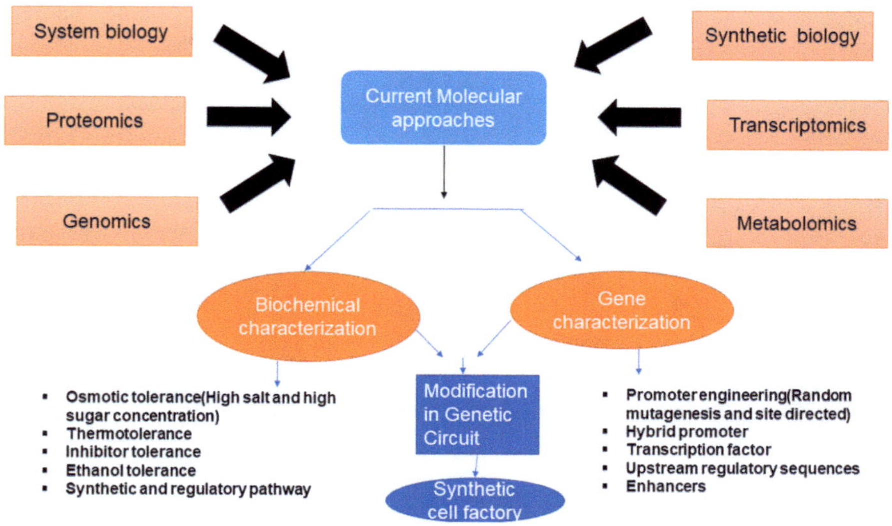

Fig. 9.4 Recent molecular approaches for the engineering of various fungi for improvement in bioethanol production

biomass to bioethanol. The most efficient industrial workforce for the production of bioethanol was documented for the utilization of *Saccharomyces cerevisiae* as a biological tool. The major limitation of *S. cerevisiae* (Argueso et al. 2009) is the inability to ferment a wide range of substrates like arabinose, xylose, and glycerol as well as could not grow at higher temperature. The current "omics" and molecular interventions like advent of synthetic biology, system biology, and metabolic engineering revolutionized the field of bioenergy production, and these efforts have already been conferred in *S. cerevisiae*. The combined strategy of various molecular approaches for the improvements in fungi (Madhavan et al. 2017) is depicted in Fig. 9.4. The successful production of homologous and heterologous products has been mediated through promoter selection and productivity has been enhanced through promoter engineering. In *S. cerevisiae*, the development of synthetic promoter leads to the dynamic utilization of this organism at commercial level. Thus, the fine and tunable control of metabolic pathways has been done through promoter engineering approaches that can be performed through engineering core promoter elements through random or site-directed mutagenesis and modifications in the upstream regulatory sequences and transcription factors. However, the major C_5 and C_6 sugars utilization pathway is depicted in Fig. 9.5 (Aristidou and Penttilä 2000).

More recently, construction of hybrid promoter may open up the possibilities to explore the promoter sequences of other organisms and their introduction in the desired host for the development of highly active synthetic promoters. The bioethanol productivity has been enhanced by the development of hybrid promoter by combining β-galactosidase (lac4) promoter of *K. lactis* with the core promoter

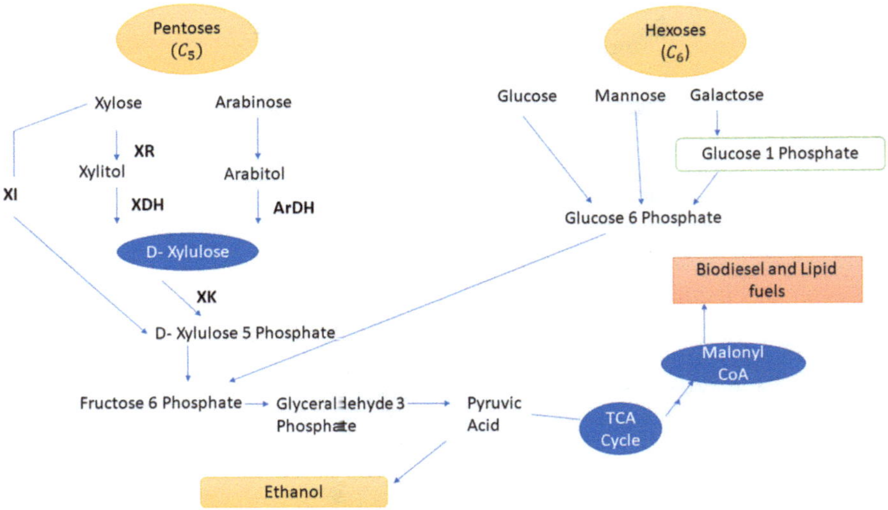

Fig. 9.5 The depiction of C_5 and C_6 sugars utilization pathway of microorganisms

sequences of cellobiohydrolase 1 (cbh1) from *Trichoderma reesei*. The advent of (Borodina and Nielsen 2014) CRISPR-Cas (Clustered Regularly Interspaced Short Palindromic Repeat) has led to paradigmatic shift in regulation and control of the expression of genes inside the yeast. This system utilizes the RNA-guided control of gene through nuclease-deficient Cas9 (dCas9) for gene silencing. The technique seems to be promising in context of generation of rapid and efficient double-stranded breaks and multiple targeted integrations in yeast. The component histone deacetylase (Sin3p homolog) is strongly bound with Mxi1 protein after the attachment of Cas9 that further down-regulate the TEF promoter of *S. cerevisiae* for enhancing bioethanol production. On the other hand, it is worth mentioning that besides the extensive utilization of conventional yeasts for the bioethanol production, non-conventional yeast systems possess remarkable credentials like thermotolerance (Edgardo et al. 2008), ethanol tolerance, genetic diversity, inhibitor tolerance, etc. The application of system biology and synthetic biology in these biological systems expands their horizons for the efficient production of bioethanol. The current examples of non-yeast systems that have been explored yet include *Kluyveromyces lactis*, *Pichia pastoris* (Soccol et al. 2010), *Hansenula polymorpha* (Ryabova et al. 2003), *Yarrowia lipolytica* (Tsigie et al. 2013), etc. It was observed that conventional yeast metabolize glucose by fermentative mode (Crabtree positive) (Azhar et al. 2017) and non-conventional yeasts metabolize glucose by oxidative pathway (Crabtree negative) (Souto-Maior et al. 2009). Therefore, research studies were exhaustively tried for the conversion of Crabtree-negative yeasts to Crabtree-positive systems for enhancing the efficiency of bioethanol production. Recently, the metabolic engineering has been successfully applied for shifting the metabolic pathway from respiratory to fermentative in *Kluyveromyces lactis*. A null mutant was constructed for the gene encoding for mitochondrial internal dehydrogenase that displayed higher ethanol productivity by depicting non-susceptibility towards oxidation of exogenous NADH.

In a similar line, the metabolically engineered strain of Crabtree-positive yeast (*Dekkera bruxellensis*) (De Souza Liberal et al. 2007) has been shown to produce high yield of ethanol. In this yeast system, the overexpression of key enzyme of fermentative pathway, i.e., alcohol dehydrogenase has been performed under the control of highly active TEF1 promoter that enhanced the ethanol productivity 1.7 times more than wild-type strain. In a similar line, the productivity of bioethanol has been significantly enhanced under various physiological stress conditions like high temperature and osmotic stress. The non-conventional yeast systems like some strains of has been *Kluyveromyces marxianus* (Ballesteros et al. 2004) depicted high temperature tolerance as compared with conventional yeast systems. The thermotolerant strains of *Kluyveromyces marxianus* NIRE-K1 and NIRE-K3 (Arora et al. 2015) were found to ferment both xylose and glucose at high temperature approximately above 45 °C and displayed high ethanol productivity of about 39.12 and 43.25 g/L, respectively. Similarly, thermotolerant variety of *Kluyveromyces* sp. *IIPE453* has been screened for enhanced production of 0.43 g/g bioethanol at 50 °C from glucose and xylose-rich bagasse hydrolyzates as a lignocellulosic feedstock. In addition to that, non-conventional yeast system including *Ogataea polymorpha* (Hasunuma and Kondo 2012) has been found to produce bioethanol by fermenting xylose at 45 °C. The osmotolerant strain *Zygosaccharomyces rouxii* was also identified and screened for producing high concentration of ethanol in the presence of 90% glucose concentrations and 3M NaCl. The resistance towards osmotic stress in terms of high sugar and salt has been attributed due to the presence of unique transporters in plasma membrane. Furthermore, the substrate utilization capacities of non-conventional yeast have also been modified through system biology and metabolic engineering approaches. The capacity of fermenting xylose and starch in the thermotolerant variety of *H. polymorpha* has been developed by the heterologous expression ofgenes *SWA2* and *GAM1* (Voronovsky et al. 2009) encoding amylolytic and xylanolytic enzymes isolated from the yeast *Schwanniomyces occidentalis*. The heterologous expression has been mediated under constitutive promoter of glyceraldehyde-3-phosphate dehydrogenase gene in the *H. polymorpha.* In a similar line, the less capacity of xylose utilization of *K. marxianus* has been enhanced by altering the pathway for the conversion of xylose to xylitol followed by ethanol at elevated temperature by replacement of native xylose reductase gene with foreign gene of xylose reductase of *P. stipitis* (Nigam 2001). Furthermore, the various promoters like *TEF1, ADH1p* (Almeida et al. 2008), *Xyl1p* (Ponniah et al. 2017), and *ADH2p* (Salusjärvi et al. 2003) have been screened and engineered for high ethanol tolerance and reducing the Crabtree effect of *Pichia stipitis*. Thus, extensive efforts (Shaw et al. 2008) have been done for eliciting the higher production of bioethanol in conventional and non-conventional yeast system; however, there is pressing need to make them suitable industrial workforce for (Sticklen 2008) commercial utilization.

7 Current Bottlenecks

The potentiality of bioethanol production from lignocellulose feedstock has been increasingly realized with respect to attaining sustainability in bioenergy demands as well as environmental safety. Despite possessing the great potential, the large-scale production of bioethanol from these magnificent resources has not yet been demonstrated as an economically sustainable option. The various technological hurdles have been realized from processing to pretreatment followed by separation through distillation process. The recalcitrant nature and structural complexity is the major impediment for the efficient production of bioethanol. This recalcitrance comes naturally in plants from the evolutionary perspective for their survival from differential environmental barriers therefore to break this natural mechanism is technologically a cumbersome task for the scientific community. The inefficiency of fermentation of xylose is another major challenge for the efficient production of bioethanol. The toxicity generated due to the production of various inhibitory molecules during various pretreatment strategies has been a technologically challenging aspect of bioethanol production Therefore, the finding of ideal pretreatment strategies is still the major concern for its commercialization. The bioconversion process through the selection of ideal microorganisms also affects the production efficiency and cost. The recent technological developments in genetic engineering approaches opens up the avenue for engineering of microbes especially in conventional and non-conventional yeast in a desired manner but still suffers from limitations for commercial applicability. Therefore, engineering microbes for the economic production of bioethanol has major concerns that include their utilization on wide substrate ranges, efficient and simple framework, and easy downstream processes. Moreover, the inhibitor tolerance, thermotolerance, fuel tolerance, control of redox balance also posits significant consideration for the bioethanol production. The non-homologous recombination pathway, as well as longer expression cassettes followed by larger cloning steps and PCR reactions also limits the industrial applicability of non-yeast systems at industrial scale. Therefore, rigorous research efforts are still required for alleviation of these challenges for the viable production of bioethanol.

8 Future Research Priorities

Inspite of the above-mentioned bottlenecks present in bioethanol production and its commercialization, the tremendous potential of bioethanol production from the lignocellulosic biomass still fascinates scientific community for the development of efficient and sustainable process technology for its industrial scale production. The continuous demand around the globe for the biofuels and bioethanol production has been increased (Limayem and Ricke 2012; Alvira et al. 2010; Sarkar et al. 2012) over time that can be represented in (Fig. 9.6a, b).

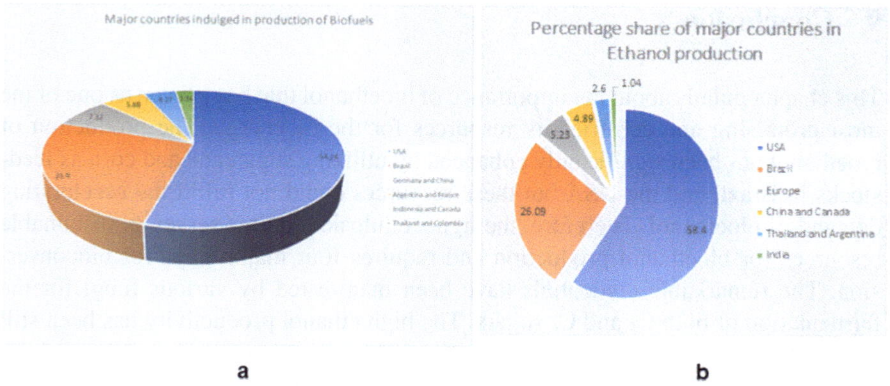

Fig. 9.6 The global share of bioenergy production (**a**) Major countries involved in biofuels production. (**b**) Major countries involved in bioethanol production

There is pressing need for identifying novel fungi that can produce cellullase enzymes of higher efficiency and specific activities. The efforts will be done for enhancing the fermentative capacity of different fungus species that has the ability to degrade biomass efficiently but less production efficiency of ethanol. Among the fungal diversity, yeast serves as the most promising workforce for the production of bioethanol. The extensive metabolic engineering of yeast through "promoter-based gene assembly and simultaneous overexpression" (PGASO) doesn't circumvent the toxicity of products during bioethanol production that leads to disruption of cell membrane due to the inhibition of ATP-producing pumps. Therefore, future studies will focus on the engineering of non-conventional, yeast that imparts better selectivity towards inhibitor tolerance, thermotolerance, and product tolerance. The system biology approaches and synthetic biology tools for improving various fungi will decrease the cost through reduction in the number of expression constructs. The transcriptomic data will help to comprehend the regulatory and rate-limiting steps in biosynthetic pathway for ethanol production. The bioinformatics tools will be helpful for making alteration in metabolic pathway by introducing novel catalytic enzymes or making knockouts for increasing biofuel production. Genome-scale metabolic models (GEMs) can be used for species correlation, metabolic flux design, identification of transcription factors and genes, and metabolic blueprint of the cell. The recent interventions of synthetic biology tools via RNA interference, codon optimization, and CRISPR-Cas technology can be utilized for enhancing the biofuel production. The construction of genetic circuits as well as finding novel metabolic pathways and their synergistic combination will be done by ligation-free assembling techniques and BioBricks softwares. Therefore, finding and constructing a genetic switch that can respond towards various environmental cues will definitely enhance the bioethanol production. The future research will also focus on the application of synergistic microbial communities that can develop the robust synthetic fungal-bacterial consortia for the efficient production of bioethanol.

9 Conclusion

This chapter entails about the importance of bioethanol that has proven as one of the most promising and eco-friendly resources for the bioenergy. The production of bioethanol has been significantly enhanced by utilizing sugarcane and corn as feedstocks in Brazil and the USA but these resources could not fulfill the accelerating demand of bioethanol. Therefore, the lignocellulosic biomass serves as sustainable resources for bioethanol production and requires four major steps for bioconversion. The remarkable credentials have been manifested by various fungi for the fermentation of both C_5 and C_6 sugars. The high ethanol productivity has been still impeded with various technological barriers. The recent interventions of molecular approaches and their successful implications in various fungi led to the paradigmatic shift in enhancing the bioethanol productivity and still thriving for more efficiency. Though the cellulosic bioethanol production plants are increasing continuously, the continued research efforts are still required for the development of viable process technology for their sustainable utilization at a profitable cost.

Acknowledgement The authors are extremely grateful for Gautam Buddha University for providing all necessary facilities and support for writing this chapter. All authors declare that they have no conflict of interest.

References

Aden A, Ruth M, Ibsen K, Jechura J, Neeves K, Sheehan J, Lukas J (2002) Lignocellulosic biomass to ethanol process design and economics utilizing co-current dilute acid prehydrolysis and enzymatic hydrolysis for corn stover (No. NREL/TP-510-32438). National renewable energy lab golden co.

Al-Hasan M (2003) Effect of ethanol–unleaded gasoline blends on engine performance and exhaust emission. Energy Convers Manag 44(9):1547–1561

Almeida JR, Röder A, Modig T, Laadan B, Lidén G, Gorwa-Grauslund MF (2008) NADH-vs NADPH-coupled reduction of 5-hydroxymethyl furfural (HMF) and its implications on product distribution in Saccharomyces cerevisiae. Appl Microbiol Biotechnol 78(6):939–945

Alvira P, Tomás-Pejó E, Ballesteros MJ, Negro MJ (2010) Pretreatment technologies for an efficient bioethanol production process based on enzymatic hydrolysis: a review. Bioresour Technol 101(13):4851–4861

Alzate CC, Toro OS (2006) Energy consumption analysis of integrated flowsheets for production of fuel ethanol from lignocellulosic biomass. Energy 31(13):2447–2459

Amin FR, Khalid H, Zhang H, u Rahman S, Zhang R, Liu G, Chen C (2017) Pretreatment methods of lignocellulosic biomass for anaerobic digestion. AMB Express 7(1):72

Anderson JE, DiCicco DM, Ginder JM, Kramer U, Leone TG, Raney-Pablo HE, Wallington TJ (2012) High octane number ethanol–gasoline blends: quantifying the potential benefits in the United States. Fuel 97:585–594

Argueso JL, Carazzolle MF, Mieczkowski PA, Duarte FM, Netto OV, Missawa SK, Dominska M (2009) Genome structure of a *Saccharomyces cerevisiae* strain widely used in bioethanol production. Genome Res 19(12):2258–2270

Aristidou A, Penttilä M (2000) Metabolic engineering applications to renewable resource utilization. Curr Opin Biotechnol 11(2):187–198

Arora R, Behera S, Sharma NK, Kumar S (2015) A new search for thermotolerant yeasts, its characterization and optimization using response surface methodology for ethanol production. Front Microbiol 6:889

Azhar SHM, Abdulla R, Jambo SA, Marbawi H, Gansau JA, Faik AAM, Rodrigues KF (2017) Yeasts in sustainable bioethanol production: a review. Biochem Biophys Rep 10:52–61

Baba Y, Tanabe T, Shirai N, Watanabe T, Honda Y, Watanabe T (2011) Pretreatment of Japanese cedar wood by white rot fungi and ethanolysis for bioethanol production. Biomass Bioenergy 35(1):320–324

Bakshi AS, Loescher ME, Sahay N (2007) U.S. Patent No. 7,273,957. Washington, DC: U.S. Patent and Trademark Office

Balat M (2011) Production of bioethanol from lignocellulosic materials via the biochemical pathway: a review. Energy Convers Manag 52(2):858–875

Ballesteros M, Oliva JM, Manzanares P, Negro MJ, Ballesteros I (2002) Ethanol production from paper material using a simultaneous saccharification and fermentation system in a fed-batch basis. World J Microbiol Biotechnol 18(6):559–561

Ballesteros M, Oliva JM, Negro MJ, Manzanares P, Ballesteros I (2004) Ethanol from lignocellulosic materials by a simultaneous saccharification and fermentation process (SFS) with *Kluyveromyces marxianus* CECT 10875. Process Biochem 39(12):1843–1848

Banerjee G, Scott-Craig JS, Walton JD (2010) Improving enzymes for biomass conversion: a basic research perspective. Bioenergy Res 3(1):82–92

Bauer MW, Driskill LE, Callen W, Snead MA, Mathur EJ, Kelly RM (1999) An endoglucanase, EglA, from the hyperthermophilic Archaeon *Pyrococcus furiosus* hydrolyzes β-1, 4 bonds in mixed-linkage (1→3),(1→4)-β-D-glucans and cellulose. J Bacteriol 181(1):284–290

Bhardwaj N, Kumar B, Agarwal K, Chaturvedi V, Verma P (2019) Purification and characterization of a thermo-acid/alkali stable xylanases from *Aspergillus oryzae* LC1 and its application in Xylo-oligosaccharides production from lignocellulosic agricultural wastes. Int J Biol Macromol 122:1191–1202

Borodina I, Nielsen J (2014) Advances in metabolic engineering of yeast *Saccharomyces cerevisiae* for production of chemicals. Biotechnol Lett 9(5):609–620

Boykoff M, Daly M, McAllister L, McNatt M, Nacu-Schmidt A, Oonk D, Pearman O (2019) Life as we know it: a review of the month's media coverage of climate change and global warming. Media and Climate Change Observatory, Center for Science and Technology Policy Research, Cooperative Institute for Research in Environmental Sciences, 29. https://doi.org/10.25810/8qh4-1973

Brandt A, Ray MJ, To TQ, Leak DJ, Murphy RJ, Welton T (2011) Ionic liquid pretreatment of lignocellulosic biomass with ionic liquid-water mixtures. Green Chem 13(9):2489–2499

Buxton DR, Russell JR (1988) Lignin constituents and cell-wall digestibility of grass and legume stems. Crop Sci 28(3):553–558

Cheng JJ, Timilsina GR (2011) Status and barriers of advanced biofuel technologies: a review. Renew Energy 36(12):3541–3549

Chovau S, Degrauwe D, Van der Bruggen B (2013) Critical analysis of techno-economic estimates for the production cost of lignocellulosic bio-ethanol. Renew Sust Energ Rev 26:307–321

Chundawat SP, Venkatesh B, Dale BE (2007) Effect of particle size based separation of milled corn stover on AFEX pretreatment and enzymatic digestibility. Biotechnol Bioeng 96(2):219–231

Dadi AP, Varanasi S, Schall CA (2006) Enhancement of cellulose saccharification kinetics using an ionic liquid pretreatment step. Biotechnol Bioeng 95(5):904–910

De Frias JA, Feng H (2013) Switchable butadiene sulfone pretreatment of Miscanthus in the presence of water. Green Chem 15(4):1067–1078

De Souza Liberal AT, Basílio ACM, do Monte Resende A, Brasileiro BTV, Da Silva-Filho EA, De Morais JOF, de Morais MA Jr (2007) Identification of *Dekkera bruxellensis* as a major contaminant yeast in continuous fuel ethanol fermentation. J Appl Microbiol 102(2):538–547

Demirbas A (2008) Biofuels sources, biofuel policy, biofuel economy and global biofuel projections. Energy Convers Manag 49(8):2106–2116
Dhyani V, Bhaskar T (2018) A comprehensive review on the pyrolysis of lignocellulosic biomass. Renew Energy 129:695–716
Dien BS, Cotta MA, Jeffries TW (2003) Bacteria engineered for fuel ethanol production: current status. Appl Microbiol Biotechnol 63(3):258–266
Dodo CM, Mamphweli S, Okoh O (2017) Bioethanol production from lignocellulosic sugarcane leaves and tops. J Energy South Afr 28(3):1–11
Edgardo A, Carolina P, Manuel F, Juanita F, Baeza J (2008) Selection of thermotolerant yeast strains *Saccharomyces cerevisiae* for bioethanol production. Enzym Microb Technol 43(2):120–123
Evcan E (2012) Bioethanol production from fungal sources using low-cost agro-industrial waste products. Master's thesis, İzmir Institute of Technology
Festucci-Buselli RA, Otoni WC, Joshi CP (2007) Structure, organization, and functions of cellulose synthase complexes in higher plants. Braz J Plant Physiol 19(1):1–13
Gamage J, Lam H, Zhang Z (2010) Bioethanol production from lignocellulosic biomass: a review. J Biobased Mater Bioenergy 4(1):3–11
García-Aparicio MP, Ballesteros M, Manzanares P, Ballesteros I, González A, Negro MJ (2007) Xylanase contribution to the efficiency of cellulose enzymatic hydrolysis of barley straw. Appl Microbiol Biotechnol 137(1–12):353–365
Gray KA, Zhao L, Emptage M (2006) Bioethanol. Curr Opin Chem Biol 10(2):141–146
Guerriero G, Hausman JF, Strauss J, Ertan H, Siddiqui KS (2016) Lignocellulosic biomass: biosynthesis, degradation, and industrial utilization. Eng Life Sci 16(1):1–16
Hahn-Hägerdal B, Galbe M, Gorwa-Grauslund MF, Lidén G, Zacchi G (2006) Bio-ethanol–the fuel of tomorrow from the residues of today. Trends Biotechnol 24(12):549–556
Hamelinck CN, Van Hooijdonk G, Faaij AP (2005) Ethanol from lignocellulosic biomass: techno-economic performance in short-, middle-and long-term. Biomass Bioenergy 28(4):384–410
Harmsen PFH, Huijgen W, Bermudez L, Bakker R (2010) Literature review of physical and chemical pretreatment processes for lignocellulosic biomass (No. 1184). UR-Food & Biobased Research, Wageningen
Hasunuma T, Kondo A (2012) Consolidated bioprocessing and simultaneous saccharification and fermentation of lignocellulose to ethanol with thermotolerant yeast strains. Process Biochem 47(9):1287–1294
Hendriks ATWM, Zeeman G (2009) Pretreatments to enhance the digestibility of lignocellulosic biomass. Bioresour Technol 100(1):10–18
Hsu TA (2018) Pretreatment of biomass. In: Handbook on bioethanol. Routledge, pp 179–212
Humbird D, Davis R, Tao L, Kinchin C, Hsu D, Aden A, Sexton D (2011) Process design and economics for biochemical conversion of lignocellulosic biomass to ethanol: dilute-acid pretreatment and enzymatic hydrolysis of corn stover (No. NREL/TP-5100-47764). National Renewable Energy Lab. (NREL), Golden, CO (United States)
Jönsson LJ, Martín C (2016) Pretreatment of lignocellulose: formation of inhibitory by-products and strategies for minimizing their effects. Bioresour Technol 199:103–112
Kang Q, Appels L, Tan T, Dewil R (2014) Bioethanol from lignocellulosic biomass: current findings determine research priorities. Sci World J 2014:298153
Kim H, Lee WH, Galazka JM, Cate JH, Jin YS (2014) Analysis of cellodextrin transporters from *Neurospora crassa* in *Saccharomyces cerevisiae* for cellobiose fermentation. Appl Microbiol Biotechnol 98(3):1087–1094
Konda NVSNM, Shi J, Singh S, Blanch HW, Simmons BA, Marcuschame DK (2014) Understanding cost drivers and economic potential of two variants of ionic liquid pretreatment for cellulosic biofuel production. Biotechnol Biofuels 7:86–99
Kumar AK, Sharma S (2017) Recent updates on different methods of pretreatment of lignocellulosic feedstocks: a review. Bioresour Bioprocess 4(1):7
Kumar R, Singh S, Singh OV (2008) Bioconversion of lignocellulosic biomass: biochemical and molecular perspectives. J Ind Microbiol Biotechnol 35(5):377–391

Kumar P, Barrett DM, Delwiche MJ, Stroeve P (2009) Methods for pretreatment of lignocellulosic biomass for efficient hydrolysis and biofuel production. Ind Eng Chem Res 48(8):3713–3729
Kutas G, Lindberg C, Steenblik R (2007) Biofuels—at what cost?: government support for ethanol and biodiesel in the European Union. International Institute for Sustainable Development, Geneva, pp 14–25
Lee J (1997) Biological conversion of lignocellulosic biomass to ethanol. J Biotechnol 56(1):1–24
Li X (2010) Bioethanol production from lignocellulosic feedstock using aqueous ammonia pretreatment and simultaneous saccharification and fermentation (SSF): process development and optimization, Graduate theses and dissertations, 11300
Limayem A, Ricke SC (2012) Lignocellulosic biomass for bioethanol production: current perspectives, potential issues and future prospects. Prog Energy Combust Sci 38(4):449–467
Liu Q, Li J, Gao R, Li J, Ma G, Tian C (2019) CLR-4, a novel conserved transcription factor for cellulase gene expression in ascomycete fungi. Mol Microbiol 111(2):373–394
Lynd LR, Van Zyl WH, McBride JE, Laser M (2005) Consolidated bioprocessing of cellulosic biomass: an update. Curr Opin Biotechnol 16(5):577–583
Madhavan A, Jose AA, Binod P, Sindhu R, Sukumaran RK, Pandey A, Castro GE (2017) Synthetic biology and metabolic engineering approaches and its impact on non-conventional yeast and biofuel production. Front Energy Res 5:8
Malça J, Freire F (2006) Renewability and life-cycle energy efficiency of bioethanol and bio-ethyl tertiary butyl ether (bioETBE): assessing the implications of allocation. Energy 31(15):3362–3380
Menegol D, Scholl AL, Dillon AJP, Camassola M (2017) Use of elephant grass (*Pennisetum purpureum*) as substrate for cellulase and xylanase production in solid-state cultivation by *Penicillium echinulatum*. Braz J Chem Eng 34(3):691–700
Meng QS, Liu CG, Zhao XQ, Bai FW (2018) Engineering *Trichoderma reesei* Rut-C30 with the overexpression of egl1 at the ace1 locus to relieve repression on cellulase production and to adjust the ratio of cellulolytic enzymes for more efficient hydrolysis of lignocellulosic biomass. J Biotechnol 285:56–63
Mood SH, Golfeshan AH, Tabatabaei M, Jouzani GS, Najafi GH, Gholami M, Ardjmand M (2013) Lignocellulosic biomass to bioethanol, a comprehensive review with a focus on pretreatment. Renew Sustain Energy Rev 27:77–93
Mora-Pale M, Meli L, Doherty TV, Linhardt RJ, Dordick JS (2011) Room temperature ionic liquids as emerging solvents for the pretreatment of lignocellulosic biomass. Biotechnol Bioeng 108(6):1229–1245
Moreno AD, Alvira P, Ibarra D, Tomás-Pejó E (2017) Production of ethanol from lignocellulosic biomass. In: Production of platform chemicals from sustainable resources. Springer, Singapore, pp 375–410
Mosier N, Wyman C, Dale B, Elander R, Lee YY, Holtzapple M, Ladisch M (2005) Features of promising technologies for pretreatment of lignocellulosic biomass. Bioresour Technol 96(6):673–686
Nagdeote DD, Deshmukh MM (2012) Experimental study of diethyl ether and ethanol additives with biodiesel-diesel blended fuel engine. Int J Emerg Technol Adv Eng 2(3):2250–2459
Nasir M, Ghazi MT (2015) Pretreatment of lignocellulosic biomass from animal manure as a means of enhancing biogas production. Eng Life Sci 15(7):733–742
Nasir Iqbal HM, Kamal S (2012) Economical bioconversion of lignocellulosic materials to value-added products. J Biotechnol Biomater 2:e112
Nguyen TY, Cai CM, Osman O, Kumar R, Wyman CE (2016) CELF pretreatment of corn stover boosts ethanol titers and yields from high solids SSF with low enzyme loadings. Green Chem 18(6):1581–1589
Nigam JN (2001) Ethanol production from wheat straw hemicellulose hydrolysate by *Pichia stipitis*. J Biotechnol 87(1):17–27
Nitsos CK, Matis KA, Triantafyllidis KS (2013) Optimization of hydrothermal pretreatment of lignocellulosic biomass in the bioethanol production process. ChemSusChem 6(1):110–122

Opoku A (2019) Biodiversity and the built environment: implications for the sustainable development goals (SDGs). Resour Conserv Recycling 141:1–7
Perez-Pimienta JA, Flores-Gómez CA, Ruiz HA, Sathitsuksanoh N, Balan V, da Costa Sousa L, Dale BE, Singh S, Simmons BA (2016) Evaluation of agave bagasse recalcitrance using AFEX™, autohydrolysis, and ionic liquid pretreatments. Bioresour Technol 211:216–223
Ponniah SK, Shang Z, Akbudak MA, Srivastava V, Manoharan M (2017) Down-regulation of hydroxycinnamoyl CoA: shikimate hydroxycinnamoyl transferase, cinnamoyl CoA reductase, and cinnamyl alcohol dehydrogenase leads to lignin reduction in rice (*Oryza sativa* L. ssp. japonica cv. Nipponbare). Plant Biotechnol Rep 11(1):17–27
Puttaswamy CT, Sagar BR, Simha U, Manjappa S, Kumar CV (2016) Production of bioethanol from lignocellulosic biomass. Indian J Adv Chem Sci 239:244
Rastogi M, Shrivastava S (2018) Current methodologies and advances in bio-ethanol production. J Biotechnol Biores 1(1):1–8
Rezania S, Din MFM, Mohamad SE, Sohaili J, Taib SM, Yusof MBM, Ahsan A (2017) Review on pretreatment methods and ethanol production from cellulosic water hyacinth. BioResources 12(1):2108–2124
Robak K, Balcerek M (2018) Review of second generation bioethanol production from residual biomass. Food Technol Biotechnol 56(2):174
Roberts SB, Gowen CM, Brooks JP, Fong SS (2010) Genome-scale metabolic analysis of *Clostridium thermocellum* for bioethanol production. BMC Syst Biol 4(1):31
Rocha-Meneses L, Raud M, Orupõld K, Kikas T (2017) Second-generation bioethanol production: a review of strategies for waste valorisation. Agron Res 15(3):830–847
Rouches E, Herpoël-Gimbert I, Steyer JP, Carrere H (2016) Improvement of anaerobic degradation by white-rot fungi pretreatment of lignocellulosic biomass: a review. Renew Sust Energ Rev 59:179–198
Ryabova OB, Chmil OM, Sibirny AA (2003) Xylose and cellobiose fermentation to ethanol by the thermotolerant methylotrophic yeast *Hansenula polymorpha*. FEMS Yeast Res 4(2):157–164
Salusjärvi L, Poutanen M, Pitkänen JP, Koivistoinen H, Aristidou A, Kalkkinen N, Penttilä M (2003) Proteome analysis of recombinant xylose-fermenting *Saccharomyces cerevisiae*. Yeast 20(4):295–314
Sánchez C (2009) Lignocellulosic residues: biodegradation and bioconversion by fungi. Biotechnol Adv 27(2):185–194
Sarkar N, Ghosh SK, Bannerjee S, Aikat K (2012) Bioethanol production from agricultural wastes: an overview. Renew Energy 37(1):19–27
Sathitsuksanoh N, Zhu Z, Ho TJ, Bai MD, Zhang YHP (2010) Bamboo saccharification through cellulose solvent-based biomass pretreatment followed by enzymatic hydrolysis at ultra-low cellulase loadings. Bioresour Technol 101(13):4926–4929
Scarlat N, Dallemand JF (2019) Future role of bioenergy. The role of bioenergy in the bioeconomy. Academic, pp 435–547
Seo JS, Chong H, Park HS, Yoon KO, Jung C, Kim JJ et al (2005) The genome sequence of the ethanologenic bacterium *Zymomonas mobilis* ZM4. Nat Biotechnol 23(1):63
Shaw AJ, Podkaminer KK, Desai SG, Bardsley JS, Rogers SR, Thorne PG, Lynd LR (2008) Metabolic engineering of a thermophilic bacterium to produce ethanol at high yield. PNAS 105(37):13769–13774
Shuai L, Questell-Santiago YM, Luterbacher JS (2016) A mild biomass pretreatment using γ-valerolactone for concentrated sugar production. Green Chem 18(4):937–943
Sindhu R, Binod P, Pandey A (2016) Biological pretreatment of lignocellulosic biomass–an overview. Bioresour Technol 199:76–82
Singh A, Pant D, Korres NE, Nizami AS, Prasad S, Murphy JD (2010) Key issues in life cycle assessment of ethanol production from lignocellulosic biomass: challenges and perspectives. Bioresour Technol 101(13):5003–5012

Singh R, Shukla A, Tiwari S, Srivastava M (2014) A review on delignification of lignocellulosic biomass for enhancement of ethanol production potential. Renew Sustain Energy Rev 32:713–728

Singh S, Simmons BA (2015) Comparison of different biomass pretreatment techniques and their impact on chemistry and structure. Front Energy Res 2(62):1–9

Singh S, Simmons BA, Vogel KP, (2009) Visualization of biomass solubilization and cellulose regeneration during ionic liquid pretreatment of switch grass. Biotech Bioeng 104(1):68–75

Soccol CR, de Souza Vandenberghe LP, Medeiros ABP, Karp SG, Buckeridge M, Ramos LP, da Silva Bon EP (2010) Bioethanol from lignocelluloses: status and perspectives in Brazil. Bioresour Technol 101(13):4820–4825

Souto-Maior AM, Runquist D, Hahn-Hägerdal B (2009) Crabtree-negative characteristics of recombinant xylose-utilizing *Saccharomyces cerevisiae*. J Biotechnol 143(2):119–123

Sticklen MB (2008) Plant genetic engineering for biofuel production: towards affordable cellulosic ethanol. Nat Rev Genet 9(6):433

Stöcker M (2008) Biofuels and biomass-to-liquid fuels in the biorefinery: catalytic conversion of lignocellulosic biomass using porous materials. Angew Chem 47(48):9200–9211

Sukumaran RK, Singhania RR, Mathew GM, Pandey A (2009) Cellulase production using biomass feed stock and its application in lignocellulose saccharification for bio-ethanol production. Renew Energy 34(2):421–424

Sun Y, Cheng J (2002) Hydrolysis of lignocellulosic materials for ethanol production: a review. Bioresour Technol 83(1):1–11

Swatloski RP, Spear SK, Holbrey JD, Rogers RD (2002) Dissolution of cellose with ionic liquids. J Am Chem Soc 124(18):4974–4975

Tsigie YA, Wu CH, Huynh LH, Ismadji S, Ju YH (2013) Bioethanol production from *Yarrowia lipolytica* Po1g biomass. Bioresour Technol 145:210–216

Van Tilbeurgh H, Tomme P, Claeyssens M, Bhikhabhai R, Pettersson G (1986) Limited proteolysis of the cellobiohydrolase I from *Trichoderma reesei*: separation of functional domains. FEBS Lett 204(2):223–227

Viñals-Verde M, Bell-García A, Michelena-Álvarez G, Ramil-Mesa M (2012) Ethanol production from lignocellulosic biomass. ICIDCA Sobre los Derivados de la Caña de Azúcar. 46(1):7–16

Voronovsky AY, Rohulya OV, Abbas CA, Sibirny AA (2009) Development of strains of the thermotolerant yeast *Hansenula polymorpha* capable of alcoholic fermentation of starch and xylan. Metab Eng 11(4–5):234–242

Wilhelm RC, Singh R, Eltis LD, Mohn WW (2019) Bacterial contributions to delignification and lignocellulose degradation in forest soils with metagenomic and quantitative stable isotope probing. ISME J 13(2):413–429

Wu Z, Li Y, Xu D, Meng H (2019) Co-pyrolysis of lignocellulosic biomass with low-quality coal: optimal design and synergistic effect from gaseous products distribution. Fuel 236:43–54

Wyman CE (1994) Ethanol from lignocellulosic biomass: technology, economics, and opportunities. Bioresour Technol 50(1):3–15

Wyman CE (2018) Ethanol production from lignocellulosic biomass: overview. In: Handbook on bioethanol. Routledge, pp 1–18

Xiong S, Martín C, Eilertsen L, Wei M, Myronycheva O, Larsson SH et al (2019) Energy-efficient substrate pasteurisation for combined production of shiitake mushroom (*Lentinula edodes*) and bioethanol. Bioresour Technol 274:65–72

Xu Q, Singh A, Himmel ME (2009) Perspectives and new directions for the production of bioethanol using consolidated bioprocessing of lignocellulose. Curr Opin Biotechnol 20(3):364–371

Xu H, Che X, Ding Y, Kong Y, Li B, Tian W (2019) Effect of crystallinity on pretreatment and enzymatic hydrolysis of lignocellulosic biomass based on multivariate analysis. Bioresour Technol 279:271–280

Zabed H, Sahu J, Suely A, Boyce AN, Faruq G (2017) Bioethanol production from renewable sources: current perspectives and technological progress. Renew Sustain Energy Rev 71:475–501. https://doi.org/10.1016/j.rser.2016.12.076

Zhao X, Zhang L, Liu D (2012) Biomass recalcitrance. Part I: the chemical compositions and physical structures affecting the enzymatic hydrolysis of lignocellulose. Biofuels Bioprod Biorefin 6(4):465–482

Zhao C, Chen S, Fang H (2018) Consolidated bioprocessing of lignocellulosic biomass to itaconic acid by metabolically engineering *Neurospora crassa*. Appl Microbiol Biotechnol 102(22):9577–9584

Zheng Y, Pan Z, Zhang R (2009) Overview of biomass pretreatment for cellulosic ethanol production. Int J Agric Biol Eng 2(3):51–68

Zheng Z, Jiang T, Zou L, Ouyang S, Zhou J, Lin X, Ouyang J (2018) Simultaneous consumption of cellobiose and xylose by *Bacillus coagulans* to circumvent glucose repression and identification of its cellobiose-assimilating operons. Biotechnol Biofuels 11(1):320

Zhu JY, Zhu W, OBryan P, Dien BS, Tian S, Gleisner R, Pan XJ (2010) Ethanol production from SPORL-pretreated lodgepole pine: preliminary evaluation of mass balance and process energy efficiency. Appl Microbiol Biotechnol 86(5):1355–1365

Chapter 10
Multiple Factors Influencing the Strategy of Lignin Mycodegradation

Gautam Anand, Sangeeta Yadav, and Dinesh Yadav

1 Introduction

The term "lignin" is derived from the Latin word "lignum," meaning wood. It is the most abundant aromatic compound on earth and is the second (after cellulose) most abundantly present natural polymer. It accounts for 15–30% of the lignocellulosic biomass (Yao et al. 2017). Of all naturally produced organic chemicals, lignin is probably the most recalcitrant. Together with hemicellulose, it forms strong adhesive for cellulose fibers (Grzegorz et al. 2017). This property of lignin is consistent with its biological functions, i.e., to give vascular plants the rigidity and to protect their structural polysaccharides from attack by other organisms. The conversion isolated cellulose or hemicellulose into monosaccharides is known to be a relatively simple process. Many cellulolytic as well as ligninolytic fungi make use of hydrolases which are capable of producing monosaccharides from polysaccharide components of wood. However, when these components are complexed with lignin, they are resistant to hydrolytic breakdown. Therefore, lignin appears to inhibit hydrolytic activity.

G. Anand
Department of Plant Pathology and Microbiology, Robert H. Smith Faculty of Agriculture, Food and Environment, Hebrew University of Jerusalem, Rehovot, Israel

Department of Biotechnology, D.D.U. Gorakhpur University, Gorakhpur, Uttar Pradesh, India

S. Yadav · D. Yadav (✉)
Department of Biotechnology, D.D.U. Gorakhpur University, Gorakhpur, Uttar Pradesh, India

R. Naraian (ed.), *Mycodegradation of Lignocelluloses*, Fungal Biology,
https://doi.org/10.1007/978-3-030-23834-6_10

2 Lignin Structure and Biosynthesis

Lignin is a complex, amorphous, three-dimensional aromatic polymer. Lignin is formed in vascular plant cell walls by the oxidative coupling of several related phenylpropanoid precursors: coniferyl alcohol, sinapyl alcohol, and *p*-coumaryl alcohol (Fig. 10.1). Because of the different monomers, lignin can be divided into three types: syringyl lignin polymerized by syringyl propane, guaiacyl lignin polymerized by guaiacyl propane, and hydroxy-phenyl lignin polymerized by hydroxy-phenyl propane. Usually, gymnosperm mainly contains guaiacyl (G) lignin; the dicotyledon mainly contains guaiacyl-syringyl (GS) lignin; the monocotyledon mainly contains guaiacyl-syringyl-hydroxy-phenyl (GSH) lignin (Wei and Song 2001). G lignin is chiefly formed through dehydrated oligomerization of coniferyl alcohol, and its structure is homogeneous. Most lignin in softwood belongs to G lignin, which is copolymerized by guaiacyl. GSH lignin is the result of the dehydrated oligomerization of coniferyl alcohol and sinapyl alcohol; the content of lignin is 17–23%. The ratio of syringyl propane to guaiacyl propane is 0.5:0.1; it also contains 7–12% ester groups. *p*-Coumaryl alcohol in it is linked to lignin in the form of ester (Gao and Tang 1996).

Peroxidases or laccases in the plant cell wall oxidize these monomers by one electron, yielding transient resonance-stabilized phenoxyradicals that then polymerize in a variety of configurations (Adler 1977; Higuchi 1990). The molecular weight of lignin is difficult to determine because lignins are highly polydisperse materials (Argyropoulos and Menachem 1997). Lignin is covalently associated with hemicelluloses in the cell wall via numerous types of linkage (Fig. 10.2). Among the most important are ether bonds between the benzylic carbon of lignin and the carbohydrate moiety, ester bonds between the benzylic carbon of lignin and uronic acid residues, and lignin-glycosidic bonds.

Recent studies show that lignin can incorporate many more monolignols than the traditional three basic units (Vanholme et al. 2008), e.g., acetylated lignin units have been identified in non-woody plants (Martínez et al. 2005). In graminaceous plants,

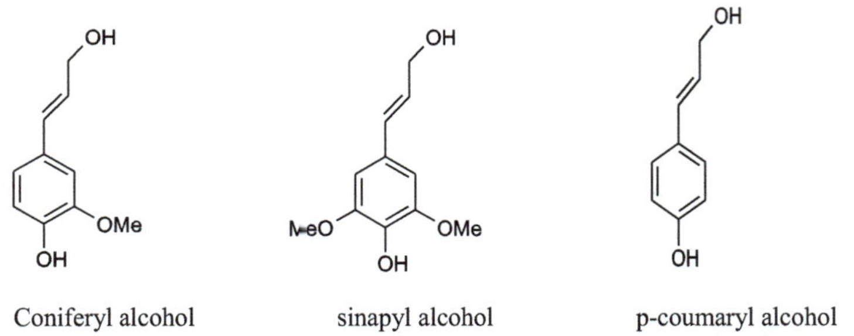

Fig. 10.1 Three precursors of lignin

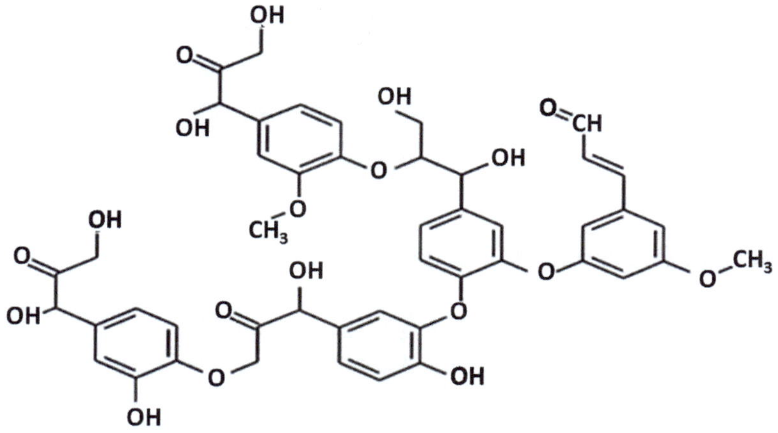

Fig. 10.2 Polymeric structure of lignin

hydroxycinnamic acid residues are frequent in the lignin and are attached to hemicelluloses via ester linkages. The matrix of lignin and hemicellulose encrusts and protects the cellulose of the plant cell wall (Jeffries 1991).

Lignin polymer contains characteristic methoxyl groups, phenolic hydroxyl groups, and some terminal aldehyde group. Most of the phenolic hydroxyl groups are bonded to neighboring phenylpropane units. Carbonyl groups and alcoholic hydroxyl groups are introduced in the final lignin structure during dehydrogenative polymerization process. Also, in some of the wood species substantial alcoholic hydroxyl groups are esterified with *p*-hydroxybenzoic acid or *p*-hydroxycinnamic acid.

Biosynthesis of monolignols is carried through the phenylpropanoid pathway. In the shikimate pathway, glucose obtained by photosynthesis is first converted to shikimic acid. This shikimic acid is converted into the final products of the pathway, i.e., phenylalanine and tyrosine through prephenic acid. Phenylalanine and tyrosine are starting components for the cinnamic acid pathway. After a set of reactions, such as deamination, hydroxylation, methylation, and reduction, three monomers of lignin are finally synthesized (Lin et al. 2003; Geng et al. 2003).

The rate-limiting step, controlling lignin synthesis and the pathway from which carbon can go into lignin biosynthesis is catalyzed by Cinnamoyl CoA reductase (CCR). Reduction of coniferaldehyde, another redox reaction in the lignin synthesis process, is catalyzed by cinnamyl alcohol dehydrogenase (CAD). CAD is located in tissues with precipitation of G lignin. Sinapyl alcohol dehydrogenase (SAD), ferulic acid 5-hydroxylase (F5H), and bispecific caffeic acid/5-hydroxyferulic acid *O*-methyltransferase (COMT) are immunolocalized in cells and tissues that have S lignin. It is believed that the last step of redox reactions of different types of lignin go through different pathways which are catalyzed by different enzymes (Lin et al. 2003). Two-step methylation reaction at the 3′ and 5′ site is an important step in lignin monomer synthesis. This step is carried out by COMT and CCoAOMT (caffeoyl CoA 3-Omethyltransferase) which are methylases on two different substrate levels.

3 Fungal Degradation of Lignocellulose

Ligninolytic fungi are not able to use lignin as their sole source of energy and carbon. Instead, they depend on the more digestible polysaccharides in lignocellulosic substrates, and the primary function of ligninolysis is to expose these polysaccharides so that they can be cleaved by fungal cellulases and hemicellulases. Oxidative coupling of lignin aromatic alcohol monomers creates a complex structure in lignin which is highly recalcitrant to degradation (Wong 2009). By linking to both hemicelluloses and cellulose, lignin acts as a barrier to any solutions or enzymes and prevents penetration of lignocellulolytic enzymes to the interior lignocellulosic structure. Not surprisingly, of the components of lignocellulosic material, lignin is the most resistant to degradation (Sánchez 2009; Himmel et al. 2007). Although lignin resists attack by most microorganisms, basidiomycetes white-rot fungi are able to degrade lignin efficiently (Wong 2009; Abbas et al. 2005). In most fungi that have been examined, ligninolysis occurs during secondary metabolism, i.e., under nutrient limitation. With this approach, the fungus avoids synthesizing and secreting metabolically expensive ligninolytic agents when substrates more accessible than lignocellulose are present. The limiting nutrient for fungal growth in most woods and soils is probably nitrogen, and most laboratory studies of ligninolytic fungi have been done in nitrogen-limited culture media (Kirk and Farrell 1987). However, a few ligninolytic fungi, e.g., some species of *Bjerkandera,* are ligninolytic even when sufficient nitrogen is present (Kaal et al. 1993).

Lignin degrading fungi produce extracellular enzymes to break lignin. These fungi are widespread and include species from the ascomycetes and basidiomycetes phyla such as white-rot and brown-rot fungi. In addition, a few anaerobic species are also found to be able to degrade cellulose in the gastrointestinal tracts of ruminant animals (Ljungdahl 2008; Yoon et al. 2007). Biomass degradation by these fungi is performed by complex mixtures of cellulases (Bayer et al. 1998), hemicellulases (Ljungdahl 2008), and ligninases (Weng et al. 2008; Sánchez 2009), reflecting the complexity of the materials. In nature, degradation of cellulosic biomass is performed by mixtures of hydrolytic enzymes collectively known as cellulases. Many microorganisms, including fungi and bacteria, have been found to be capable of degrading cellulose and other plant cell wall fibers contributing significantly to the decay of lignocellulosic residues in nature. The degradation of lignin by different fungi is briefly discussed:

4 White-Rot Fungi

White-rot fungi are a heterogeneous group of fungi classified into the basidiomycota. They are most efficient organisms capable of degrading lignin and the related litter. More than 90% of all wood-rotting basidiomycetes are of the white-rot type (Gilbertson 1980). Different white-rot fungi vary considerably in the relative rates at which they attack lignin. Some of them simultaneously attack lignin, hemicellulose, and cellulose, whereas some other white-rot fungi preferentially work on lignin in a selective manner. Their purpose is to decompose the lignin in wood so that they can gain access to the cellulose and hemicelluloses that are embedded in the lignin matrix. Usually, syringyl (S) units of lignin are preferentially degraded, whereas guaiacyl (G) units are more resistant to degradation. Many white-rot fungi colonize cell lumina and cause cell wall erosion. Eroded zones coalesce as decay progresses and large voids filled with mycelium are formed. This type of rot is referred as non-selective or simultaneous rot. Calcium oxalate and MnO_2 accumulate when the decay proceeds (Blanchette 1995). Some white-rot fungi degrade lignin in woody plant cell walls relatively to a higher extent than cellulose, and they are called selective white-rot fungi. In nature, they may cause white-pocket or white-mottled types of rot, e.g., *Phellinus nigrolimitatus* (Blanchette 1995). There are also fungi, e.g., the tree pathogen *Heterobasidion annosum*, that are able to produce both types of attack in the same wood (Eriksson et al. 1990). Typical examples of such fungi are *Ganoderma applanatum* and *Heterobasidion annosum*. Because fungi selectively degrading lignin are considered the most promising fungi for applications in the pulp and paper industry, the search among these fungi has attained a considerable interest. However, the ratio lignins, hemicelluloses, and celluloses decayed by a selected fungus can differ enormously, and even different strains of the same species, e.g., of *Phanerochaete chrysosporium* and *Ceriporiopsis subvermispora*, may behave differently on the same kind of wood. Several screening studies to find suitable fungi for biopulping of wood or straw have revealed fungi that, under certain conditions, degrade lignin preferentially to cellulose. Such lignin-selective fungi are, e.g., *P. chrysosporium*, *C. subvermispora* (Otjen et al. 1987; Eriksson et al. 1990), *Pycnoporus cinnabarinus* (Ander and Eriksson 1977), *Phlebia radiata* (Ander and Eriksson 1977), *Phlebia tremellosa (syn. Merulius tremellosus)* (Ander and Eriksson 1977; Eriksson et al. 1990), *Phellinus pini* (Eriksson et al. 1990), and *Dichomitus squalens* (Eriksson et al. 1990).

White-rot fungi produce a number of extracellular enzymes that directly attack lignin, cellulose, and hemicellulose of the plant cell wall to decompose it. These enzymes include laccases and peroxidases, such as lignin peroxidase (LiP), manganese peroxidase (MnP), and versatile peroxidase (VP). Laccases and peroxidase enzymes can also cause lignin degradation through low molecular weight-free radicals such as OH, depolymerize the phenolic and non-phenolic lignin polymer, and mineralize the insoluble lignin. Basic research on lignin degradation, e.g., its mechanisms, physiology, enzymology, and molecular biology, has been mainly carried out with the corticoid fungus *P. chrysosporium* (Kirk and Farrell 1987; Eriksson et al. 1990). After more and taxonomically different fungi had been studied in more detail, it was revealed that both the physiological conditions for lignin degradation and the enzyme systems expressed are fungus-specific and differ from those found in *P. chrysosporium*. Differences may be connected to the taxonomic position and/or ecology of the fungi, e.g., substrate specialization (hardwood, softwood, or certain wood species, heartwood or sapwood), the stage of degradation, etc.

5 Brown Rot

Brown-rot fungi are basidiomycetes that degrade wood to yield brown, shrunken specimens that typically exhibit a pattern of cubical cracks and easily disintegrate upon handling. Only a small proportion, roughly 70%, of all wood decay basidiomycete species falls into this group, which occurs most frequently on gymnosperm wood. In contrast to white-rot fungi, brown-rot fungi, such as *Postia placenta*, *Laetiporus portentosus*, *Piptoporus betulinus*, and *Gloeophyllum trabeum*, can degrade wood carbohydrates, but not oxidized lignin. Brown-rot fungi mainly decompose the cellulose and hemicellulose components in wood, but they can also modify the lignin to a limited extent (Eriksson et al. 1990). They have been much less investigated than white-rot fungi in spite of their enormous economic importance in the destruction of wood. Lignin degradation by brown-rot fungi mainly involves non-enzymatic oxidation reactions producing hydroxyl radicals via Fenton chemistry (Kirk et al. 1991; Kerem et al. 1998, 1999). Brown-rot fungi partially oxidize lignin via aromatic ring demethylation. During this process, the phenolic hydroxyl content of the reaction mixture increases due to partial oxidation and partially due to the addition of new carboxyl and carbonyl groups (Hatakka and Hammel 2011).

Brown-rotted wood is dark, shrink, and typically broken into brick-shaped or cubical fragments that easily break down into brown powder (Blanchette 1995). The brown color indicates the presence of modified lignin in wood. During brown rot, the hemicelluloses in wood are degraded most rapidly, after which virtually all of the cellulose is removed, leaving behind a complex, aromatic ring-containing polymer derived from the original lignin. To some extent, brown-rot fungi have similar degradative capabilities and pathways as white-rot fungi. Both wood decay mechanisms rely on radical formation, low pH, and the production of organic acids. They cause increased alkali solubility of lignin, and the decay is enhanced by high oxygen tension, all of which indicate a crucial involvement of radicals, especially in the early stages of decay (Kirk 1975)

6 Factors Affecting Lignin Degradation

Lignin degradation, owing to its complex structure and features, is a complicated process. Apart from type of fungi, many different environmental factors influence lignin degradation. The environmental factors affect growth and metabolism of fungi thereby influencing lignin degradation. Moisture content, temperature, pH, aeration, carbon, and nitrogen sources are important parameters that affect fungal growth (Yang et al. 1980; Leisola et al. 1984). Among different microbes, white-rot fungi are the most efficient organisms for lignin degradation. *Basidiomycetes* govern the list among different white-rot fungi. Expression of *P. chrysosporium* genes are strongly influenced by nitrogen and carbon limitation. Regulatory elements present in the promoter regions of genes encoding lignolytic enzymes play an important role in transcriptional activation (Cohen et al. 2001). Under aerobic environment *P. chrysosporium* can reduce 1 g of different separated lignins in 2 days. The fungus is able to degrade lignin efficiently under 30% CO_2 and 10% O_2 concentrations. Addition of 0.12% nitrogen has been shown to increase lignin degradation. The fungus uses lignin, hemicelluloses, and cellulose as substrate. The lignin reduction happens at the end of primary growth by cooperation of other metabolism-like nitrogen (Kang et al. 2013; Leisola et al. 1984; Yang et al. 1980; Hatakka and Hammel 2011; Hatakka 2005). The fungi use oxidation reactions which decrease methoxy, phenolic, and aliphatic content of lignin. These reactions also cleave aromatic rings and form new carbonyl groups. When it comes to mixed microbial population, soft-rot fungi can efficiently degrade lignin. They are better adapted to variations in temperature, pH, and oxygen content (Hamed 2013; Gupta and Verma 2015; Gupta et al. 2016).

pH plays a prominent role in the growth of fungi and it is very difficult to control an optimum pH when the fungi growing on a solid-state substrate. Production of lignolytic gene is influenced by the initial pH of the medium. At pH 4.0–5.0, most of the white-rot fungi grow optimally and also the substrate acidity decreases their growth (Patel and Gupte 2016).

The optimum temperature during biological pre-treatment varies with the type of fungi used. Most of the white-rot ascomycetes fungi grow best around 39 °C while the white-rot basidiomycetes grow optimally around 25 and 30 °C. The heat generated during fungal growth develops temperature gradients in solid-state media. The accumulated heat can inhibit fungal growth and metabolism. Variations in temperature employed are due to the differences in fungal strain, physiology, and type of substrate (Rouches et al. 2016).

Inoculum concentrations have significant role in lignin degradation. The time required for the colonization of the substrate is affected by the type and amount of inoculums. Fungal spores are the commonly used inoculum. Mostly, it is found that larger quantity of inoculum leads to shorter time for colonization of the substrate (Van Kuijk et al. 2015).

High substrate concentrations have to be used for biological pre-treatment to make the process economically viable. When dry substrate is used, it is found that concentration of inhibitors increases and this may influence the degradation process. Hence, pre-treatment should be carried in conditions to minimize the generation as well as accumulation of inhibitory compounds. Initial moisture content is essential for the establishment of microbial growth in the biomass. Initial moisture content critically affects the fungal growth and enzyme production and significantly affects lignin degradation (Rouches et al. 2016).

Aeration also has significant role in fungal degradation of lignin. Aeration has many functions which are including oxygenation, CO_2 removal, heat dispersion, humidity conservation, and also dispersal of volatile combinations produced during metabolism. It also affects production of enzymes that are required during lignin metabolism. Since lignin degradation is an oxidative process, oxygen availability is important for ligninase activity of white-rot fungi. However, high oxygen content has also been shown to slow down lignin degradation process in *P. chrysosporium*. High aeration could improve delignification rate and hence controlled aeration is essential for improvement of biological pre-treatment (Salvachua et al. 2011; Leisola et al. 1984).

7 Steps in Lignolysis

Lignin degradation by the fungi involves both depolymerization and cleavage of aromatic ring. Extracellular enzyme secreted by fungi oxidizes lignin in the following steps:

1. Oxidation of β–*O*–4 linkages to arylglycerol compounds
2. Aromatic rings cleavage by the α–ketoadipate pathway
3. Formation of cyclic carbonate structures.

8 Enzymes Involved in Lignin Degradation

Lignin does not contain hydrolyzable linkages, which means that the enzymes must be oxidative. Ligninases can be classified as either phenol oxidases (laccase) or heme peroxidases, lignin peroxidase (LiP), manganese peroxidase (MnP), and versatile peroxidase (VP) (Martínez et al. 2005). In general, laccases use molecular oxygen as electron acceptors while peroxidases use hydrogen peroxide as a co-substrate. These enzymes can directly attack lignin, cellulose, and hemicellulose of the plant cell wall to decompose it. Some of these enzymes and their isoenzymes gene have been investigated and characterized. For example, the genome of *P. chrysosporium* contains ten LiP and five MnP genes (Martinez et al. 2004). Four laccase genes have been recognized in the plant pathogenic fungus *Rhizoctonia solani* (Wahleithner et al. 1996). Although LiP is able to oxidize the non-phenolic part of lignin (which forms 80–90% of lignin composition), it is absent from many

Table 10.1 Features of fungal lignolytic enzymes

Enzyme	Prosthetic group	Substrate	Mediators	Optimum pH	Reaction
Laccase	Four Cu atoms	Ortho and para-diphenols, aminophenol, polyphenols, polyamines, lignin, aryl diamines	Hydroxybenzotriazole, syringaldazine, phenols, aniline	2.0–10.0	4 benzenediol + O_2 = 4 benzosemiquinone + 2 H_2O
Lignin peroxidase	Heme	Halogenated phenolic compounds, polycyclic aromatic compounds	Veratryl alcohol	1.0–5.0	1,2-bis(3,4-dimethoxyphenyl)propane-1,3-diol + H_2O_2 = 3,4-dimethoxybenzaldehyde + 1-(3,4-dimethoxyphenyl) ethane-1,2-diol + H_2O
Manganese peroxidase	Heme	Lignin and other phenolic compounds	Thiols, unsaturated fatty acids, organic acids as chelators, Mn^{3+}	2.5–6.8	2Mn(II) + $2H^+$ + H_2O_2 = 2Mn(III) + $2H_2O$
Versatile peroxidases	Heme	Phenolic and non-phenolic aromatic compounds	Veratryl alcohol, compounds similar to LiP and MnP mediators	3.0–5.0	Donor + H_2O_2 = oxidized Donor + $2H_2O$

lignin degrading fungi (Wang et al. 2008). Also, oxidative ligninolytic enzymes are too large to penetrate into the wood cell wall micropores (Srebotnik et al. 1988). Hence, prior to the enzymatic attack, low-molecular weight diffusible reactive oxidative compounds initiate changes to the lignin structure (Srebotnik et al. 1988; Tanaka et al. 1999). A comparative list of features of fungal lignolytic enzymes is given in Table 10.1.

8.1 *Laccase*

Laccase is a copper-containing oxidase that uses molecular oxygen to oxidize various aromatic and nonaromatic compounds through a free radical catalyzed reaction mechanism. Most white-rot fungi typically produce laccase (Bollag and Leonowicz 1984), and the enzyme is common also in higher plants and in other fungi. Laccases couple the electron reduction of dioxygen into two molecules of water with the oxidation of a vast variety of substrates, such as phenols, arylamines, anilines, thiols, and lignins. Four copper ions in their catalytic center mediate the redox process. The oxidation reactions catalyzed by laccases lead to the formation of free radicals which act as intermediate substrates for the enzymes (Thurston 1994; Messerschmidt and Huber 1990; Ferraroni et al. 2007). These mediators can leave the enzyme site and react with a broad range of high-redox potential substrates and thus create non-enzymatic routes of oxidative polymerizing or depolymerizing reactions. Laccases have been found and studied in white-rot fungi, such as *Pleurotus ostreatus* D1 (Pozdnyakova et al. 2006), *Lentinus tigrinus* (Ferraroni et al. 2007), *Cerrena unicolor* strain 137 (Michniewicz et al. 2006), *Trametes* sp. strain AH28-2 (Xiao et al. 2003), *Trametes pubescens* (Shleev et al. 2007), and *Cyathus bulleri* (Salony et al. 2006). Also, ascomycetes such as *Melanocarpus albomyces* (Hakulinen et al. 2006), *Magnaporthe grisea* (Iyer and Chattoo 2003), *Myrothecium verrucaria* 24G-4 (Sulistyaningdyah et al. 2004), and *Neurospora crassa* (Germann et al. 1988) were also found to produce laccases.

8.2 Lignin Peroxidase

Lignin peroxidase (LiP, EC 1.11.1.14) is a glycosylated enzyme containing heme protein with an iron protoporphyrin prosthetic group that requires hydrogen peroxide (H_2O_2) to catalyze the oxidation of non-phenolic lignin units and mineralize the recalcitrant aromatic compounds. LiPs are heme-containing glycoproteins and play a central role in the biodegradation of the cell wall constituent, lignin (Piontek et al. 2001). LiPs catalyze the H_2O_2-dependent oxidative depolymerization of a variety of non-phenolic lignin compounds (diarylpropane), β-*O*-4 non-phenolic lignin model compounds and a wide range of phenolic compounds such as guaiacol, vanillyl alcohol, catechol, syringic acid, and acteosyringone (Wong 2009). Lignin oxidation takes place by electron transfer, non-catalytic cleavages of various bonds and opening up of aromatic ring. Typically for LiP activity, the amino acid residue needed is a tryptophan and is assumed to long-range electron transfer from a protein radical at the surface of the enzyme, which would act as the substrate oxidizer, to the heme cofactor (Ruiz-Dueñas and Martínez 2009). This helps in oxidization of bulky lignin substrate. They have been found in many white-rot fungi such as *Phanerochaete chrysosporium, Trametes versicolor*, and *Bjerkandera* sp. (Kirk and Farrell 1987; Kaal et al. 1993; Orth et al. 1993). LiPs resemble other peroxidases such as the classical, extensively studied enzyme from horseradish, in that they contain ferric heme and operate via a typical peroxidase catalytic cycle (Kirk and Farrell 1987). However, LiPs are more powerful oxidants than typical peroxidases are, and consequently oxidize not only the usual peroxidase substrates such as phenols and anilines, but also a variety of non-phenolic lignin structures and other aromatic ethers that resemble the basic structural unit of lignin. Unlike the other peroxidases like MnP, LiP is able to oxidize non-phenolic aromatic substrates and does not require the participation of mediators due to its unusually high redox potential (Wong 2009).

8.3 *Manganese Peroxidases*

Manganese peroxidases (MnPs) are extracellular glycoproteins and are secreted in multiple isoforms which contain one molecule of heme as iron protoporphyrin (Asgher et al. 2008). MnPs occur in most white-rot fungi, and are similar to conventional peroxidases, except that manganese (Mn) is essential for the catalysis. The enzyme MnP plays an important role during the initial stages of lignin degradation (Perez and Jeffries 1990). Compared to laccase, MnP causes greater degradation of phenolic lignin due to its higher redox potential with the eventual release of carbon dioxide (Ten Have and Teunissen 2001). MnP catalyzes the peroxide-dependent oxidation of Mn(II) to Mn(III), which is then released from the enzyme surface in complex with oxalate or with other chelators. Chelated Mn (III) complex acts as a reactive low molecular weight, diffusible redox-mediator of phenolic substrates including simple phenols, amines, dyes, phenolic lignin substructures and dimers (Wesenberg et al. 2003; Asgher et al. 2008). In addition, organic acids such as lactate and malonate can chelate Mn^{3+} ion. Since their discovery in *P. chrysosporium,* they have been found in many basidiomycetes such as *Lenzites betulinus*, *Phanerochaete flavido-alba*, and *Panus tigrinus* (Hoshino et al. 2002; de la Rubia et al. 2002; Lisov et al. 2003).

8.4 *Versatile Peroxidases*

Versatile peroxidase (VP), as the name suggests, has catalytic properties of both LiP and MnP. VPs are glycoproteins with hybrid properties capable of oxidizing typical substrates of other basidiomycetes peroxidases including Mn(II) and also veratryl alcohol (VA), MnP and the typical LiP substrate, respectively (Wesenberg et al. 2003; Asgher et al. 2008). VP was first purified from the genera of fungi *Bjerkandera* and was found to transform lignin even without an external mediator (Moreira et al. 2007). VPs form an attractive ligninolytic enzyme group due to their dual oxidative ability to oxidize Mn(II) and also phenolic and non-phenolic aromatic compounds. Versatile peroxidase possesses a hybrid molecular architecture with several binding sites including Mn^{2+} and is able to oxidize Mn^{2+} like MnP and LiP. However, unlike MnP, VP has the dual ability to oxidize Mn^{2+} in the independent oxidation of simple amines and phenolic monomers. VP can also oxidize a variety of substrates (with high and low redox potentials) including Mn^{2+}, phenolic and non-phenolic lignin dimers, and aromatic alcohols (Perez-Boada et al. 2005; Camarero et al. 1999). This makes versatile peroxidase a better catalyst than MnP and LiP. Similar to the MnP mechanism, Mn(III) is released from VPs and acts as a diffusible oxidizer of phenolic lignin and free phenol substrates. Like other members of heme peroxidases, heme is buried in the interior of the protein and has access to the outer medium through two channels (Heinfling et al. 1998).

9 Measurement of Lignin Degradation

The delignification of woody tissues by fungi can be assessed by microscopy (Blanchette 1991) with selective staining (Srebotnik and Messner 1994); however, these techniques are relatively complex and semiquantitative. Microscopy is useful only when the substrate contains lignified cell walls.

The removal of lignin from lignocellulosic material can be measured by the simple procedure of extracting its low molecular weight components, weighing the leftover woody tissues, degrading the remaining polysaccharide component with strong acid, and then reweighing the leftover lignin, which is referred to as Klason lignin. Measurement of Klason lignin is simple and useful if interfering substances are not present. However, this method is not reliable if it is used on plant tissues that contain high molecular weight components that cannot be are not removed by extraction and acid treatment. Interfering substances may be proteins and tannins (Theander and Westerlund 1993). On the other hand, the Klason procedure is also not accurate when woody tissues contain polymers which are acid-soluble and consequently lost during the hydrolysis of polysaccharides.

Another approach to study fungal ligninolysis is to assess whether the fungus degrades a simpler molecule rather than growth substrate itself being delignified. Substrates of this type can be incorporated into the natural lignocellulosic growth medium, e.g., wood or litter, or they can be used as probes in defined liquid growth media (Kirk 1975, 1978; Srebotnik and Messner 1994). C^{14}-labelled synthetic lignins are most frequently used in this type study. Carbon dioxide produced during degradation of the radiolabelled polymer can be trapped in alkali and determined by scintillation counting. However, this method is expensive and requires expert facilities for radioisotope study.

Several chemical procedures have also been introduced for the estimation of lignolysis. For example, pulverized wood samples can be treated with acetyl bromide in acetic acid, and the absorbance of the resulting solution is measured at 280 nm. Comparison of absorbance with absorbance obtained from known lignin standards can be used to determine the extent of lignolysis. Methods of this type are subject to interference from other components, but can be useful for the comparison of closely related lignocellulosic samples (Theander and Westerlund 1993).

10 Importance of Biodegradation of Lignin

Of all naturally produced organic chemicals, lignin is probably the most recalcitrant. Lignin is the most abundant aromatic compound on earth and is second only to cellulose in its contribution to living terrestrial biomass. Lignin degradation can thus play a major role in improving earth's biofuel resources and also serve as an alternative to harsh technologies used in the paper and pulp industry. When plants die or drop litter, lignified organic carbon is incorporated into the top layer of the soil. This recalcitrant material has to be broken down and recycled by microorganisms to maintain the earth's carbon cycle. If this process fails, imagine the amount of carbon which is irreversibly sequestered as lignocellulose.

Lignin biodegradation has diverse effects on soil quality. Microbial lignolysis facilitates the formation of humus by promoting the release of aromatic humus precursors such as flavonoids, terpenes, and tannins from the litter (Hudson 1992). Undegraded lignocellulose, has adverse effect on soil fertility by supporting high microbial load. Increasing population of microbes in soil may not only produce phytotoxins but also compete with crop plants for soil nitrogen and other nutrient. By breaking down the most refractory component of litter, ligninolysis thus contributes to the removal of conditions that inhibit crop productivity.

In paper and textile industries, wood chips are often treated with chemical or mechanical forces to separate their fibers. In chemical pulping, strong chemicals, such as soda, and high pressure are applied to remove lignin. The resulting pulp has high strength but the process is polluting and the overall yield is relatively low (40–50%). Mechanical pulping uses mechanical forces to separate wood fibers. In this process, yields are higher (up to 95%) and results in paper with good printing properties. However, the process requires a lot of electrical energy for refining, the fibers have poor strength properties, and high content of lignin causes a tendency to yellowing. Mechanical pulping can be combined with biological pulping to save refining energy (Akhtar et al. 1998). Also fungal treatment of fibers as a substitute for chemical treatment can not only make the process eco-friendly but high yielding as well by removing the hemicellulose content of the fibers.

References

Abbas A, Koc H, Liu F, Tien M (2005) Fungal degradation of wood: initial proteomic analysis of extracellular proteins of *Phanerochaete chrysosporium* grown on oak substrate. Curr Genet 47(1):49–56

Adler E (1977) Lignin chemistry—past, present and future. Wood Sci Technol 11(3):169–218

Akhtar M, Blanchette RA, Myers GC, Kirk TK (1998) An overview of biomechanical pulping research. In: young RA, Akhtar M (eds.) Environmentally friendly technologies for the pulp and paper industry. Wiley, New York, pp 309–419

Ander P, Eriksson KE (1977) Selective degradation of wood components by white-rot fungi. Physiol Plant 41(4):239–248

Argyropoulos DS, Menachem SB (1997) Lignin. In: Biotechnology in the pulp and paper industry. Springer, Berlin, Heidelberg, pp 127–158

Asgher M, Bhatti HN, Ashraf M, Legge RL (2008) Recent developments in biodegradation of industrial pollutants by white rot fungi and their enzyme system. Biodegradation 19(6):771

Bayer EA, Chanzy H, Lamed R, Shoham Y (1998) Cellulose, cellulases and cellulosomes. Curr Opin Struct Biol 8(5):548–557

Blanchette RA (1991) Delignification by wood-decay fungi. Annu Rev Phytopathol 29(1):381–403

Blanchette RA (1995) Degradation of the lignocellulose complex in wood. Can J Bot 73(S1):999–1010

Bollag JM, Leonowicz A (1984) Comparative studies of extracellular fungal laccases. Appl Environ Microbiol 48(4):849–854

Camarero S, Sarkar S, Ruiz-Dueñas FJ, Martínez MJ, Martínez ÁT (1999) Description of a versatile peroxidase involved in the natural degradation of lignin that has both manganese peroxidase and lignin peroxidase substrate interaction sites. J Biol Chem 274(15):10324–10330

Cohen R, Hadar Y, Yarden O (2001) Transcript and activity levels of different *Pleurotus ostreatus* peroxidases are differentially affected by Mn^{2+}. Environ Microbiol 3(5):312–322

de la Rubia T, Linares A, Pérez J, Muñoz-Dorado J, Romera J, Martínez J (2002) Characterization of manganese-dependent peroxidase isoenzymes from the ligninolytic fungus *Phanerochaete flavido-alba*. Res Microbiol 153(3):547–554

Eriksson KEL, Blanchette RA, Ander P (1990) Morphological aspects of wood degradation by fungi and bacteria. In: Microbial and enzymatic degradation of wood and wood components. Springer, Berlin, Heidelberg, pp 1–87

Ferraroni M, Myasoedova NM, Schmatchenko V, Leontievsky AA, Golovleva LA, Scozzafava A, Briganti F (2007) Crystal structure of a blue laccase from *Lentinus tigrinus*: evidences for intermediates in the molecular oxygen reductive splitting by multicopper oxidases. BMC Struct Biol 7(1):60

Gao J, Tang LG (1996) Cellulose science. Science Press, Beijing

Geng S, Xu C, Li Y (2003) Advance in biosynthesis of lignin and its regulation. Acta Botan Boreali-Occiden Sin 23(1):171–181

Germann UA, Müller G, Hunziker PE, Lerch K (1988) Characterization of two allelic forms of *Neurospora crassa* laccase. Amino-and carboxyl-terminal processing of a precursor. J Biol Chem 263(2):885–896

Gilbertson RL (1980) Wood-rotting fungi of North America. Mycologia 72(1):1–49

Grzegorz J, Anna P, Justyna S, Urszula S-B, Anna J-W, Andrzej P (2017) Lignin degradation: microorganism, enzymes involved, genome analysis and evolution. FEMS Microbiol Rev 41(6):941–962

Gupta A, Verma JP (2015) Sustainable bio-ethanol production from agro-residues: a review. Renew Sust Energ Rev 41:550–567

Gupta VK, Kubicek CP, Berrin JG, Wilson DW, Couturier M, Berlin A, Edivaldo Filho XF, Ezeji T (2016) Fungal enzymes for bio-products from sustainable and waste biomass. Trends Biochem Sci 41(7):633–645

Hakulinen N, Kruus K, Koivula A, Rouvinen J (2006) A crystallographic and spectroscopic study on the effect of X-ray radiation on the crystal structure of *Melanocarpus albomyces* laccase. Biochem Biophys Res Commun 350(4):929–934

Hamed SAM (2013) In-vitro studies on wood degradation in soil by soft-rot fungi: *Aspergillus niger* and *Penicillium chrysogenum*. Int Biodeter Biodegrad 78:98–102

Hatakka A (2005) Biodegradation of lignin. Biopolym Online 15:1

Hatakka A, Hammel KE (2011) Fungal biodegradation of lignocelluloses. In: Industrial applications. Springer, Berlin, Heidelberg, pp 319–340

Heinfling A, Ruiz-Dueñas FJ, Martínez MJ, Bergbauer M, Szewzyk U, Martínez AT (1998) A study on reducing substrates of manganese-oxidizing peroxidases from *Pleurotus eryngii* and *Bjerkandera adusta*. FEBS Lett 428(3):141–146

Higuchi T (1990) Lignin biochemistry: biosynthesis and biodegradation. Wood Sci Technol 24(1):23–63

Himmel ME, Ding SY, Johnson DK, Adney WS, Nimlos MR, Brady JW, Foust TD (2007) Biomass recalcitrance: engineering plants and enzymes for biofuels production. Science 315(5813):804–807

Hoshino F, Kajino T, Sugiyama H, Asami O, Takahashi H (2002) Thermally stable and hydrogen peroxide tolerant manganese peroxidase (MnP) from Lenzites betulinus. FEBS Lett 530(1–3):249–252

Hudson HJ (1992) Fungal biology. Cambridge University Press, Cambridge, pp 106–170

Iyer G, Chattoo BB (2003) Purification and characterization of laccase from the rice blast fungus, *Magnaporthe grisea*. FEMS Microbiol Lett 227(1):121–126

Jeffries TW (1991) Biodegradation of lignin-carbohydrate complexes. In: Physiology of biodegradative microorganisms. Springer, Dordrecht, pp 163–176

Kaal EE, de Jong ED, Field JA (1993) Stimulation of ligninolytic peroxidase activity by nitrogen nutrients in the white rot fungus *Bjerkandera* sp. strain BOS55. Appl Environ Microbiol 59(12):4031–4036

Kang S, Li X, Fan J, Chang J (2013) Hydrothermal conversion of lignin: a review. Renew Sust Energ Rev 27:546–558

Kerem Z, Bao W, Hammel KE (1998) Rapid polyether cleavage via extracellular one-electron oxidation by a brown-rot basidiomycete. Proc Natl Acad Sci 95(18):10373–10377

Kerem Z, Jensen KA, Hammel KE (1999) Biodegradative mechanism of the brown rot basidiomycete *Gloeophyllum trabeum*: evidence for an extracellular hydroquinone-driven fenton reaction. FEBS Lett 446(1):49–54

Kirk TK (1975) Effects of a brown-rot fungus, *Lenzites trabea*, on lignin in spruce wood. Holzforschung Int J Biol Chem Phys Technol Wood 29(3):99–107

Kirk TK, Farrell RL (1987) Enzymatic "combustion": the microbial degradation of lignin. Annu Rev Microbiol 41:465–505

Kirk TK, Schultz E, Connors WJ, Lorenz LF, Zeikus JG (1978) Influence of culture parameters on lignin metabolism by *Phanerochaete chrysosporium*. Arch Microbiol 117(3):277–285

Kirk TK, Ibach R, Mozuch MD, Conner AH, Highley L (1991) Characteristics of cotton cellulose depolymerized by a brown-rot fungus, by acid, or by chemical oxidants. Holzforschung-Int J Biol Chem Phys Technol Wood 45(4):239–244

Leisola MS, Ulmer DC, Fiechter A (1984) Factors affecting lignin degradation in lignocellulose by *Phanerochaete chrysosporium*. Arch Microbiol 137(2):171–175

Lin Z, Ma Q, Xu Y (2003) Lignin biosynthesis and its molecular regulation. Prog Nat Sci 13(5):321–328

Lisov AV, Leontievsky AA, Golovleva LA (2003) Hybrid Mn-peroxidase from the ligninolytic fungus *Panus tigrinus* 8/18. Isolation, substrate specificity, and catalytic cycle. Biochemistry (Moscow) 68(9):1027–1035

Ljungdahl LG (2008) The cellulase/hemicellulase system of the anaerobic fungus *Orpinomyces* PC-2 and aspects of its applied use. Ann N Y Acad Sci 1125(1):308–321

Martinez D, Larrondo LF, Putnam N, Gelpke MDS, Huang K, Chapman J, Helfenbein KG, Ramaiya P, Detter JC, Larimer F, Coutinho PM (2004) Genome sequence of the lignocellulose degrading fungus *Phanerochaete chrysosporium* strain RP78. Nat Biotechnol 22(6):695

Martínez ÁT, Speranza M, Ruiz-Dueñas FJ, Ferreira P, Camarero S, Guillén F, Martínez MJ, Gutiérrez Suárez A, Río Andrade JCD (2005) Biodegradation of lignocellulosics: microbial, chemical, and enzymatic aspects of the fungal attack of lignin. Int Microbiol 8:195–204

Messerschmidt A, Huber R (1990) The blue oxidases, ascorbate oxidase, laccase and ceruloplasmin modelling and structural relationships. Eur J Biochem 187(2):341–352

Michniewicz A, Ullrich R, Ledakowicz S, Hofrichter M (2006) The white-rot fungus *Cerrena unicolor* strain 137 produces two laccase isoforms with different physico-chemical and catalytic properties. Appl Microbiol Biotechnol 69(6):682–688

Moreira PR, Almeida-Vara E, Malcata FX, Duarte JC (2007) Lignin transformation by a versatile peroxidase from a novel *Bjerkandera* sp. strain. Int Biodeter Biodegrad 59(3):234–238

Orth AB, Royse DJ, Tien MING (1993) Ubiquity of lignin-degrading peroxidases among various wood-degrading fungi. Appl Environ Microbiol 59(12):4017–4023

Otjen L, Blanchette R, Effland M, Leatham G (1987) Assessment of 30 white rot basidiomycetes for selective lignin degradation. Holzforschung 41(6):343–349

Patel H, Gupte A (2016) Optimization of different culture conditions for enhanced laccase production and its purification from *Tricholoma giganteum* AGHP. Bioresour Bioprocess 3(1):11

Perez J, Jeffries TW (1990) Mineralization of 14C-ring-labeled synthetic lignin correlates with the production of lignin peroxidase, not of manganese peroxidase or laccase. Appl Environ Microbiol 56(6):1806–1812

Perez-Boada M, Ruiz-Duenas FJ, Pogni R, Basosi R, Choinowski T, Martínez MJ, Piontek K, Martínez AT (2005) Versatile peroxidase oxidation of high redox potential aromatic com-

pounds: site-directed mutagenesis, spectroscopic and crystallographic investigation of three long-range electron transfer pathways. J Mol Biol 354(2):385–402
Piontek K, Smith AT, Blodig W (2001) Lignin peroxidase structure and function. Biochem Soc Trans 29:111–116
Pozdnyakova NN, Turkovskaya OV, Yudina EN, Rodakiewicz-Nowak Y (2006) Yellow laccase from the fungus *Pleurotus ostreatus* D1: purification and characterization. Appl Biochem Microbiol 42(1):56–61
Rouches E, Herpoël-Gimbert I, Steyer JP, Carrere H (2016) Improvement of anaerobic degradation by white-rot fungi pretreatment of lignocellulosic biomass: a review. Renew Sust Energ Rev 59:179–198
Ruiz-Dueñas FJ, Martínez ÁT (2009) Microbial degradation of lignin: how a bulky recalcitrant polymer is efficiently recycled in nature and how we can take advantage of this. Microb Biotechnol 2(2):164–177
Salony, Mishra S, Bisaria VS (2006) Production and characterization of laccase from *Cyathus bulleri* and its use in decolourization of recalcitrant textile dyes. Appl Microbiol Biotechnol 71(5):646–653
Salvachua D, Prieto A, Lopez-Abelairas M, Lu-Chau T, Martinez AT, Martinez MJ. (2011) Fungal pretreatment: An alternative in second-generation ethanol from wheat straw. Bioresour Technol. 102(16):7500–7506
Sánchez C (2009) Lignocellulosic residues: biodegradation and bioconversion by fungi. Biotechnol Adv 27(2):185–194
Shleev S, Nikitina O, Christenson A, Reimann CT, Yaropolov AI, Ruzgas T, Gorton L (2007) Characterization of two new multiforms of *Trametes pubescens* laccase. Bioorg Chem 35(1):35–49
Srebotnik E, Messner K (1994) A simple method that uses differential staining and light microscopy to assess the selectivity of wood delignification by white rot fungi. Appl Environ Microbiol 60(4):1383–1386
Srebotnik E, Messner K, Foisner R (1988) Penetrability of white rot-degraded pine wood by the lignin peroxidase of *Phanerochaete chrysosporium*. Appl Environ Microbiol 54(11):2608–2614
Sulistyaningdyah WT, Ogawa J, Tanaka H, Maeda C, Shimizu S (2004) Characterization of alkaliphilic laccase activity in the culture supernatant of *Myrothecium verrucaria* 24G-4 in comparison with bilirubin oxidase. FEMS Microbiol Lett 230(2):209–214
Tanaka H, Itakura S, Enoki A (1999) Hydroxyl radical generation by an extracellular low-molecular-weight substance and phenol oxidase activity during wood degradation by the white-rot basidiomycete *Trametes versicolor*. J Biotechnol 75(1):57–70
Ten Have R, Teunissen PJ (2001) Oxidative mechanisms involved in lignin degradation by white-rot fungi. Chem Rev 101(11):3397–3414
Theander O, Westerlund E (1993) Quantitative analysis of cell wall components. In: Forage cell wall structure and digestibility, (forage cell walls). ASA-CSSA-SSSA, 677 S. Segoe Rd., Madison, WI 53711, USA, 83–104
Thurston CF (1994) The structure and function of fungal laccases. Microbiology 140(1):19–26
Vanholme R, Morreel K, Ralph J, Boerjan W (2008) Lignin engineering. Curr Opin Plant Biol 11(3):278–285
Van Kuijk SJA, Sonnenberg ASM, Baars JJP, Hendriks WH, Cone JW (2015) Fungal treated lignocellulosic biomass as ruminant feed ingredient: a review. Biotechnol Adv 33(1):191–202
Wahleithner JA, Xu F, Brown KM, Brown SH, Golightly EJ, Halkier T, Kauppinen S, Pederson A, Schneider P (1996) The identification and characterization of four laccases from the plant pathogenic fungus *Rhizoctonia solani*. Curr Genet 29(4):395–403
Wang L, Yan W, Chen J, Huang F, Gao P (2008) Function of the iron-binding chelator produced by *Coriolus versicolor* in lignin biodegradation. Sci China Ser C Life Sci 51(3):214
Wei J, Song Y (2001) Recent advances in study of lignin biosynthesis and manipulation. Acta Bot Sin 43(8):771–779
Weng JK, Li X, Bonawitz ND, Chapple C (2008) Emerging strategies of lignin engineering and degradation for cellulosic biofuel production. Curr Opin Biotechnol 19(2):166–172

Wesenberg D, Kyriakides I, Agathos SN (2003) White-rot fungi and their enzymes for the treatment of industrial dye effluents. Biotechnol Adv 22(1–2):161–187

Wong DW (2009) Structure and action mechanism of ligninolytic enzymes. Appl Biochem Biotechnol 157(2):174–209

Xiao Y, Tu X, Wang J, Zhang M, Cheng Q, Zeng W, Shi Y (2003) Purification, molecular characterization and reactivity with aromatic compounds of a laccase from basidiomycete *Trametes* sp. strain AH28-2. Appl Microbiol Biotechnol 60(6):700–707

Yang HH, Effland MJ, Kirk TK (1980) Factors influencing fungal degradation of lignin in a representative lignocellulosic, thermomechanical pulp. Biotechnol Bioeng 22(1):65–77

Yao L, Yong-Chao L, Hong-Qin H, Feng-Jin X, Xian-Yong W, Xing F (2017) Structural characterization of lignin and its degradation products with spectroscopic methods. J Spectrosc. https://doi.org/10.1155/2017/8951658

Yoon JJ, Cha CJ, Kim YS, Son DW, Kim YK (2007) The brown-rot basidiomycete *Fomitopsis palustris* has the endo-glucanases capable of degrading microcrystalline cellulose. J Microbiol Biotechnol 17(5):800–805

Chapter 11
Application and Biodegradation of Lignocellulosic Biomass

M. P. Singh, Sonam Agarwal, Ankita Kushwaha, and Vivek K. Chaturvedi

1 Introduction

Lignocellulosic wastes are generated in large quantity throughout the world as a result of crop residues and these wastes pose an environmental threat because of their disposal problems. To get rid of this bulk, these wastes are generally burnt or left in the field to rot, leading to unhygienic conditions as well as waste of resources. Burning of these wastes not only results in oxygen-deficient environment but also causes several respiratory diseases like asthma, tuberculosis, etc. as well as poor visibility at night (Singh and Gautam 2004; Agarwal et al. 2016). These wastes are composed of lignin, cellulose, and hemicelluloses (Jeznabadi et al. 2016; Agarwal et al. 2018). Because of the composition of these wastes, they are difficult to degrade and are very slowly degraded by ruminants, as lignin polymers provide physical and chemical barrier thereby preventing the access of hydrolytic enzymes, such as cellulases and hemicellulases to their substrates, which are responsible for the degradation of cellulose and hemicelluloses of the substrate, respectively (Cohen et al. 2002; Chaturvedi et al. 2019). Lignin is the second most abundant plant polymer which provides resistance against microbial attack and also gives strength in combination with cellulosic fibers. Lignin forms a matrix around the cellulose, because of which there is slow microbial depolymerization of cellulose, which otherwise can be used for several industrial purposes; therefore, it is necessary to first degrade lignin for the global carbon cycle (Gold and Alic 1993).

The problem related to safe disposal of lignocellulosic wastes and their strict nature towards degradation can be solved by using these wastes for the cultivation of edible mushroom species, especially white rot fungus (WRF), which with its potential enzymatic system convert these wastes into value-added products like nutrient-rich food, improved animal fodder, plant fertilizer, biofuels, chemicals,

M. P. Singh (✉) · S. Agarwal · A. Kushwaha · V. K. Chaturvedi
Centre of Biotechnology, University of Allahabad, Prayagraj, Uttar Pradesh, India

R. Naraian (ed.), *Mycodegradation of Lignocelluloses*, Fungal Biology,
https://doi.org/10.1007/978-3-030-23834-6_11

cheap carbon source for fermentation, etc. (Philippoussis et al. 2000; Poppe 2000; Kuhad et al. 2007). Mushrooms produce ligninolytic and hydrolytic enzymes which help in degrading these wastes which are rich source of carbohydrates and therefore, mushrooms can be cultivated on large variety of these wastes like pasteurized wheat straw, paddy straw, sawdust, etc. (Kabirifard et al. 2012; Kulshreshtha et al. 2013; Mikiashvili et al. 2006; Chaturvedi et al. 2018). The enzymes secreted by WRF, i.e., cellulases, hemicellulases, and ligninases, are nonspecific in nature and therefore apart from playing a role in degradation they are also used for bioremediation (Adebayo and Martinez-Carrera 2015). These enzymes have many industrial applications including chemicals, fuel, food, brewery and wine, animal feed, textile and laundry, pulp and paper, and agriculture (Kuhad et al. 2007).

The chemical composition of lignocelluloses makes them a suitable substrate for other industrial purposes as well. Several technologies are available that are used for the bioconversion of lignocelluloses to value-added products. These technologies are gaining interest with the belief that they will provide us with improved quality products with low price (Muralilal and Pereira 2006). They can be used in different industries like pulp and paper, chemical, textile, food, feed and fertilizer, biosensor, bioconversion, etc. The present chapter focuses on the lignocellulosic production, their composition, their degradation through mushroom cultivation making use of its enzymes and applications of lignocelluloses in various industries.

2 Chemical Composition of Lignocellulosic Biomass

From the last several years, agro-industrial biomass comprising of lignocellulosic waste is an inexpensive, renewable, abundant in quantity and provides a unique natural resource for large-scale and cost-effective bio-energy collection (Anwar et al. 2014).In general, the composition of lignocelluloses highly depends on its source whether it is derived from the hardwood, softwood, or grasses. Table 11.1 shows the typical chemical compositions of components present in various lignocellulosic materials. Lignocellulosic biomass is mainly composed of three components such as cellulose, hemicellulose, and lignin together with small amounts of other components like acetyl groups, minerals, and phenolic substituents (Isikgor and Becer 2015).

3 Lignocellulose Production

Large amounts of lignocellulosic wastes are generated as a result of agriculture, horticulture, forests, food and wood processing industries. About 4664.63 million tonnes of agricultural waste is produced as a part of crop residues (Rai and Ahlawat 2005). With expanding agricultural production, there is an increase in the quantities

Table 11.1 Chemical compositions of components present in various lignocellulosic materials

S. n.	Lignocellulosic biomass	Cellulose (%)	Lignin (%)	Hemicellulose (%)	Reference
1.	Newspaper	40–55	18–30	25–40	Howard et al. (2003)
2.	Corn stover	37–42	19	26	Zhu et al. (2005)
3.	Banana waste	13.2	14	14.8	John et al. (2006)
4.	Sugarcane bagasse	42	20	25	Kim and Day (2011)
5.	Sponge gourd fibers	66.59	15.46	17.44	Guimaraes et al. (2009)
6.	Nut shells	25–30	30–40	25–30	Howard et al. (2003)
7.	Softwood stems	45–50	25–35	25–35	Malherbe and Cloete (2002)
8.	Grasses	25–40	10–30	25–50	Malherbe and Cloete (2002)
9.	Barley straw	33–40	8–17	20–35	Pedersen and Meyer (2010)
10.	Sweet sorghum	45	21	27	Kim and Day (2011)
11.	Rice straw	32–41	10–18	15–24	Van Dyk and Pletschke (2012)
12.	Corn cobs	45	15	35	Prasad et al. (2007)
13.	Bagasse	54.87	23.33	16.52	Guimaraes et al. (2009)
14.	Energy crops	43–45	19–12	24–31	Prasad et al. (2007)
15.	Wheat straw	31–44	16–24	22–24	Szulczyk et al. (2010)

of livestock waste, agricultural crop residues, and agro-industrial by-products (Obi et al. 2016). Annually, there is an average rate of 5–10% increase in agricultural wastes (Wang et al. 2016). It has been found that after the processing of one ton of the main product about 1.5 tons of crop residues are generated. According to a survey by Govt. of India, it is estimated that every year about 500 metric tons of crop residue is generated (Mahawar et al. 2015). About 100 million tons of paddy straw alone contributes to crop residue (Tewari and Pandey 2002), adding to this paddy stubble, the leftover lower portion of paddy crop, being heavier, denser, and more bulky in nature as compared to paddy straw accounts for 5–10% of entire crop residue (Singh and Gautam 2004). Dry biomass net productivity as a result of photosynthesis by plants is around 155.2 billion tons per year of which two third is produced on land and one third ends up in the ocean (Singh and Gautam 2004). About 10^{12} metric tonnes of plant waste is produced each year in the form of cellulose (Wilson and Irwin 1999). These wastes generally consist of wood residues, waste paper, grasses, agricultural residues (like straw, stalks and bagasse), domestic wastes (lignocellulosic garbage and sewage), and municipal solid wastes (Rodríguez et al. 2008).

4 Lignocellulose Degradation

The structure of lignocellulose is very complex and resistant to enzymatic hydrolysis. The lignin layer of the lignocellulose makes their degradation tough and generally, the rate of decay is directly proportional to the lignin content in the plant material (Cohen et al. 2002). Earlier methods used for degradation are either costly or not much efficient, like the use of alkaline and acid hydrolysis in which weak acids help to remove the lignin but cause problem for cellulose hydrolysis and on the other hand, in strong acid treatment physiological conditions have to be maintained like temperature and pH which also requires expensive instruments (Kuhad and Singh 2007). Therefore, biological delignification is one of the most promising methods to improve their digestion or degradation (Streeter et al. 1982; Kamra and Zadražil 1986).

Cultivation of edible mushrooms on these wastes is value-added process, as the aim is not just to degrade lignin but also the yield of another product of commercial use like human and animal nutrient-rich food (Zhang et al. 2002). It serves as one of the most efficient biological ways of recycling of wastes with another advantage of using spent compost, rich in lignin content, for different purposes like biofuels production, as a fertilizer, etc. (Madan et al. 1987; Murarilal and Pereira 2006).

4.1 Biodegradation Through Mushroom Cultivation

Mushroom cultivation on lignocellulose waste is carried out from long time with varying degrees of success. The slow rate of degradation of lignocellulose is because of the presence of lignin and therefore white rot fungus is best suitable for degradation with the help of its lignin modifying enzymes (LDEs). They convert the lignin polymer to CO_2 and H_2O, after which the cellulose and hemicellulose can easily be attacked by cellulases and hemicellulases, respectively, of the WRF. They can colonize on a wide variety of lignocellulosic wastes like wheat straw, paddy straw, cotton, peanut shells, poplar sawdust, oak sawdust, corncobs, olive press-cake, sorghum straw, pearl millet stalks, rye straw, maize cobs, banana waste, maize straw, hard wood chips, sugarcane bagasse, rice and wheat bran, etc. (Zervakis et al. 2001; Adebayo and Martinez-Carrera 2015; Davari et al. 2012; Orts et al. 2008). These fungi with their phenol-targeting redox enzymes selectively degrade lignin and related compounds (Ntougias et al. 2012). Higher fungi with its enzyme convert the high molecular weight compound to low molecular weight compound, which can be further used for the fungal nutrition (Elisashvili et al. 2003). The enzymes and the exploitation of substrate vary with species, strains, and cultivation method (Zadrazil and Dube 1992). It has been observed that fungus helps in converting the cellulose and other plant carbohydrates to acetic, propionic, and butyric acid in large amounts, which can further be used by ruminant animals as energy and carbon source. These

WRF also can convert the lignocellulose wastes anaerobically in liquid digesters (Ezeji et al. 2006; Martín et al. 2006; Alborés et al. 2006).

During the cultivation of mushrooms, it was reported that the fiber, hemicellulose, and cellulose content of the spent cultivation substrate decreases, whereas the protein content in the spent substrate increases (Koutrotsios et al. 2014). It has been observed that due to large amount of cellulose, which is the most biodegraded component of the lignocellulose waste, followed by hemicelluloses, which is found in less abundant and lastly lignin which is present in the smallest quantity (Salmones et al. 2005). The cultivation of mushrooms are used for various purposes like biodegradation of hazardous compounds, biological detoxification of toxic agro-industrial wastes, biotransformation of agro-industrial residues to mushroom food and animal feed as well as value-added products like biologically active metabolites, enzymes, flavor compounds, etc. (Philippoussis 2009; Philippoussis and Diamantopoulou 2011). During the cultivation of mushroom, care should be taken to check that the wastes do not contain chemical components and toxic substances, which otherwise can affect the mycelium growth and also can cause human health issue (Stamets 2011). The supplementation of substrate improves the nutrient requirements for mushroom production. The carbon and nitrogen content in the substrate have significant effect on mushroom production and it has been observed that during the mushroom cultivation carbon content decreases whereas, nitrogen and ash content in the substrate increases.

The enzymes present in WRF that help in the degradation of these wastes are cellulases, hemicellulases, and ligninases which degrade the cellulose, hemicellulose, and lignin part of the waste, respectively.

4.2 *Enzymatic Studies of Cellulases*

Cellulases are enzymes that degrade the cellulose part of the lignocellulosic biomass. Cellulose is a homopolysaccharide consisting of β-D-glucopyranose units, which are linked by β-(1→4)-glucosidic bonds. Both nonreducing and reducing ends are present in cellulose. The repetitive unit of cellulose is called as cellobiose, which can be converted into glucose residues. Fungi are the most studied microorganisms in relation to cellulose degradation by cellulolytic enzymes. The cellulases are further divided into three groups: endo-β-1,4-glucanases (EC 3.2.1.4), exo-β-1,4 glucanases I and II (EC 2.1.1.91), and β-glucosidase (EC-3.2.1.21) (Adebayo and Martinez-Carrera 2015).

The endo-β-1,4-glucanases part of the enzyme randomly catalyze the internal bonds of cellulose chain, i.e., they split the β-1,4-glucosidic linkages. Exo-β-1,4 glucanases split the cellobiose or glucose from the cellulose chain, in which exo-β-1,4 glucanases I attack at the reducing end of the chain whereas exo-β-1,4 glucanases II attack the nonreducing end of the chain. Moreover, β-glucosidase attack the cellobiose and other water-soluble cellodextrin and release the glucose monomer units from them (Eriksson et al. 1990; Adebayo and Martinez-Carrera 2015). There are two other enzymes as well that are involved in the degradation of cellulose; these are the oxidative enzymes, first one is cellulose: quinone oxidoreductase (cellobiose dehydrogenase) which reduces quinones and phenoxy radicals with the oxidation of cellobiose to cellobiono-δ-lactone, and the second one is cellobiose oxidase, which with the help of molecular oxygen oxidizes cellobiose and higher cellodextrins to their corresponding onic acids (Eriksson et al. 1990).

4.3 Enzymatic Studies of Hemicellulases

The hemicellulases degrade the hemicellulose part of the lignocellulosic biomass. Hemicelluloses are the heterogeneous polymers that are built up by pentoses (D-xylose, L-araninose), hexoses (D-glucose, D-galactose), sugar acids (ferulic acids and 4-*O*-methyl-D-glucuronic acid) and acetyl group (Adebayo and Martinez-Carrera 2015). Fungi secrete some hemicellulolytic enzymes that degrade the hemicellulose; these are either extracellular cell wall bound or intracellular (Dekker 1985; Kuhad et al. 2007). The hemicellulases include α-D-glucuronidase (EC 3.2.1), endo-β-1,4-xylanase (EC 3.2.1.8), α-L-arabinofuranosidase (EC 3.2.1.55), α-D-galactosidase, β-D-xylosidase (EC 3.21.37), feruloyl esterase, and acetylxylan esterase (EC 3.1.1.6) (Biely 1985; Adebayo and Martinez-Carrera 2015).The α-D-glucuronidase attack 4-*O*-methyl-D-glucuronic acid, endo-β-1,4-xylanase degrade the xylan chain, whereas α-L-arabinofuranosidase attack the L-araninose chain end. After this ferulic acid and acetyl groups are removed by feruloyl esterase and acetylxylan esterase, respectively. The α-D-galactosidase reduces xylan to xylobiose and this xylobiose is further degraded to D-xylose, a monomer unit of xylobiose, by β-D-xylosidase (Adebayo and Martinez-Carrera 2015).

4.4 Enzymatic Studies of Ligninases

WRF secrete various ligninases, which very efficiently degrade the lignin component of the lignocellulose biomass. Lignin is a complex oxyphenol propanoid polymer monomeric unit linked together with different types of bonds. It is synthesized by one-electron oxidation of precursors: *p*-coumaryl, coniferyl, and sinapyl alcohol generate phenoxy radicals and then non-enzymatic polymerization occurs which forms high molecular weight, heterogeneous, three-dimensional polymers (Garg and Modi 1999). The main enzymes of ligninases are laccase, lignin peroxidase (LiP), manganese peroxidase (MnP), and versatile peroxidase (VP). The LDEs are oxido-reductive enzymes that degrade the lignin by a series of redox reactions. They oxidize the aromatic compounds until the aromatic ring is cleaved, which is then degraded by other enzymes (Husain 2006). They have broad range of specificity, high redox potential, extracellular in nature, and are expressed in nutrient-deficient environment; therefore, they are suitable for lignin degradation (Reddy

1995). Laccase is a benzenediol: oxygen oxidoreductase multi-copper enzyme that uses molecular oxygen to oxidize various aromatic and nonaromatic compounds by radical catalyzed reaction mechanism (Claus 2004; Baldrian 2006). Laccase is monomeric, dimeric, or tetrameric glycoprotein with four copper atoms classified into three types (T1, T2 and T3) present in its catalytic center, which mediate the redox process. T1 carries out the oxidation of substrate that leads to the formation of free radicals, which further acts as intermediate substrate for the enzymes (Yoshitake et al. 1993; Ferraroni et al. 2007; Adebayo and Martinez-Carrera 2015). Laccase catalyzes demethoxylation reactions of terminal phenolic units. It degrades β-dimers and β-*O*-4 dimers by α-oxidation of alkyl-aryl cleavage and C_α-C_β cleavage (Youn et al. 1995). Lignin mediator system carries out various physiological functions like lignolysis, lignin synthesis, morphogenesis, pathogenesis, and detoxification (Mayer and Staples 2002). LiP is a heme containing glycoprotein that catalyzes the oxidation of non-phenolic aromatic lignin compounds (Wong 2009). It catalyzes the H_2O_2-dependent oxidation depolymerization of a wide range of phenolic and non-phenolic compounds, polycyclic aromatic hydrocarbons, etc. by a series of non-enzymatic reactions in a multi-step electron transfer process which then forms intermediate radicals like phenoxy radicals and veratryl alcohol radical cations. These radicals undergo different reactions, which finally results in the oxidation of lignin and cleavage of the aromatic ring structures (Ten Have and Teunissen 2001; Umezawa and Higuchi 1987; Wong 2009). LiP cleaves β-*O*-4 ether bonds and C_α-C_β bonds in dimeric lignin model compounds. It also catalyzes decarboxylation of phenylacetic acids, oxidation of aromatic C_α-alcohols to C_α-oxo compounds, hydroxylation and quinone formation (Haemmerli et al. 1987; Gupta et al. 2017). MnP is a glycosylated glycoprotein containing an iron protoporphyrin IX (heme) prosthetic group is secreted in multiple isoforms (Nie et al. 1999; Glenn and Gold 1985; Asgher et al. 2008). It is involved in the oxidation of phenols and phenolic ring structures (Kuhad et al. 2007). MnP is a H_2O_2-dependent enzyme that catalyzes the oxidation of Mn (II) to Mn (III) which results in oxidation of various phenolic compounds. MnP oxidizes Mn^{2+} to Mn^{3+}, which is then stabilized by chelators like oxalic acid (Glenn et al. 1986; Wariishi et al. 1992). Chelated Mn (III) complex acts as a reactive low molecular weight, diffusible redox-mediator of phenolic substrates. For oxidation of non-phenolic compounds, Mn (III) must form a reactive radical in the presence of other mediators like oxalate and malonate (Wong 2009; Wesenberg et al. 2003).Versatile peroxidase is also a glycoprotein that combines the properties of LiP and MnP (Adebayo and Martinez-Carrera 2015). It is an attractive ligninolytic enzyme because they oxidize Mn (II) as well as phenolic and non-phenolic aromatic compounds (Wesenberg et al. 2003). It is superior to LiP and MnP because of the hybrid molecular structure which provides multiple binding sites for the substrates (Camarero et al. 1999). It contains two channels; the first channel is similar to LiP and in the second channel oxidation of Mn (II) to Mn (III) takes place (Ruiz-Dueñas et al. 1999). Other enzymes apart from these, which are involved in lignin degradation, are aryl-alcohol oxidase (AAO), aryl-alcohol dehydrogenases (AAD), quinone reductases (QR) and cellobiose dehydrogenase (CDH) (Adebayo and Martinez-Carrera 2015).

5 Biotechnological Application of Lignocellulose

5.1 Biofuel

Over the years, indiscriminate use of fossil fuels has resulted not only in concerns about their environmental impact, but also their inevitable exhaustion. One possible solution to these problems is the replacement of materials from these non-renewable sources with renewable ones. The use of lignocellulosic biomass and agro-industrial residues from various sources is of fundamental importance for the construction of various sectors of the bioeconomy based on the innovation of biofuels and bioproducts. The bioconversion of lignocellulosic biomass into fermentable sugars for the production of bioethanol has been considered a promising alternative to increase the production of ethanol to meet global demand (Kuila and Sharma 2017; Maurya et al. 2016).

Currently, the first generation processes for the production of ethanol from corn. Only a small part of the corn plant, i.e., the corn kernels are taken from the corn plant and only the starch, which represents about 50% of the dry kernel mass, is transformed into ethanol. Two types of second-generation processes are under development. The first type uses enzymes and yeast fermentation to convert the plant cellulose into ethanol while the second type uses pyrolysis to convert the whole plant to either a liquid bio-oil. Second-generation processes can also be used with plants such as grasses, wood or agricultural waste material such as straw. The principal advantage of second-generation biofuel over fossil fuel is that biofuels are produced from renewable and sustainable resource. The combustion of biofuel releases almost zero or very little net carbon dioxide emission to the atmosphere and has a positive impact on environmental protection, especially the global climate change. Bioethanol is produced from corn and sugarcane, depending on the climatic conditions whereas the feedstock used for the production of bioethanol mainly produced from sugarcane in the tropical regions such as Brazil and Columbia whereas it is produced from corn in the areas such as the USA, European regions, and China. Ethanol fuel has a "gasoline gallon equivalency" (GGE) value of 1.5, i.e., to replace the energy of 1 volume of gasoline, 1.5 times the volume of ethanol is needed. Bioethanol consists of gasoline ethanol mixtures such as E15 (15% Ethanol and 85% Gasoline) and E85 (85% Ethanol and 15% Gasoline) which was partially replaced by gasoline. One of the major applications of lignocellulosic biomass is the production of second-generation bioethanol. Lignocellulosic biomass such as agricultural residues, woods, and grasses are abundant in most land areas of the world and their generation does not have to compete for arable land against food and feed production. They are abundant almost all over the world and have high contents of cellulose and hemicelluloses; therefore, they are used for the production of bioethanol. According to the International Energy Agency, cellulosic ethanol could allow ethanol fuels to play a much bigger role in the future (www.world energyoutlook.org). There is difficulty in directly converting lignocellulosic materials to ethanol; hence, a three-step process is required for the bioconversion:

5.2 Pretreatment/Delignification

The purpose of pretreatment is the primary step to make biomass accessible by breaking the seal of lignin through the process of delignification, by removing hemicelluloses, or by disrupting the structure of crystalline structure of cellulose and increase the porosity of the material, so the hydrolytic enzymes can access their substrates (cellulose and hemicellulose) in the following enzymatic hydrolysis. This pretreatment including the physical, chemical (alkali and acid hydrolysis), and biological processes for lignocellulosic materials. Among these three pretreatment processes, biological pretreatment is probably the most economical and environmentally friendly pretreatment technology for the lignocellulosic materials because fungi such as brown, white, and soft-rot fungi degrade lignin and hemicellulose in lignocellulosic materials.

Brown rot fungi mainly attack cellulose, while white and soft rots attack both cellulose and lignin. White rot fungi are the most effective basidiomycetes for biological pretreatment of lignocellulosic materials. Biological pretreatment using metabolites of white rot fungi in nature for ethanol production from biomass is a promising technology due to several advantages such as its being an eco-friendly and economically viable strategy for enhancing enzymatic scarification rate. Since no chemicals are used in this process, there is no need for recycling of chemicals and it does not release toxic compounds into the environment (Sindhu et al. 2016). Potumarthi et al. (2013) studied simultaneous pretreatment and saccharification of rice husk by *Phanerochaete chrysosporium*. Effective delignification was carried out by growing the fungus on rice husk and the pretreated biomass was subjected to enzymatic hydrolysis. This method avoids operational costs associated with washing and the removal of inhibitors during conventional pretreatment.

5.3 Fermentation

The three major components of lignocelluloses viz. cellulose have the highest conversion rate to ethanol which is 85–90%, hemicelluloses 30–85%, and lignin 0%. The main products of hemicelluloses hydrolysis include hexoses and pentoses. The cellulose and hemicellulose can be readily fermented to ethanol by microorganism whereas lignin is difficult to be converted to ethanol easily but it can be used as a fuel in the fractional distillation in ethanol purification. Ethanol has lower pressure than gasoline, thereby reducing smog formation by lowering evaporative emissions. It also has a high octane value, high heat of vaporization, and other favorable fuel properties.

5.4 Animal Feed

Due to high lignin and low protein contents, the feed value of lignocelluloses is quite low. Hence, it is desirable to utilize some processes for enhancing the protein content of the biomass. Kamara and Zadrazil (1988) reported that white rot fungi can utilize lignocellulosic wastes to make them palatable animal fodder by increasing the protein level, an increase in the digestibility of the lignocellulosic material and an improvement in the dry product palatability. Solid state fermentation (SSF) of lignocellulosic residues by white rot fungi improves the availability of cellulose to ruminants (Mahesh and Mohini 2013). White rot fungi such as *P. chrysosporium*, *Cyathus* sp., *Coriolus hirsutus*, *Dichomitus squalens*, and *Pleurotus* spp. are the extensively studied organisms for solid state bioconversion of lignocellulosic materials. SSF is a process in which solid substrates are decomposed by known cultures of microorganisms (mainly fungi, which can grow on and through the substrate) under standard conditions (Zadrazil and Reiniger 1988) and makes these lignocellulosic substrates to more digestible and nutritious feedstuffs for ruminants. Digestibility of the lignocellulosic materials such as straw is dependent on the depolymerization of its structural carbohydrates (Mahesh and Mohini 2013). Species of *Pleurotus* are most studied white rot fungi for improving digestibility and nutritional quality of straw (Kundu 2005; Streeter et al. 1982; Kakkar et al. 1990). Up to 75% of wheat straw in vitro digestibility is done by using *C. stercoreus* or *D. squaleris* (Agosin and Odier 1985).

5.5 Fertilizers

Generation of the equivalent amount of spent mushroom compost from the metric tonnes of mushroom are produced from the mushroom industry. Fermor et al. (2000) reported that the disposal of these spent composts is bulky, and charges and taxes on landfill are implemented in the developed countries. Spent mushroom compost consisting of degraded cellulose, hemicelluloses, and lignin having high contents of mineral, phosphate with high porosity, is a good soil fertilizer for inducing the seed germination and as a prospective bioremediating agent (Chiu et al. 2000) because this spent substrates can be a source of extracellular and intracellular lignocellulolytic enzymes (Fermor et al. 2000). In addition to providing balanced carbon and nitrogen sources for plant growth and germination, this spent compost undergoes further conversion in the soil to humus. According to Ball and Jackson (1995), the composition, activity, and stability of enzymes found from compost extract suggest that they could further be used for the preparation of animal fodder. This compost actually speeds up emission of methane biogas.

5.6 *High Value-Added Bioproducts/Chemicals*

A number of high-value bioproducts such as organic acids, amino acids, vitamins, and fungal polysaccharides such as xanthans are produced by fermentation using glucose as the initial substrate could be manufactured from lignocellulosic residues (Sánchez 2009). Ribbons (1987) reported that several potential value-added products could be derived from lignin such as vanillin and gallic acid which can be used as herbicide, antifoaming agents, antimicrobial agents and drugs using white rot fungi *P. chrysosporium*. A method developed for the production of furfural from lignocellulosic materials like birch and bagasse biomass which is also the best raw material for levoglucosan production (Gravitis et al. 2004).

The acid hydrolysis of lignocellulosic materials such as xylose (hemicellulose) solutions and this xylose further utilized for making xylitol which is used as a sweetener and a teeth hardening agent (Parajó et al. 1998).

6 Conclusion

As a result of agriculture, food and wood processing industries, large amount of lignocellulosic waste is generated. These wastes are made up of complex compounds and therefore degrade very slowly thereby causing several environmental problems and unhygienic conditions. One of the best utilization and environment-friendly disposals of these wastes is the cultivation of mushroom on these wastes as they are rich source of nutrients. The mushrooms secrete many extracellular and intracellular enzymes which help in the breakdown of complex compounds of wastes into low molecular weight compounds, which are then taken up by mushrooms for their growth and nourishment. Apart from this, these wastes have found their importance in different industries as well, where they serve as the main source for production of several bioproducts, chemicals as well as biofuel. The wastes generated after mushroom cultivation, that is called as spent mushroom compost, being rich in nutrients can further be used as animal feed and fertilizers, thereby solving the problem of waste produced after mushroom cultivation which is in large quantities. So, keeping in view these many solutions, one can make best use of the lignocellulosic wastes which are generally burnt in the field leading to oxygen-deficient environment which ultimately causes several respiratory diseases.

References

Adebayo EA, Martinez-Carrera D (2015) Oyster mushrooms (*Pleurotus*) are useful for utilizing lignocellulosic biomass. Afr J Biotechnol 14:52–67

Agarwal S, Vaseem H, Kushwaha A, Gupta KK, Maurya S, Chaturvedi VK, Ravi P, Singh MP (2016) Yield, biological efficiency and nutritional value of *Pleurotus sajor-caju* cultivated on floral and agro-waste. Cell Mol Biol 62:1–5

Agarwal S, Gupta KK, Chaturvedi VK, Kushwaha A, Chaurasia PK, Singh MP (2018) The potential application of peroxidase enzyme for the treatment of industry wastes. In: Research advancements in pharmaceutical, nutritional, and industrial enzymology. IGI Global, pp 278–293

Agosin E, Odier E (1985) Solid-state fermentation, lignin degradation and resulting digestibility of wheat straw fermented by selected white-rot fungi. Appl Microbiol Biotechnol 21:397–403

Alborés S, Pianzzola MJ, Soubes M, Cerdeiras MP (2006) Biodegradation of agroindustrial wastes by *Pleurotus* spp. for its use as ruminant feed. Electron J Biotechnol 9:52–67

Anwar Z, Gulfraz M, Irshad M (2014) Agro-industrial lignocellulosic biomass a key to unlock the future bio-energy: a brief review. J Radiat Res Appl Sci 7:163–173

Asgher M, Bhatti HN, Ashraf M, Legge RL (2008) Recent developments in biodegradation of industrial pollutants by white rot fungi and their enzyme system. Biodegradation 19:771

Ball AS, Jackson AM (1995) The recovery of lignocellulose-degrading enzymes from spent mushroom compost. Bioresour Technol 54:311–314

Baldrian P (2006) Fungal laccases–occurrence and properties. FEMS Microbiol Rev 30:215–242

Biely P (1985) Microbial xylanolytic systems. Trends Biotechnol 3:286–290

Camarero S, Sarkar S, Ruiz-Dueñas FJ, Martínez MJ, Martínez ÁT (1999) Description of a versatile peroxidase involved in the natural degradation of lignin that has both manganese peroxidase and lignin peroxidase substrate interaction sites. J Biol Chem 274:10324–10330

Chaturvedi VK, Agarwal S, Gupta KK, Ramteke PW, Singh MP (2018) Medicinal mushroom: boon for therapeutic applications. 3 Biotech 8:334

Chaturvedi VK, Singh A, Singh VK, Singh MP (2019) Cancer nanotechnology: a new revolution for cancer diagnosis and therapy. Curr Drug Met 20:416–429

Chiu SW, Law SC, Ching ML, Cheung KW, Chen MJ (2000) Themes for mushroom exploitation in the 21st century: sustainability, waste management, and conservation. J Gen Appl Microbiol 46:269–282

Claus H (2004) Laccases: structure, reactions, distribution. Micron 35:93–96

Cohen R, Persky L, Hadar Y (2002) Biotechnological applications and potential of wood-degrading mushrooms of the genus *Pleurotus*. Appl Microbiol Biotechnol 58:582–594

Davari M, Sharma SN, Mirzakhani M (2012) Residual influence of organic materials, crop residues, and biofertilizers on performance of succeeding mung bean in an organic rice-based cropping system. Int J Recycling Org Waste Agric 1(1):14

Dekker RF (1985) Biodegradation of the hemicelluloses. Biodegrad Wood Com 1985:505–533

Elisashvili V, Chichua D, Kachlishvili E, Tsiklauri N, Khardziani T (2003) Lignocellulolytic enzyme activity during growth and fruiting of the edible and medicinal mushroom *Pleurotus ostreatus* (Jacq.: Fr.) Kumm. (Agaricomycetideae). Int J Med Mush 5:193–198

Eriksson KE, Blanchette RA, Ander P (1990) Morphological aspects of wood degradation by fungi and bacteria. In: Microbial and enzymatic degradation of wood and wood components. Springer, Berlin, Heidelberg, pp 1–87

Ezeji TC, Qureshi N, Karcher P, Blaschek HP (2006) Production of butanol from corn. Chemical industries, vol 112. Marcel Dekker, New York, p 99

Fermor T, Watts N, Duncombe T, Brooks R, McCarthy A, Semple K, Reid B (2000) Bioremediation: use of composts and composting technologies. Mush Sci 15:833–839

Ferraroni M, Myasoedova NM, Schmatchenko V, Leontievsky AA, Golovleva LA, Scozzafava A, Briganti F (2007) Crystal structure of a blue laccase from *Lentinus tigrinus*: evidences for intermediates in the molecular oxygen reductive splitting by multicopper oxidases. BMC Struct Biol 7:60

Garg SK, Modi DR (1999) Decolorization of pulp-paper mill effluents by white-rot fungi. Crit Rev Biotechnol 19:85–112

Glenn JK, Gold MH (1985) Purification and characterization of an extracellular Mn (II)-dependent peroxidase from the lignin-degrading basidiomycete, *Phanerochaete chrysosporium*. Arch Biochem Biophys 242:329–341

Glenn JK, Akileswaran L, Gold MH (1986) Mn (II) oxidation is the principal function of the extracellular Mn-peroxidase from *Phanerochaete chrysosporium*. Arch Biochem Biophys 251:688–696

Gold MH, Alic MA (1993) Molecular biology of the lignin-degrading basidiomycete *Phanerochaete chrysosporium*. Microbiol Rev 57:605–622

Gravitis J, Zandersons J, Vedernikov N, Kruma I, Ozols-Kalnins V (2004) Clustering of bio-products technologies for zero emissions and eco-efficiency. Ind Crops Prod 20:169–180

Guimaraes JL, Frollini E, Da Silva CG, Wypych F, Satyanarayana KG (2009) Characterization of banana, sugarcane bagasse and sponge gourd fibers of Brazil. Ind Crops Prod 30:407–415

Gupta KK, Agarwal S, Kushwaha A, Maurya S, Chaturvedi VK, Pathak RK, Verma V, Singh MP (2017) Oyster mushroom: a rich source of antioxidants. Incredible world of biotechnology. Nova Science Publishers, Inc, New York. ISBN: 978-1-53611-282-5

Haemmerli SD, Schoemaker HE, Schmidt HW, Leisola MS (1987) Oxidation of veratryl alcohol by the lignin peroxidase of *Phanerochaete chrysosporium* Involvement of activated oxygen. FEBS Lett 220:149–154

Howard RL, Abotsi EL, Van Rensburg EJ, Howard S (2003) Lignocellulose biotechnology: issues of bioconversion and enzyme production. Afr J Biotechnol 2:602–619

Husain Q (2006) Potential applications of the oxidoreductive enzymes in the decolorization and detoxification of textile and other synthetic dyes from polluted water: a review. Crit Rev Biotechnol 26:201–221

Isikgor FH, Becer CR (2015) Lignocellulosic biomass: a sustainable platform for the production of bio-based chemicals and polymers. Polym Chem 6:4497–4559

Jeznabadi EK, Jafarpour M, Eghbalsaied S (2016) King oyster mushroom production using various sources of agricultural wastes in Iran. Int J Recycling Org Waste Agric 5:17–24

John F, Monsalve G, Medina PLV, Ruiz CAA (2006) Ethanol production of banana shell and cassava starch. Dyna 73:21–27

Kabirifard A, Fazaeli H, Kafilzadeh F (2012) Comparison of growth rate of four *Pleurotus* fungi species on wheat straw and date palm leaf substrates. Mala J Anim Sci 15:55–63

Kakkar VK, Garcha HS, Dhanda S, Makkar GS (1990) Mushroom harvested spent straw as feed for buffaloes. Indian J Anim Nutr 7:267–272

Kamara DN, Zadrazil F (1988) Microbiological improvement of lignocellulosic in animal feed production. Treat Ligno White Rot Fungi 1988:56–60

Kamra DN, Zadražil F (1986) Influence of gaseous phase, light and substrate pretreatment on fruit-body formation, lignin degradation and in vitro digestibility of wheat straw fermented with *Pleurotus* spp. Agric Wastes 18:1–7

Kim M, Day DF (2011) Composition of sugar cane, energy cane, and sweet sorghum suitable for ethanol production at Louisiana sugar mills. J Ind Microbiol Biotechnol 38:803–807

Koutrotsios G, Mountzouris KC, Chatzipavlidis I, Zervakis GI (2014) Bioconversion of lignocellulosic residues by *Agrocybe cylindracea* and *Pleurotus ostreatus* mushroom fungi–assessment of their effect on the final product and spent substrate properties. Food Chem 161:127–135

Kuhad RC, Singh A (eds) (2007) Lignocellulose biotechnology: future prospects. IK International Pvt Ltd, New Delhi

Kuhad RC, Kuhar S, Kapoor M, Sharma KK, Singh A (2007) Lignocellulolytic microorganisms, their enzymes and possible biotechnologies based on lignocellulolytic microorganisms and their enzymes. lignocellulose biotechnology: future prospects. IK International Pvt Ltd, New Delhi, pp 3–22

Kulshreshtha S, Mathur N, Bhatnagar P, Kulshreshtha S (2013) Cultivation of *Pleurotus citrinopileatus* on handmade paper and cardboard industrial wastes. Ind Crops Prod 41:340–346

Kundu SS (2005) Roughage processing technology. Satish Serial Pub. House, New Delhi

Kuila A, Sharma V (eds) (2017) Lignocellulosic biomass production and industrial applications. Wiley, pp 1–276

Madan M, Vasudevan P, Sharma S (1987) Cultivation of *Pleurotus sajor-caju* on different wastes. Biol Wastes 22:241–250

Mahawar N, Goyal P, Lakhiwal S, Jain S (2015) Agro waste: a new eco-friendly energy resource. Int Res J 4:1–5

Mahesh MS, Mohini M (2013) Biological treatment of crop residues for ruminant feeding: a review. Afr J Biotechnol 12(27):4221–4231

Malherbe S, Cloete TE (2002) Lignocellulose biodegradation: fundamentals and applications. Rev Environ Sci Biotechnol 1:105–114

Martín C, López Y, Plasencia Y, Hernández E (2006) Characterisation of agricultural and agro-industrial residues as raw materials for ethanol production. Chem Biochem Eng Q 20:443–447

Maurya S, Bhardwaj AK, Gupta KK, Agarwal S, Kushwaha A (2016) Green synthesis of silver nanoparticles using *Pleurotus* and its bactericidal activity. Cell Mol Biol 62:131

Mayer AM, Staples RC (2002) Laccase: new functions for an old enzyme. Phytochemistry 60:551–565

Mikiashvili N, Wasser SP, Nevo E, Elisashvili V (2006) Effects of carbon and nitrogen sources on *Pleurotus ostreatus* ligninolytic enzyme activity. World J Microbiol Biotechnol 22:999–1002

Murarilal, Pereira BMJ (2006) Integrated biosystems for lignocellulosic waste management. In: Kuhad RC (ed) Lignocellulose biotechnology: present and future prospects. I.K. International Pvt. Ltd, pp 72–82

Nie G, Reading NS, Aust SD (1999) Relative stability of recombinant versus native peroxidases from *Phanerochaete chrysosporium*. Arch Biochem Biophys 365:328–334

Ntougias S, Baldrian P, Ehaliotis C, Nerud F, Antoniou T, Merhautová V, Zervakis GI (2012) Biodegradation and detoxification of olive mill wastewater by selected strains of the mushroom genera *Ganoderma* and *Pleurotus*. Chemosphere 88:620–626

Obi FO, Ugwuishiwu BO, Nwakaire JN (2016) Agricultural waste concept, generation, utilization and management. Niger J Technol 35:957–964

Orts WJ, Holtman KM, Seiber JN (2008) Agricultural chemistry and bioenergy. J Agric Food Chem 56:3892–3899

Parajó JC, Dominguez H, Domínguez J (1998) Biotechnological production of xylitol. Part 3: operation in culture media made from lignocellulose hydrolysates. Bioresour Technol 66:25–40

Pedersen M, Meyer AS (2010) Lignocellulose pretreatment severity–relating pH to biomatrix opening. New Biotechnol 27:739–750

Philippoussis AN (2009) Production of mushrooms using agro-industrial residues as substrates. In: Biotechnology for agro-industrial residues utilisation. Springer, Dordrecht, pp 163–196

Philippoussis AN, Diamantopoulou PA (2011) Agro-food industry wastes and agricultural residues conversion into high value products by mushroom cultivation. In: Proceedings of the 7th international conference on mushroom biology and mushroom products (ICMBMP7), France, 4–7 Oct 2011

Philippoussis A, Diamantopoulou P, Zervakis G, Ioannidou S (2000) Potential for the cultivation of exotic mushroom species by exploitation of Mediterranean agricultural wastes. Mush Sci 15:523–530

Poppe J (2000) Use of agricultural waste materials in the cultivation of mushrooms. Mush Sci 15:3–23

Potumarthi R, Baadhe RR, Nayak P, Jetty A (2013) Simultaneous pretreatment and saccharification of rice husk by *Phanerochete chrysosporium* for improved production of reducing sugars. Bioresour Technol 128:113–117

Prasad S, Singh A, Joshi HC (2007) Ethanol as an alternative fuel from agricultural, industrial and urban residues. Resour Conserv Recycling 50:1–39

Rai RD, Ahlawat OP (2005) Recycling of agro-wastes for protein production through mushroom cultivation. Fungi Diver Biotechnol 2005:149–179

Reddy CA (1995) The potential for white-rot fungi in the treatment of pollutants. Curr Opin Biotechnol 6:320–328

Ribbons DW (1987) Chemicals from lignin. Philos Trans R Soc Lond A 321:485–494

Rodríguez G, Lama A, Rodríguez R, Jiménez A, Guillén R, Fernández-Bolaños J (2008) Olive stone an attractive source of bioactive and valuable compounds. Bioresour Technol 99:5261–5269

Ruiz-Dueñas FJ, Guillén F, Camarero S, Pérez-Boada M, Martínez MJ, Martínez ÁT (1999) Regulation of peroxidase transcript levels in liquid cultures of the ligninolytic fungus *Pleurotus eryngii*. Appl Environ Microbiol 65:4458–4463

Salmones D, Mata G, Waliszewski KN (2005) Comparative culturing of *Pleurotus* spp. on coffee pulp and wheat straw: biomass production and substrate biodegradation. Bioresour Technol 96:537–544

Sánchez C (2009) Lignocellulosic residues: biodegradation and bioconversion by fungi. Biotechnol Adv 27:185–194

Sindhu R, Binod P, Pandey A (2016) Biological pretreatment of lignocellulosic biomass–an overview. Bioresour Technol 199:76–82

Singh MP, Gautam NC (2004) An overview of lignocellulose biotechnology. Recent Adv Biotechnol 2004:3–20

Stamets P (2011) Growing gourmet and medicinal mushrooms. Ten Speed Press, 13 July

Streeter CL, Conway KE, Horn GW, Mader TL (1982) Nutritional evaluation of wheat straw incubated with the edible mushroom, *Pleurotus ostreatus*. J Anim Sci Biotechnol 54:183–188

Szulczyk KR, McCarl BA, Cornforth G (2010) Market penetration of ethanol. Renew Sustain Energy Rev 14:394–403

Ten Have R, Teunissen PJ (2001) Oxidative mechanisms involved in lignin degradation by white-rot fungi. Chem Rev 101:3397–3414

Tewari RP, Pandey M (2002) Sizeable income generating venture. The Hindu Survey of Indian Agriculture 2002:155–167

Umezawa T, Higuchi T (1987) Mechanism of aromatic ring cleavage of β-O-4 lignin substructure models by lignin peroxidase. FEBS Lett 218:255–260

Van Dyk JS, Pletschke BI (2012) A review of lignocellulose bioconversion using enzymatic hydrolysis and synergistic cooperation between enzymes factors affecting enzymes, conversion and synergy. Biotechnol Adv 30:1458–1480

Wang B, Dong F, Chen M, Zhu J, Tan J, Fu X, Wang Y, Chen S (2016) Advances in recycling and utilization of agricultural wastes in China: based on environmental risk, crucial pathways, influencing factors, policy mechanism. Procedia Environ Sci 31:12–17

Wariishi H, Valli K, Gold MH (1992) Manganese (II) oxidation by manganese peroxidase from the basidiomycete *Phanerochaete chrysosporium*. Kinetic mechanism and role of chelators. J Biol Chem 267:23688–23695

Wesenberg D, Kyriakides I, Agathos SN (2003) White-rot fungi and their enzymes for the treatment of industrial dye effluents. Biotechnol Adv 22:161–187

Wilson DB, Irwin DC (1999) Genetics and properties of cellulases. In: Recent progress in bioconversion of lignocellulosics. Springer, Berlin, Heidelberg, pp 1–21

Wong DW (2009) Structure and action mechanism of ligninolytic enzymes. Appl Biochem Biotechnol 157:174–209

Yoshitake K, Katayama Y, Nakamura M, Iimura Y, Kawai S, Morohoshi N (1993) N-linked carbohydrate chains protect laccase III from proteolysis in *Coriolus versicolor*. Microbiology 139:179–185

Youn HD, Hah YC, Kang SO (1995) Role of laccase in lignin degradation by white-rot fungi. FEMS Microbiol Lett 132:183–188

Zadrazil F, Reiniger P (1988) Treatment of lignocellulosics with white rot fungi. In: Elsevier applied science 1–225

Zadrazil F, Dube HC (1992) The oyster mushroom–importance and prospects. Mushroom Res 1

Zervakis G, Philippoussis A, Ioannidou S, Diamantopoulou P (2001) Mycelium growth kinetics and optimal temperature conditions for the cultivation of edible mushroom species on lignocellulosic substrates. Folia Microbiol 46:231

Zhang R, Li X, Fadel JG (2002) Oyster mushroom cultivation with rice and wheat straw. Bioresour Technol 82:277–284

Zhu Y, Lee YY, Elander RT (2005) Optimization of dilute-acid pretreatment of corn stover using a high-solids percolation reactor. Appl Biochem Biotechnol 124:1045–1054

Chapter 12
Myco-Nanotechnological Approach for Improved Degradation of Lignocellulosic Waste: Its Future Aspect

Abhishek K. Bhardwaj, Manish Kumar Gupta, and R. Naraian

1 Introduction

Huge biomass of naturally produced lignocellulosic waste during agricultural practices is a big problem during its disposal in developing country. Lignocellulose is a complex carbohydrate polymer that is bonded to strong bonds to give it a highly stable structure and other hand intricate nature of lignin like polysaccharide of plant cell wall is slow or resistant to enzymatic degradation in normal environmental condition (Bilandzija et al. 2018). The farmers most often dispose huge amount of wastes through the process of negligent burning in open, which release huge amount of air pollutants like dense white or black smoke and ash (Perlack 2005; Alftrén and Hobley 2014). However, the lignocellulosic biomass has enormous potential to contribute global renewable energy, chemical, and materials in a sustainable manner (Kumar et al. 2008; Limayem and Ricke 2012; Saratale and Oh 2012; Fapyane and Ferapontova 2017). The technologies need to develop the value-added product such as (sugar, organic acids, surfactants, glues, solvents or beverage softeners, etc.) from huge amount of agriculture and forests, which are prominent sources of carbohydrates (Mamilla et al. 2019).

Many researchers reported that cellulolytic enzymes are found in insect, bacteria, fungi, and plant which can be utilized into various industrial sectors for conversion of cellulose into free glucose (Goyal et al. 1991; Jahangeer et al. 2005; Sharma et al. 2007). Due to wide demand for hydrolyzed cellulose products; various industrial sectors employing cellulolytic enzymes has attracted greater attention since past decades (Tischer and Wedekind 1999). Cellulases are the combination of

A. K. Bhardwaj
Department of Environmental Science, Faculty of Science, Veer Bahadur Singh Purvanchal University, Jaunpur, Uttar Pradesh, India

M. K. Gupta · R. Naraian (✉)
Department of Biotechnology, Faculty of Science, Veer Bahadur Singh Purvanchal University, Jaunpur, Uttar Pradesh, India

R. Naraian (ed.), *Mycodegradation of Lignocelluloses*, Fungal Biology,
https://doi.org/10.1007/978-3-030-23834-6_12

intricate enzymes which catalyze the hydrolysis of cellulose due to the synergistic actions of the exoglucanase, endoglucanase, and β-glucosidase enzymes. These three important components can break the cross-linked bonds present in cellulose and generate monomers of glucose for further fermentation. Although cellulases have many applications in medicine, energy and industry, most of them have low enzyme stability at different media, P^H, and temperatures (Dutta and Saha 2019). In addition, these applications are limited due to lack of reusability (Woodley 2013; Guzik et al. 2014). Various different methods can be used, such as protein engineering, chemical modification, and immobilization to enhance enzyme stability (Rodrigues et al. 2013; Barbosa et al. 2014; Rueda et al. 2016; Vaghari et al. 2016). In view of enhanced capability and stability of cellulose, a group of researchers introduced the concept of nanosupport immobilized by lignocellulosic enzymes (Blanchette et al. 2012; Hwang and Gu 2013; Nawaz et al. 2016; Wan et al. 2016). Surface immobilization, self-immobilization, and entrapment can be divided into three major categories of immobilization of enzyme onto NPs form a complexes. These complexes are made up of immobilized enzyme on to insoluble support (NPs) which enhances the properties of enzyme like bio-catalysis, thermostability, and reusability. The previous reports are suggesting among all the nanosupports, the magnetic nanosupports having a better choice due to less harmful, high surface area and reusable nature (Bilal et al. 2018). These properties can be employed for a wide range of industrial processes such as paper and cotton manufacturing units, fuel and food industries, clarification and extraction process of wine, juice and brewery, detergents, animal feed additives, agriculture, and research (Fig. 12.1). In the industrial processing or production yield, nanomaterials modification with enzymes has the potential to improve the economic feasibility of the entire process (Khoshnevisan et al. 2017). The chapter discusses recent and up-to-date information regarding to the role of various immobilized nanomaterials and nanocomposites on lignocellulose decomposition as well as on the mechanism of enhancing the lignocelluloses decomposition. The properties of various metal and magnetic nanocomposites with immobilized enzymes such as enzyme stability, activity, reusability, recovery and lignocellulose conversion ability in different fuels, materials, and value-added chemicals are also discussed.

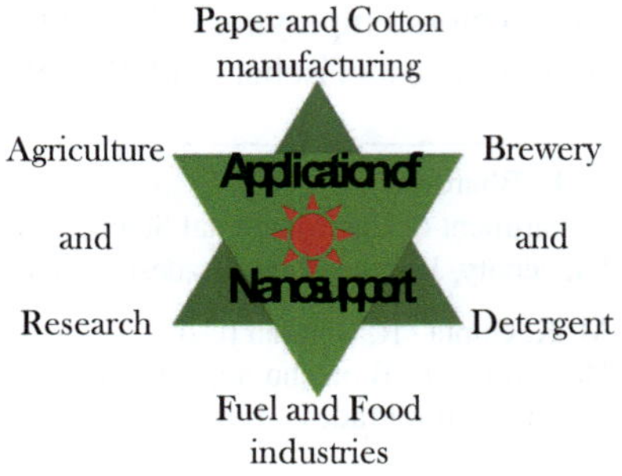

Fig. 12.1 Multiple applications of nanosupport (✷) immobilized with lignocellulosic enzymes

2 Complexity and Recalcitrant Nature of Lignocelluloses

The plant cell wall basically composed of three major biopolymers (a) cellulose (~30–50% by weight), which is the main component, (b) hemicellulose (~20–45% by weight as xylans), and (c) non-polysaccharide aromatic polymer lignin (~15–35) (Bacic et al. 1988; Metzger and Hüttermann 2009). The biomass fractionation is a very complex and lengthy process of these materials. High amount of sugars can be produced from cellulose and hemicellulose, using acid as the catalyst, or enzymatic hydrolysis and reformed into bio-ethanol by a fermentation procedure (Walker and Wilson 1991; Refaat 2012). Most physical and chemical pretreatment using acid, alkali, processes require special instrument and consume a lot of energy and generate inhibitors, which will affect enzymatic hydrolysis and fermentation (Mood et al. 2013). Many researchers reported that biological pretreatments employing bacteria, fungus (white, brown, and soft-rot fungi), deuteromycetes, and ascomycetes can enhance the hydrolysis productivity because of generating low inhibitors and limited energy utilization (Taha et al. 2015; Zhang et al. 2016). The lignocellulases are the constitutive building block of plant material composed of carbohydrate polymers like majority of cellulose (>50%), hemicelluloses, and aromatic component: lignin (~20%) as principal material. The lignocellulosic material includes not only agriculture and forest residue but also considers animal excreta, paper mill sludge, and municipal solid waste (MSW). Generation of more than 50 billion tons of lignocellulosic biomass produced annually worldwide and creates environmental problems due to its larger volume and mismanagement. Scientific communities are currently focused over such kind of abandoned renewable resources of the earth. But researchers facing various hurdles because recalcitrant nature of lignocelluloses such as: (a) considerable lignin content; (b) cellulose protection by lignin; (c) cellulose covering by hemicelluloses; (d) high degree of polymerization and crystallinity; (e) strong fiber strength (Agbor et al. 2011). The fractional segregation is the major difficulty of complex hierarchy structure of lignocellulosic biomass, in which cellulose, hemicellulose, and lignin are affected by many physical, chemical, structural, and compositional factors (Seemala et al. 2018). Lignin is a complex, amorphous hetero-biopolymer, water insoluble of phenylpropane units joined together by carbon-carbon and aryl-ether a linkage which provides frame support, impermeability, and protect from microbial attack. The chief cause of recalcitrant nature of lignocellulose is due to the presence of lignin (Grabber 2005). Generally, the softness of vegetation or wood reflects the high quantity of lignin in comparison to hardwood. The phenylpropane units of lignin are the principal bottleneck in degradation of lignocellulosic biomass (Mussatto 2016). The accessibility of cellulose is influenced by the presence of hemicelluloses which commonly coated with hemicellulose branches with short lateral chains consisting of different sugars (hexoses, pentoses, and acetylated sugars) (Yang et al. 2018). Hemicellulose can easily be hydrolyzed by diluted acid, alkali, and enzyme in comparison to cellulose under mild conditions. During the degradation, high thermochemical sensitivity of hemicelluloses produces several co-products of furfurals which inhibit the fermentation process (Chen et al. 2018). However, during the synthesis of nanocellulose co-product does not influence (Agbor et al. 2011; Hu et al.

2016). For this ground, diverse kind of pretreatment practices usually compromise to abandoned lignin and removal of hemicelluloses during maximum cellulose recovery for further nanocellulose production. Pretreatment of lignocellulosic biomass strategies are dependent on a variety of parameters like pH, temperature, treatment time, and different catalyst. This diversity of effects is pretreatment severity during degradation of biomass (Pedersen and Meyer 2010). The different kind of pretreatment techniques can be summarized as physical (milling and grinding), chemical (acid, alkaline, oxidizing agents, and organic solvent), biological, and some other combined techniques (wet oxidation, steam/auto hydrolysis, microwave, and hydrothermal) (Kumar et al. 2009; Stern et al. 2014; Bhardwaj et al. 2017a, 2018). Physical pretreatment consisting chipping, milling, grinding, and thermal process are safer than chemical methods, but less efficient, and require high energy (Stern et al. 2014). Chemical pretreatments are efficient, low cost biomass disintegration with less pretreatment severity, which may cause ecological threat. In case of biological methods, enzymes play a significant role in mild biomass pretreatments due to low stability of enzymes against pH and high temperatures (Van Dyk and Pletschke 2012).

3 Cellulases Action

Cellulase is an enzyme produced by different taxonomic groups of microorganism like fungi, bacteria, and protozoans. Cellulase hydrolyzes β–1, 4-D-glucan linkages present in the structure of cellulose and produce glucose, shorter polysaccharides and cello-oligosaccharides. Cellobiohydrolase, endo-β–1,4-glucanase, and β-glucosidase are the three major types of cellulose enzymes. These three enzymes disintegrate cellulose with several other enzymes which attack synergistically to change crystalline cellulose to glucose (Béguin and Aubert 1994). Generally, different cellulases are characterized by two domain structures including catalytic domain and cellulose binding domain or carbohydrate binding module connected through a linker peptide. Catalytic or core domain contains catalytic site while cellulose binding domains assist in binding of enzyme with cellulose, EG are regarded as the break down the crystalline structure of the cellulosic substrate (Husain et al. 2011). Tables 12.1 and 12.2 represents the interaction mode; adsorption and covalently with different nanomaterials. These cellulases immobilized on the surface of the NPs or embedded inside the polymeric nanospheres exhibit high catalytic efficiency and immobilization yield.

4 Nanoparticles (NPs) to Enhance Degradation of Lignocelluloses

Currently, scientists are looking a wide scope in the field of nanobiotechnology due to unique nature of nanoscale materials for lignocellulosic degradation. Nanoparticles (NPs) have exclusive physicochemical properties, large surface to volume ratio,

Table 12.1 Nanohybrids (adsorption immobilization mode) employed for minimization of steric hindrance during cellulase activity

Name of support	Enzyme	Properties improvement	Reference
Silver NPs	Cellulase	Efficient in cellulose hydrolysis	Salunke et al. (2015)
Silver NPs and gold NPs	Cellulase	Heat stability and reusable up to sixth times	Mishra and Sardar (2015)
MnO_2 NPs	Cellulase	Stability in a broad range of pH and temperatures, reusability, and cellulose stability	Cherian et al. (2015)
Functionalized MWCNTs	Cellulase	High binding efficiency and loading reusability	Mubarak et al. (2014)
Fe_3O_4@SiO_2 NPs	Cellulase	Catalytic efficiency, temperature optima, reusability, half life, and thermal stability	Li et al. (2014); Harmoko et al. (2016); Tao et al. (2016)
Magnetic NPs	Cellulase	Effective bioethenol production using *Sesbania aculeate* biomass	Baskar et al. (2016)
Magnetic NPs and silica-coated magnetic NPs	Cellulase	Recycle up to ten step, recovery yield up to 75%	Roth et al. (2016)
Zinc-doped magnetic NPs	Cellulase from *Trichoderma reesei*	Immobilization of enzyme 94%, thermostability up to 80 °C, hydrolysis recorded 93%	Abraham et al. (2014)
Nickel–cobaltite	Cellulase from *Aspergillus fumigatus*	Thermostable for 7 h at 80 °C, and better hydrolysis performance	Srivastava et al. (2014)
3-(2-aminoethyl aminopropyl)-trimethoxysilane-magnetic Nanosphere	Cellulase	High temperature, broader pH stability, and 87% activity recovery	Zhang et al. (2015)
Fe_3O_4 NPs and Fe_3O_4 Alginate nanocomposites	Cellulase from *Aspergillus fumigatus*	Cellulase activity increased 35% and 40%, thermally stable up to 8 h at 70 °C, and retained 56% of its relative activity; in comparison to 19% for sample	Srivastava et al. (2015)
Magnetic gold mesoporous silica NPs core shell	Cellulase	Thermal stability and facilitated its long time storage (58% enzyme maintained 9 h after catalytic activity)	Poorakbar et al. (2018)

high catalytic efficiency, strong adsorption ability as well as high surface reactivity which can be easily harnessed to modify for the purpose of enzyme immobilization and surface capping. Such a nanotechnological approach can play an important role to improve thermostability of cellulose enzymes (Srivastava et al. 2014). Nanoparticles demonstrate the potential of immobilized enzymes to reduce the steric hindrances in comparison to bulk support with enzymes (Ansari et al. 2011; Khan et al. 2011) (Fig. 12.2).

Table 12.2 Nanohybrids (covalent immobilization mode) employed for minimization of steric hindrance during cellulase activity

Name of support	Enzyme	Properties improvement	Reference
Chitosan-coated magnetic NPs	Cellulase from *Trichoderma reesei*	80% activity after 15 cycle CMC hydrolysis	Sánchez-Ramírez et al. (2017)
TiO_2 NPs	Cellulase from *Aspergillus niger*	Strongly bound enzyme has better stability	Ahmad and Sardar (2014)
β-Cyclodextrin-coated Fe_3O_4 NPs	Cellulase from *Aspergillus niger*	Stability, reusability makes immobilized cellulase is 20-fold better than free cellulose	Huang et al. (2015)
Chitosan-coated magnetic NPs	Cellulase	50% cellulose immobilized after 10 cycles and stable at 50 °C	Zang et al. (2014)
Fe_3O_4 magnetic NPs	Cellulase	Storage stability 60% of activity after 6 cycle	Jordan and Theegala (2014)
Fe_3O_4 magnetic NPs	Cellulase	Wide range of pH and temperature resistant	Xu et al. (2011)
Fe_3O_4@silica core-shell magnetic NPs	Cellulase	Improved heat and operational stability, and reusability	Jafari Khorshidi et al. (2016)
Amine functionalized cobalt ferrite magnetic NPs	Cellulase	The immobilized cellulase retains >64% of the activity after 6 cycles at high pH and temperature	Bohara et al. (2016)
Fe_3O_4 NPs entrapped within cross-linked epoxy polymer	Cellulase	Thermal stability and reusability	Hosseini et al. (2018)
PEGylated graphene oxide	Cellulase *Aspergillus oryzae*	Ionic liquid stability enhanced 30 times and hydrolysis rate reached 87%	Xu et al. (2016)
Attapulgite@ chitosan nanocomposite	Cellulase	Enhanced heat stability and effective hydrolysis of wheat straw	Yang et al. (2016a)
Magnetic core-shell metal-organic framework	Cellulase	Enhanced pH stability, thermal stability, and catalytic efficiency	Hosseini et al. (2018)

CMC carboxy methyl cellulose; *PEG* Poly ethylene glycol

5 Immobilization of Lignolytic Enzymes on Nanoparticles

Significant technical hurdles were noted during biomass conversion because most enzymes are easily denatured due to small changes in pH, temperature and lose their activity (Mateo et al. 2007). Cellulases are strongly influenced by hydrolyzed product (Andrić et al. 2010) and a chunk amount of enzyme is consumed by slow hydrolysis before being recovered from the reaction system. Therefore, icebreaking research is needed to solve these problems because it has become a bottleneck pro-

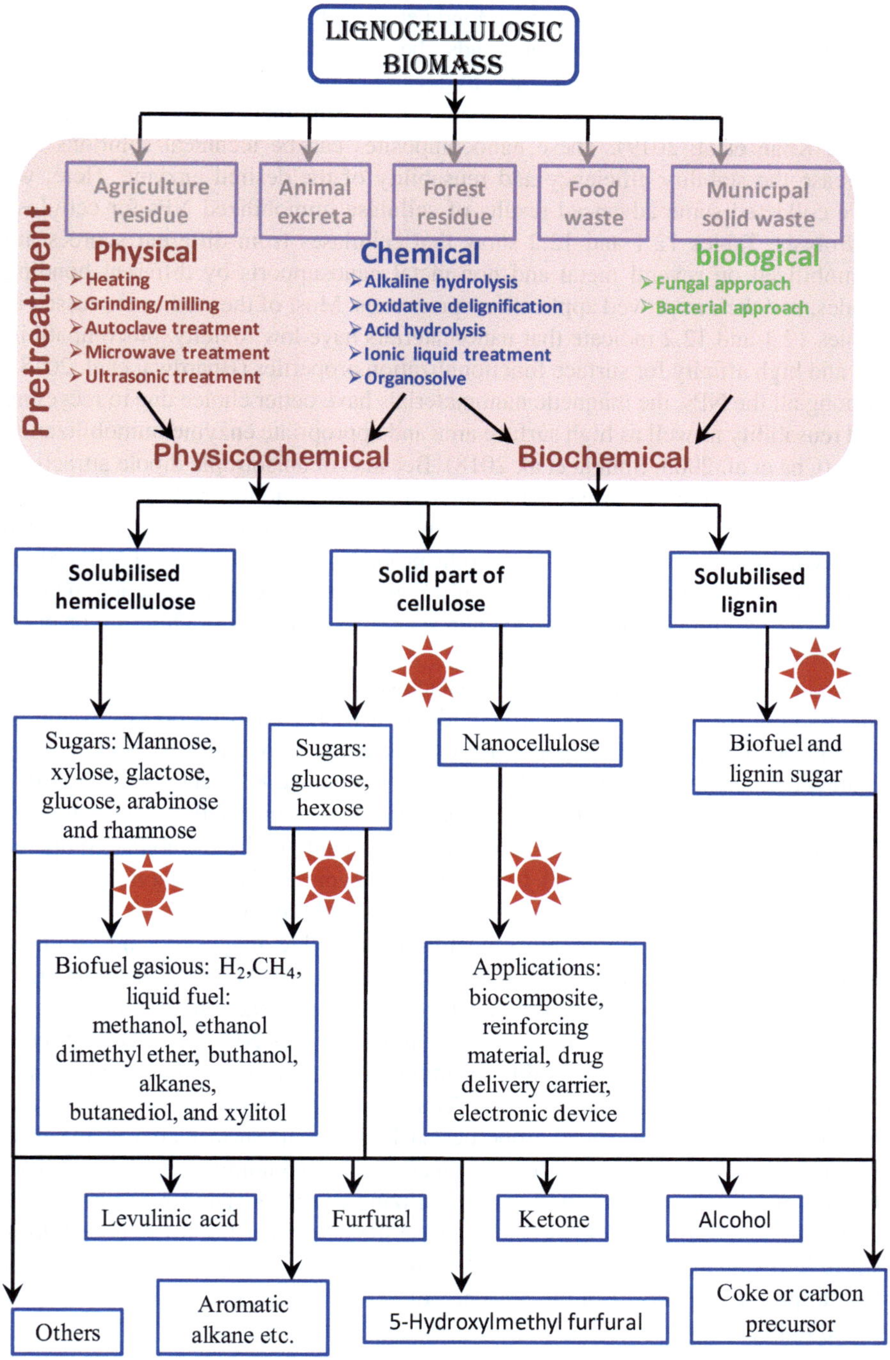

Fig. 12.2 Sketch diagram illustrating fate of lignocellulosic biomass in the form of intermediates and metabolites with the employment of NPs

cess from an economic perspective (Honda et al. 2015). In advanced nanotechnology, nanohybrid materials are used as adsorbent, covalent, embedded, cross-linked, and encapsulated cellulase aggregates as a result of immobilized enzymes on nanosupports (Nery and Kubota 2016; Iype et al. 2017; Sulaiman et al. 2017; Carli et al. 2019; Khan et al. 2019). These nanocomposites can be technical solutions that increase the stability efficiency and reusability of the desired enzyme. Here, we have collected some advanced results of cellulase immobilized NPs for cellulose hydrolysis. Tables 12.1 and 12.2 show that cellulases from different sources are immobilized on several metal and non-metal nanosupports by different bonding modes, and their improved applications/properties. Most of the studies expressed in Tables 12.1 and 12.2 indicate that nanomaterials have low toxicity, biocompatibility, and high affinity for surface functionalization properties (Bhardwaj et al. 2018). Among all the NPs, the magnetic nanomaterials have better choice due to recycling and reusability as well as high surface area and appropriate enzyme immobilization sites (Cha et al. 2005; Shukla et al. 2018). Because of anisotropic dipole attraction, hydrophobic nature and large surface area; uncoated nanomaterials are commonly unstable, which leads to the formation of aggregates, ultimately NPs may oxidized in presence of oxygen and lost their specific properties (Atacan et al. 2017; Bhardwaj et al. 2017b; Shukla et al. 2017). Therefore, an appropriate surface fictionalization is required with molecule as amino-salines (Zhang et al. 2015), chitosan, or polymers to overcome such limitations.

Khan et al. (2016), reported that β-galactosidase isolated from *Aspergillus oryzae,* immobilized onto graphene-iron oxide ($Gr@Fe_3O_4$) nanocomposites by simple adsorption. The prepared magnetic nanosupports are providing excellent stability against pH, temperature, and the bound enzyme retained 83% activity after eighth successive reuse (Khan et al. 2016). β-galactosidase is immobilized on bulk ZnO and nano-ZnO, and ZnO nanoparticles have significant yield and stability compared to bulk ZnO (Husain et al. 2011) Immobilized enzymes on NPs supports are similar to their use in industrial, environmental, clinical, and analytical fields. Magnetite NPs (Fe_3O_4) have excellent biocompatibility, reusable magnetic, and cellulose immobilization (Zang et al. 2014; Chiaradia et al. 2016). Magnetic nanomaterials are used as nanosupports that ultimately recover from the reaction mixture by applying magnetic flux after the reaction is complete. Metal NPs (MNPs) have been shown to be potentially compatible in immobilizing various cellulolytic enzymes (Zhang et al. 2015; Khan et al. 2016). Thus, nanobiotechnology and nanocarriers have inherent physicochemical properties such as specific surface area, shape, size, surface adjustability, mass transfer resistance, and high binding, demonstrating useful and efficient nanoscaffolds for immobilizing enzymes (Dinçer and Telefoncu 2007). In addition, it can be used to improve the deconstruction of lignocellulosic waste into valuable metabolites. A schematic mechanism of crystalline nanocellulose to product (glucose) in the presence of cellulase-based nanosupport has proposed in Fig. 12.3.

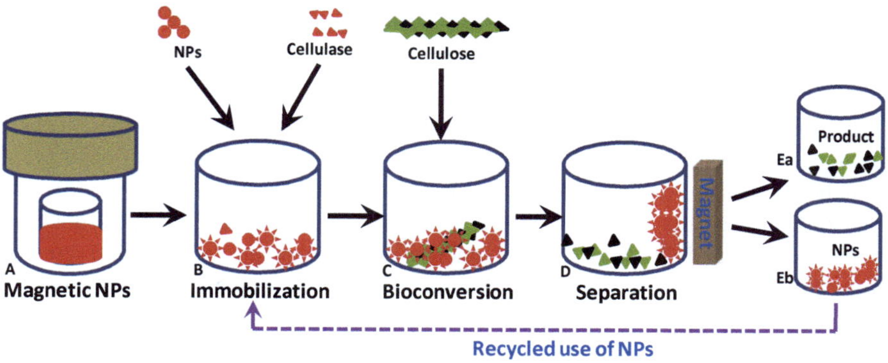

Fig. 12.3 A schematic diagram of enhanced product formation (glucose) from a crystalline cellulose using nanosupports (A) independent magnetic NPs (B) immobilization of cellulolytic enzymes on NPs (C) bioconversion of cellulose into simple sugars (D) magnetic separation of product and NPs (Ea) separated cellulosic product (Eb) separated NPs complex ready to reuse

6 Characterization of NPs

It is important to understand that how enzymatic system will interact with nanomaterials along with nanomaterials formation and stabilization. The particle size is inversely proportional to surface area and volume ratio, which allows tunable interaction between the materials and its surroundings (Reineck et al. 2019). Thus, the size and surface area of nanomaterials are basic characteristics which highly influenced enzymatic stability, yield efficiency, surface immobilization of enzyme and fate in environment. The size, shape, surface, and structural characteristics of the material are important factors because it provides important information for understanding the behavior of the interaction between the support and the enzyme (Campbell et al. 2014). Therefore, the established interaction can be evaluated and effectively adjusted in a large-scale process. To this end, several techniques are used to characterize the nanosupport (nanomaterials) to obtain morphological, structural, quantitative, and qualitative information that can help to further manipulate the appropriate interactions. The major techniques used in the characterization of the characteristic information of the support and enzyme have accounted below.

6.1 *UV-Visible Spectroscopy*

The electromagnetic interaction of material can be explored about shape, size, density, and concentration. Generally, the UV-VIS spectra measure the wavelength ranging between 200 and 800 nm during different stages of nanomaterials synthesis. The acquired spectra depict the characteristic peaks, comparative analysis of shape and size along with the concentration of samples (Maurya et al. 2016; Bhardwaj et al. 2018).

6.2 Fourier-Transform Infrared Spectroscopy (FTIR)

FTIR spectroscopy is a very potent technique used for the characterization of enzymatic matrix. The infrared radiation interacts with the modified nanosupport surface along with different enzymes. This interaction leads to transmission or absorption which corresponds specifically to the chemical bonds present in enzyme vibration mode. Typically, the vibration frequencies are measured as the wave numbers in between 4000 and 480 cm^{-1}. The characteristic FTIR spectra of NPs and enzyme immobilized NPs can produce characteristic vibration frequency which provides clear-cut information about interaction pattern with each other (Baby et al. 2010).

6.3 X-Ray Diffraction (XRD)

The use of XRD in the field of enzyme loaded and engineered nanomaterials study is extensively increasing due to its broad range of application. XRD is the most powerful technique to determine the crystallinity and phase composition of nanosupport. The Bragg's equation can be described by the reflection of collimated X-rays in case of material phase determination. The beam is incident on the crystal face of the sample to be characterized. XRD-based wide-angle elastic scattering, usually used for ordered materials only. X-ray beam passes through the sample and is scattered or diffracted by atoms in the X-ray path are studied. Interference caused by scattering mutual observation using Bragg's law and properly positioned X-rays the crystal structure characteristics of the detector and material was determined (Epp 2016). In this method, X-ray source is used for getting diffraction pattern, which can plot intensity verses 2θ using inbuilt software in XRD machine, consequently some peaks are observed against different nanosupport (Yang et al. 2016b). Finally, the characteristic peaks can be matched using Joint Committee on Powder Diffraction Standards (JCPDS) software. The particles size can also be calculated using Debye Scherrer's formula given as below

$$D = \frac{K\lambda}{\beta \cos\theta}$$

where the constant $K = 0.97$, $\lambda = 0.6255$ Å is the wavelength of X-ray line used and β is the full width at half maximum.

6.4 X-Ray Photoelectron Spectroscopy (XPS)

This is one of the highly acceptable techniques used for the surface elemental composition analysis of nanosuport modified by enzymes, known as X-ray photoelectron spectroscopy (XPS). The XPS spectrum provides an elementary analysis of the nanosupports based on the photoelectric effect permits the characterization of the sample up to the 10 nm depth. The characteristic peaks are observed at different region and voltage (electron volts) of sample, which may represent a particular element over the nanosupport surface to determine the information about immobilized enzymes (Wilson and Langell 2014). Almost all elements of the materials can be recognized using XPS spectra excluding the lighter atoms like hydrogen and helium. The low energy X-ray beam bombarded over samples induces photoelectron emission through the excitation of electrons (Wang et al. 2012). The emission is further captured by the detector to analyze on the basis of emission kinetic energies which is the characteristics of the particular atom (Xue et al. 2011). The quantitative and qualitative analysis of function of kinetic energy also provides the ideas referring to the valance and the chemical bonds available in enzyme-modified nanosupports.

6.5 Transmission Electron Microscopy (TEM)

Transmission electron microscopy (TEM) is used for getting clear information characterizing shape, size, and structure of nanosupport. TEM is the preferred method for particle size analysis and particle size distribution of NP samples because it is capable of imaging particles from 1 nm to a few microns. This technique is also used to observe the morphology of NPs. Because of its ultra-high resolution capability, it is possible to study particle surfaces particle size or even lattice structure. The image is formed just like simple light microscope but in this technique the electron source is used to illuminate NPs with higher resolution (Zhu et al. 2017). The colloidal and diluted sample is put over a grid, this thin sample kept between path of focused electron beam and after the transmission of electron fall over the fluorescent screen and gives black and white picture (Shukla et al. 2018). The sample atom containing with high density of electron and low density leads to forming dark contrasts and clear region, respectively. Moreover, the enzymatic immobilization is responsible for change in the characteristic SEAD pattern (Khan et al. 2016). Selected area (electron) diffraction can also be analyzed to elucidate the crystalline structure of nanosupport (He and Gao 2010).

6.6 Scanning Electron Microscopy (SEM)

This technique is use for viewing surface morphology, topography of nanosupport modified with enzyme. This technique differ from TEM, sample atom excited by the electron beam and generate electron from the sample surface known as secondary electron instead of electron beam transmit through the sample. The conducting sample surface is responsible for generating secondary electron, which leads to provide information about material surface information with less than a nanometer resolution. Field emission SEM has much better resolution than normal SEM (Bhardwaj et al. 2017a).

Scanning electron microscopy (SEM) is one of the most widely used instrumental methods for the analysis and study of nano- and micro-scale imaging of solid materials. One of the reasons why SEM is preferred for particle size analysis is that its resolution is 10 nm, or 100 Å (Goldstein 2012). The updated versions of these instruments can achieve resolution about 2.5 nm (25 Å). Electron beam acceleration is performed by a high voltage system (20 kV) and the electron beam narrows after passing through the aperture and the electromagnetic lens. The beam scans the surface of sample with the aid of a scanning coil. After interaction of incoming electron beam with the specimen; it emits beam of secondary electrons, which are detected by the detectors, fitted suitably in vacuum chamber and consequently, image of specimen is generated by the signal from the beam area (Hafner 2007). The instrument can also be combined with other related energy dispersive X-ray microanalysis techniques (EDX, EDS, EDAX), used to determine the composition or orientation of a single crystal or feature. Typically, tungsten or field emission guns are used as a source of electron generation. To keep the tip free from contaminants and oxides, field launch guns require ultra-high vacuum conditions (10^{-10} to 10^{-11} Torr) (Pathak and Thassu 2016). The most appropriate and real environmental sample can be analyzed without taking care of moisture by environmental SEM; however, resolution will decrease. The inbuilt EDS analysis tool is also used for the mapping and elemental composition of composites.

6.7 *Vibrating Sample Magnetometer (VSM)*

The purpose behind the use of vibrating sample magnetometer (VSM) is to measure the induced magnetization of the magnetic nanosupport with and without enzyme. Nanosupport is mechanically vibrated in a uniform magnetic field. This field magnetize sample by aligning the magnetic domains, or the individual magnetic spins, with the field if the sample is magnetic in nature (Foner 1959; Shukla et al. 2017). The stronger is the constant field, the larger will be the magnetization, which creates a net magnetic dipole moment (Wu et al. 2012). This dipole moment of the nanosupport will create a magnetic field around the sample and sometimes called the magnetic stray field. As the sample is swipe from about −15,000 to 15,000(H) g at room temperature, this magnetic stray field will change as a function of time and can be sensed by a set of pick-up coils (Le Goff and Gallet 2004; Shukla et al. 2018). The alternating magnetic field creates an electric field in pick-up coils according to Faraday's law of electromagnetic induction (Su et al. 2015). This current is proportional to the magnetization of the sample. Therefore, the greater the magnetization, the greater would be the induced current (Mahdavi et al. 2013).

6.8 Circular Dichroism Spectroscopy (CDS)

The CSD technique is the most powerful technique widely used in the characterization of surface proteins. The CSD work based on the principle of differential absorption of light circularly polarized by optically active molecules and allows determining its absolute configurations. As light passes through a stereogenic center present in the sample, there will be differential absorbance of two circulating components to right or to left. The technique is non-destructive, easy to operate, fast, and requires low sample quantities and can be used under room temperature conditions. The temperature may change the amino acid or protein structure; this conformational change can be studied using CDS measurements. The characteristic peak of α-helix secondary structure of enzyme gets at 220 nm. This technique provides the information about the quantification of the secondary structure of protein in the form of the α-helical, β-strand, and other structure using the far UV range of the EM spectrum between 170 and 260 nm. The aromatic residues of the protein molecule can also be observed in UV range between 260 and 300 nm. The peptide bond shows two transition found in the far UV n-π* at ~220 nm and a π-π* transition at ~190 nm when ground state electrons to an excited state, in protein molecules during absorption of UV. The sum of all the signals from surface protein is determining the characteristic signal of particular immobilized enzyme. Thus, the technique has wide potential to investigate structural change takes place during the immobilization of the enzyme over the support.

7 Conclusion and Future Prospects

The global need for renewable materials as an alternative to nonrenewable energy sources have best opportunity in the development of a environmentally benign technology. The immobilization cellulolytic enzyme over different kinds of NPs leads to enhance yield and stability of enzymes. The sufficient evidence has been discussed with the help of several findings of the work in this chapter. It is observed that among all the nanomaterials the magnetic NPs immobilized enzymes provide better response than free enzyme and enhance the stability against various chemical and physical denaturants. Several NP conjugates have found remarkable stability, high yields, and better recovery to obtain a range of value-added products, especially different fuels and materials. Their application is not limited to biomass conversion; it can be used in the fields of biodiesel, medicine, sensors, cosmetics, industrial production, and bioremediation. Despite the development of smart nanohybrid materials, it still relies on magnetic materials to improve the efficiency and recovery of nanocarriers during biomass conversion. Of course, when we overcome this goal, it is indeed a real step towards sustainable development, which helps to reduce dependency over nonrenewable resources. Lignocelluloses rich in agricultural waste materials would be frequently use to produce a large number of industrial products such as organic acids, ethanol, and other industrially important compounds. The application of such kind of materials is not confined and the nanomatrix could be designed for a powerful bio-recognition probe in the field of biosensing when used in different enzyme systems. This nanocomposite with a range of fascinating functionality can also explore field of drug design, drug delivery, and biosensing applications.

References

Abraham RE, Verma ML, Barrow CJ, Puri M (2014) Suitability of magnetic nanoparticle immobilised cellulases in enhancing enzymatic saccharification of pretreated hemp biomass. Biotechnol Biofuels 7:90

Agbor VB, Cicek N, Sparling R, Berlin A, Levin DB (2011) Biomass pretreatment: fundamentals toward application. Biotechnol Adv 29:675–685

Ahmad R, Sardar M (2014) Immobilization of cellulase on TiO_2 nanoparticles by physical and covalent methods: a comparative study. Ind J Biochem Biophy 51:314–320

Alftrén J, Hobley TJ (2014) Immobilization of cellulase mixtures on magnetic particles for hydrolysis of lignocellulose and ease of recycling. Biomass Bioenergy 65:72–78

Andrić P, Meyer AS, Jensen PA, Dam-Johansen K (2010) Effect and modeling of glucose inhibition and in situ glucose removal during enzymatic hydrolysis of pretreated wheat straw. Appl Biochem Biotechnol 160:280

Ansari SA, Husain Q, Qayyum S, Azam A (2011) Designing and surface modification of zinc oxide nanoparticles for biomedical applications. Food Chem Toxicol 49:2107–2115

Atacan K, Çakıroğlu B, Özacar M (2017) Covalent immobilization of trypsin onto modified magnetite nanoparticles and its application for casein digestion. Int J Biol Macromol 97:148–155

Baby TT, Aravind SJ, Arockiadoss T, Rakhi R, Ramaprabhu S (2010) Metal decorated graphene nanosheets as immobilization matrix for amperometric glucose biosensor. Sensors Actuators B Chem 145:71–77

Bacic A, Harris PJ, Stone BA (1988) Structure and function of plant cell walls. In: The biochemistry of plants, vol 14. pp 297–371

Barbosa O, Ortiz C, Berenguer-Murcia Á, Torres R, Rodrigues RC, Fernandez-Lafuente R (2014) Glutaraldehyde in bio-catalysts design: a useful crosslinker and a versatile tool in enzyme immobilization. RSC Adv 4:1583–1600

Baskar G, Kumar RN, Melvin XH, Aiswarya R, Soumya S (2016) Sesbania aculeate biomass hydrolysis using magnetic nanobiocomposite of cellulase for bioethanol production. Renew Energy 98:23–28

Béguin P, Aubert JP (1994) The biological degradation of cellulose. FEMS Microbiol Rev 13:25–58

Bhardwaj AK, Shukla A, Mishra RK, Singh S, Mishra V, Uttam K, Singh MP, Sharma S, Gopal R (2017a) Power and time dependent microwave assisted fabrication of silver nanoparticles decorated cotton (SNDC) fibers for bacterial decontamination. Front Microbiol 8:330

Bhardwaj AK, Shukla A, Singh SC, Uttam KN, Nath G, Gopal R (2017b) Green synthesis of Cu_2O hollow microspheres. Adv Mat Proceed 2(2):132–138

Bhardwaj AK, Shukla A, Maurya S, Singh SC, Uttam KN, Sundaram S, Singh MP, Gopal R (2018) Direct sunlight enabled photo-biochemical synthesis of silver nanoparticles and their Bactericidal Efficacy: photon energy as key for size and distribution control. J Photochem Photobiol B Biol 188:42–49

Bilal M, Zhao Y, Rasheed T, Iqbal HM (2018) Magnetic nanoparticles as versatile carriers for enzymes immobilization: a review. Int J Biol Macromol 120:2530–2544

Bilandzija N, Voca N, Jelcic B, Jurisic V, Matin A, Grubor M, Kricka T (2018) Evaluation of Croatian agricultural solid biomass energy potential. Renew Sust Energ Rev 93:225–230

Blanchette C, Lacayo CI, Fischer NO, Hwang M, Thelen MP (2012) Enhanced cellulose degradation using cellulase-nanosphere complexes. PLoS One 7:e42116

Bohara RA, Thorat ND, Pawar SH (2016) Immobilization of cellulase on functionalized cobalt ferrite nanoparticles. Korean J Chem Eng 33:216–222

Campbell AS, Dong C, Meng F, Hardinger J, Perhinschi G, Wu N, Dinu CZ (2014) Enzyme catalytic efficiency: a function of bio–nano interface reactions. ACS Appl Mater Interfaces 6:5393–5403

Carli S, de Campos Carneiro LAB, Ward RJ, Meleiro LP (2019) Immobilization of a β-glucosidase and an endoglucanase in ferromagnetic nanoparticles: a study of synergistic effects. Protein Expr Purif 160:28–35

Cha T, Guo A, Zhu XY (2005) Enzymatic activity on a chip: the critical role of protein orientation. Proteomics 5:416–419

Chen D, Gao A, Cen K, Zhang J, Cao X, Ma Z (2018) Investigation of biomass torrefaction based on three major components: hemicellulose, cellulose, and lignin. Energy Convers Manag 169:228–237

Cherian E, Dharmendirakumar M, Baskar G (2015) Immobilization of cellulase onto MnO_2 nanoparticles for bioethanol production by enhanced hydrolysis of agricultural waste. Chin J Catal 36:1223–1229

Chiaradia V, Valério A, de Oliveira D, Araújo PH, Sayer C (2016) Simultaneous single-step immobilization of *Candida antarctica* lipase B and incorporation of magnetic nanoparticles on poly (urea-urethane) nanoparticles by interfacial miniemulsion polymerization. J Mol Catal B Enzym 131:31–35

Dinçer A, Telefoncu A (2007) Improving the stability of cellulase by immobilization on modified polyvinyl alcohol coated chitosan beads. J Mol Catal B Enzym 45:10–14

Dutta N, Saha MK (2019) Nanoparticle-induced enzyme pretreatment method for increased glucose production from lignocellulosic biomass under cold conditions. J Sci Food Agric 99:767–780

Epp J (2016) X-ray diffraction (XRD) techniques for materials characterization. In: Materials characterization using Nondestructive Evaluation (NDE) methods. Elsevier, pp 81–124

Fapyane D, Ferapontova EE (2017) Electrochemical assay for a total cellulase activity with improved sensitivity. Anal Chem 89:3959–3965

Foner S (1959) Versatile and sensitive vibrating-sample magnetometer. Rev Sci Instrum 30:548–557

Goldstein J (2012) Practical scanning electron microscopy: electron and ion microprobe analysis. Springer Science & Business Media, Berlin

Goyal A, Ghosh B, Eveleigh D (1991) Characteristics of fungal cellulases. Bioresour Technol 36:37–50

Grabber JH (2005) How do lignin composition, structure, and cross-linking affect degradability? A review of cell wall model studies. Crop Sci 45:820–831

Guzik U, Hupert-Kocurek K, Wojcieszyńska D (2014) Immobilization as a strategy for improving enzyme properties-application to oxidoreductases. Molecules 19:8995–9018

Hafner B (2007) Scanning electron microscopy primer. Characterization Facility, University of Minnesota-Twin Cities, pp 1–29

Harmoko C, Sucipto KI, Retnoningtyas ES, Hartono SB (2016) Vinyl functionalized cubic mesoporous silica nanoparticles as supporting materials to enhance cellulase enzyme stability. ARPN J Eng Appl Sci 11:2981–2992

He H, Gao C (2010) Supraparamagnetic, conductive, and processable multifunctional graphene nanosheets coated with high-density Fe_3O_4 nanoparticles. ACS Appl Mater Interfaces 2:3201–3210

Honda T, Tanaka T, Yoshino T (2015) Stoichiometrically controlled immobilization of multiple enzymes on magnetic nanoparticles by the magnetosome display system for efficient cellulose hydrolysis. Biomacromolecules 16:3863–3868

Hosseini SH, Hosseini SA, Zohreh N, Yaghoubi M, Pourjavadi A (2018) Covalent immobilization of cellulase using magnetic poly (ionic liquid) support: improvement of the enzyme activity and stability. J Agric Food Chem 66:789–798

Hu Y, Liang B, Fang L, Ma G, Yang G, Zhu Q, Chen S, Ye X (2016) Antifouling zwitterionic coating via electrochemically mediated atom transfer radical polymerization on enzyme-based glucose sensors for long-time stability in 37 °C serum. Langmuir 32:11763–11770

Huang P-J, Chang K-L, Hsieh J-F, Chen S-T (2015) Catalysis of rice straw hydrolysis by the combination of immobilized cellulase from *Aspergillus niger* on β-cyclodextrin-Fe_3O_4 nanoparticles and ionic liquid. Biomed Res Int 2015:409103

Husain Q, Ansari SA, Alam F, Azam A (2011) Immobilization of *Aspergillus oryzae* β galactosidase on zinc oxide nanoparticles via simple adsorption mechanism. Int J Biol Macromol 49:37–43

Hwang ET, Gu MB (2013) Enzyme stabilization by nano/microsized hybrid materials. Eng Life Sci 13:49–61

Iype T, Thomas J, Mohan S, Johnson KK, George LE, Ambattu LA, Bhati A, Ailsworth K, Menon B, Rayabandla SM (2017) A novel method for immobilization of proteins via entrapment of magnetic nanoparticles through epoxy cross-linking. Anal Biochem 519:42–50

Jafari Khorshidi K, Lenjannezhadian H, Jamalan M, Zeinali M (2016) Preparation and characterization of nanomagnetic cross-linked cellulase aggregates for cellulose bioconversion. J Chem Technol Biotechnol 91:539–546

Jahangeer S, Khan N, Jahangeer S, Sohail M, Shahzad S, Ahmad A, Khan SA (2005) Screening and characterization of fungal cellulases isolated from the native environmental source. Pak J Bot 37:739

Jordan J, Theegala C (2014) Probing the limitations for recycling cellulase enzymes immobilized on iron oxide (Fe_3O_4) nanoparticles. Biomass Convers Biorefin 4:25–33

Khan MJ, Qayyum S, Alam F, Husain Q (2011) Effect of tin oxide nanoparticle binding on the structure and activity of α-amylase from *Bacillus amyloliquefaciens*. Nanotechnology 22:455708

Khan M, Husain Q, Naqvi AH (2016) Graphene based magnetic nanocomposites as versatile carriers for high yield immobilization and stabilization of β-galactosidase. RSC Adv 6:53493–53503

Khan M, Husain Q, Ahmad N (2019) Elucidating the binding efficacy of β-galactosidase on polyaniline–chitosan nanocomposite and polyaniline–chitosan–silver nanocomposite: activity and molecular docking insights. J Chem Technol Biotechnol 94:837–849

Khoshnevisan K, Vakhshiteh F, Barkhi M, Baharifar H, Poor-Akbar E, Zari N, Stamatis H, Bordbar A-K (2017) Immobilization of cellulase enzyme onto magnetic nanoparticles: applications and recent advances. Mol Catal 442:66–73

Kumar R, Singh S, Singh OV (2008) Bioconversion of lignocellulosic biomass: biochemical and molecular perspectives. J Ind Microbiol Biotechnol 35:377–391

Kumar P, Barrett DM, Delwiche MJ, Stroeve P (2009) Methods for pretreatment of lignocellulosic biomass for efficient hydrolysis and biofuel production. Ind Eng Chem Res 48:3713–3729

Le Goff M, Gallet Y (2004) A new three-axis vibrating sample magnetometer for continuous high-temperature magnetization measurements: applications to paleo-and archeo-intensity determinations. Earth Planet Sci Lett 229:31–43

Li Y, Wang X-Y, Zhang R-Z, Zhang X-Y, Liu W, Xu X-M, Zhang Y-W (2014) Molecular imprinting and immobilization of cellulase onto magnetic Fe_3O_4@ SiO_2 nanoparticles. J Nanosci Nanotechnol 14:2931–2936

Limayem A, Ricke SC (2012) Lignocellulosic biomass for bioethanol production: current perspectives, potential issues and future prospects. Prog Energy Combust Sci 38:449–467

Mahdavi M, Namvar F, Ahmad M, Mohamad R (2013) Green biosynthesis and characterization of magnetic iron oxide (Fe_3O_4) nanoparticles using seaweed (*Sargassum muticum*) aqueous extract. Molecules 18:5954–5964

Mamilla JL, Novak U, Grilc M, Likozar B (2019) Natural deep eutectic solvents (DES) for fractionation of waste lignocellulosic biomass and its cascade conversion to value-added bio-based chemicals. Biomass Bioenergy 120:417–425

Mateo C, Palomo JM, Fernandez-Lorente G, Guisan JM, Fernandez-Lafuente R (2007) Improvement of enzyme activity, stability and selectivity via immobilization techniques. Enzym Microb Technol 40:1451–1463

Maurya S, Bhardwaj A, Gupta K, Agarwal S, Kushwaha A (2016) Green synthesis of silver nanoparticles using *Pleurotus* and its bactericidal activity. Cell Mol Biol 62:131

Metzger JO, Hüttermann A (2009) Sustainable global energy supply based on lignocellulosic biomass from afforestation of degraded areas. Naturwissenschaften 96:279–288

Mishra A, Sardar M (2015) Cellulase assisted synthesis of nano-silver and gold: application as immobilization matrix for biocatalysis. Int J Biol Macromol 77:105–113

Mood SH, Golfeshan AH, Tabatabaei M, Jouzani GS, Najafi GH, Gholami M, Ardjmand M (2013) Lignocellulosic biomass to bioethanol, a comprehensive review with a focus on pretreatment. Renew Sust Energ Rev 27:77–93

Mubarak N, Wong J, Tan K, Sahu J, Abdullah E, Jayakumar N, Ganesan P (2014) Immobilization of cellulase enzyme on functionalized multiwall carbon nanotubes. J Mol Catal B Enzym 107:124–131

Mussatto SI (2016) Biomass fractionation technologies for a lignocellulosic feedstock based biorefinery. Elsevier, Amsterdam

Nawaz MA, Karim A, Bibi Z, Rehman HU, Aman A, Hussain D, Ullah M, Qader SAU (2016) Maltase entrapment approach as an efficient alternative to increase the stability and recycling efficiency of free enzyme within agarose matrix. J Taiwan Inst Chem Eng 64:31–38

Nery EW, Kubota LT (2016) Evaluation of enzyme immobilization methods for paper-based devices—a glucose oxidase study. J Pharm Biomed Anal 117:551–559

Pathak Y, Thassu D (2016) Drug delivery nanoparticles formulation and characterization. Vol. 191. CRC Press

Pedersen M, Meyer AS (2010) Lignocellulose pretreatment severity–relating pH to biomatrix opening. New Biotechnol 27:739–750

Perlack RD (2005) Biomass as feedstock for a bioenergy and bioproducts industry: the technical feasibility of a billion-ton annual supply. Oak Ridge National Laboratory, Oak Ridge

Poorakbar E, Shafiee A, Saboury AA, Rad BL, Khoshnevisan K, Ma'mani L, Derakhshankhah H, Ganjali MR, Hosseini M (2018) Synthesis of magnetic gold mesoporous silica nanoparticles core shell for cellulase enzyme immobilization: improvement of enzymatic activity and thermal stability. Process Biochem 71:92–100

Refaat, A. A. (2012) 5.13-biofuels from waste materials, in Comprehensive Renewable Energy, S. Ali, Ed., pp. 217–261, Elsevier, Oxford, UK

Reineck P, Lin Y, Gibson BC, Dickey MD, Greentree AD, Maksymov IS (2019) UV plasmonic properties of colloidal liquid-metal eutectic gallium-indium alloy nanoparticles. Sci Rep 9:5345

Rodrigues RC, Ortiz C, Berenguer-Murcia Á, Torres R, Fernández-Lafuente R (2013) Modifying enzyme activity and selectivity by immobilization. Chem Soc Rev 42:6290–6307

Roth HC, Schwaminger SP, Peng F, Berensmeier S (2016) Immobilization of cellulase on magnetic nanocarriers. Chemistry Open 5:183–187

Rueda N, dos Santos JC, Ortiz C, Torres R, Barbosa O, Rodrigues RC, Berenguer-Murcia Á, Fernandez-Lafuente R (2016) Chemical modification in the design of immobilized enzyme biocatalysts: drawbacks and opportunities. Chem Rec 16:1436–1455

Salunke BK, Sawant SS, Kang TK, Seo DY, Cha Y, Moon SA, Alkotaini B, Sathiyamoorthi E, Kim BS (2015) Potential of biosynthesized silver nanoparticles as nanocatalyst for enhanced degradation of cellulose by cellulase. J Nanomater 16, 18

Sánchez-Ramírez J, Martínez-Hernández JL, Segura-Ceniceros P, López G, Saade H, Medina-Morales MA, Ramos-González R, Aguilar CN, Ilyina A (2017) Cellulases immobilization on chitosan-coated magnetic nanoparticles: application for *Agave atrovirens* lignocellulosic biomass hydrolysis. Bioprocess Biosyst Eng 40:9–22

Saratale GD, Oh SE (2012) Lignocellulosics to ethanol: the future of the chemical and energy industry. Afr J Biotechnol 11:1002–1013

Seemala B, Meng X, Parikh A, Nagane N, Kumar R, Wyman CE, Ragauskas A, Christopher P, Cai CM (2018) Hybrid catalytic biorefining of hardwood biomass to methylated furans and depolymerized technical lignin. ACS Sustain Chem Eng 6:10587–10594

Sharma A, Qiang Y, Antony J, Meyer D, Kornacki P, Paszczynski A (2007) Dramatic increase in stability and longevity of enzymes attached to monodispersive iron nanoparticles. IEEE Trans Magn 43:2418–2420

Shukla A, Bhardwaj AK, Pandey B, Singh S, Uttam K, Shah J, Kotnala R, Gopal R (2017) Laser synthesized magnetically recyclable titanium ferrite nanoparticles for photodegradation of dyes. J Mater Sci Mater Electron 28(20):15380–15386

Shukla A, Bhardwaj AK, Singh S, Uttam K, Gautam N, Himanshu A, Shah J, Kotnala R, Gopal R (2018) Microwave assisted scalable synthesis of titanium ferrite nanomaterials. J Appl Phys 123:161411

Srivastava N, Rawat R, Sharma R, Oberoi HS, Srivastava M, Singh J (2014) Effect of nickel–cobaltite nanoparticles on production and thermostability of cellulases from newly isolated thermotolerant *Aspergillus fumigatus* NS (class: Eurotiomycetes). Appl Biochem Biotechnol 174:1092–1103

Srivastava N, Singh J, Ramteke PW, Mishra P, Srivastava M (2015) Improved production of reducing sugars from rice straw using crude cellulase activated with Fe_3O_4/Alginate nanocomposite. Bioresour Technol 183:262–266

Stern J, Anbar M, Moraïs S, Lamed R, Bayer EA (2014) Insights into enhanced thermostability of a cellulosomal enzyme. Carbohydr Res 389:78–84

Su X, Li X, Li J, Liu M, Lei F, Tan X, Li P, Luo W (2015) Synthesis and characterization of core–shell magnetic molecularly imprinted polymers for solid-phase extraction and determination of Rhodamine B in food. Food Chem 171:292–297

Sulaiman S, Cieh NL, Mokhtar MN, Naim MN, Kamal SMM (2017) Covalent immobilization of cyclodextrin glucanotransferase on kenaf cellulose nanofiber and its application in ultrafiltration membrane system. Process Biochem 55:85–95

Taha M, Shahsavari E, Al-Hothaly K, Mouradov A, Smith AT, Ball AS, Adetutu EM (2015) Enhanced biological straw saccharification through coculturing of lignocellulose-degrading microorganisms. Appl Biochem Biotechnol 175:3709–3728

Tao Q-L, Li Y, Shi Y, Liu R-J, Zhang Y-W, Guo J (2016) Application of molecular imprinted magnetic Fe_3O_4@ SiO_2 nanoparticles for selective immobilization of cellulase. J Nanosci Nanotechnol 16:6055–6060

Tischer W, Wedekind F (1999) Immobilized enzymes: methods and applications. In: Biocatalysis—from discovery to application. Springer, pp 95–126

Vaghari H, Jafarizadeh-Malmiri H, Mohammadlou M, Berenjian A, Anarjan N, Jafari N, Nasiri S (2016) Application of magnetic nanoparticles in smart enzyme immobilization. Biotechnol Lett 38:223–233

Van Dyk J, Pletschke B (2012) A review of lignocellulose bioconversion using enzymatic hydrolysis and synergistic cooperation between enzymes—factors affecting enzymes, conversion and synergy. Biotechnol Adv 30:1458–1480

Walker L, Wilson D (1991) Enzymatic hydrolysis of cellulose: an overview. Bioresour Technol 36:3–14

Wan L, Chen Q, Liu J, Yang X, Huang J, Li L, Guo X, Zhang J, Wang K (2016) Programmable self-assembly of DNA–protein hybrid hydrogel for enzyme encapsulation with enhanced biological stability. Biomacromolecules 17:1543–1550

Wang WP, Yang H, Xian T, Jiang JL (2012) XPS and magnetic properties of $CoFe_2O_4$ nanoparticles synthesized by a polyacrylamide gel route. Mater Trans 53:1586–1589

Wilson D, Langell M (2014) XPS analysis of oleylamine/oleic acid capped Fe_3O_4 nanoparticles as a function of temperature. Appl Surf Sci 303:6–13

Woodley JM (2013) Protein engineering of enzymes for process applications. Curr Opin Chem Biol 17:310–316

Wu S, He Q, Zhou C, Qi X, Huang X, Yin Z, Yang Y, Zhang H (2012) Synthesis of Fe_3O_4 and Pt nanoparticles on reduced graphene oxide and their use as a recyclable catalyst. Nanoscale 4:2478–2483

Xu J, Huo S, Yuan Z, Zhang Y, Xu H, Guo Y, Liang C, Zhuang X (2011) Characterization of direct cellulase immobilization with superparamagnetic nanoparticles. Biocatal Biotransform 29:71–76

Xu J, Sheng Z, Wang X, Liu X, Xia J, Xiong P, He B (2016) Enhancement in ionic liquid tolerance of cellulase immobilized on PEGylated graphene oxide nanosheets: application in saccharification of lignocellulose. Bioresour Technol 200:1060–1064

Xue Y, Chen H, Yu D, Wang S, Yardeni M, Dai Q, Guo M, Liu Y, Lu F, Qu J (2011) Oxidizing metal ions with graphene oxide: the in situ formation of magnetic nanoparticles on self-reduced graphene sheets for multifunctional applications. Chem Commun 47:11689–11691

Yang C, Mo H, Zang L, Chen J, Wang Z, Qiu J (2016a) Surface functionalized natural inorganic nanorod for highly efficient cellulase immobilization. RSC Adv 6:76855–76860

Yang D, Wang X, Shi J, Wang X, Zhang S, Han P, Jiang Z (2016b) In situ synthesized rGO–Fe_3O_4 nanocomposites as enzyme immobilization support for achieving high activity recovery and easy recycling. Biochem Eng J 105:273–280

Yang Y, Deines T, Zhang M, Zhang K, Wang D (2018) Supercritical CO_2 pretreatment of cellulosic biomass for biofuel production: effects of biomass particle size. In: ASME 2018 13th international manufacturing science and engineering conference. American Society of Mechanical Engineers, pp V002T004A018

Zang L, Qiu J, Wu X, Zhang W, Sakai E, Wei Y (2014) Preparation of magnetic chitosan nanoparticles as support for cellulase immobilization. Ind Eng Chem Res 53:3448–3454

Zhang W, Qiu J, Feng H, Zang L, Sakai E (2015) Increase in stability of cellulase immobilized on functionalized magnetic nanospheres. J Magn Magn Mater 375:117–123

Zhang L, Vilà N, Klein T, Kohring G-W, Mazurenko I, Walcarius A, Etienne M (2016) Immobilization of cysteine-tagged proteins on electrode surfaces by thiol–ene click chemistry. ACS Appl Mater Interfaces 8:17591–17598

Zhu Y, Ciston J, Zheng B, Miao X, Czarnik C, Pan Y, Sougrat R, Lai Z, Hsiung C-E, Yao K (2017) Unravelling surface and interfacial structures of a metal–organic framework by transmission electron microscopy. Nat Mater 16:532

Index

R. Naraian (ed.), *Mycodegradation of Lignocelluloses*, Fungal Biology,
https://doi.org/10.1007/978-3-030-23834-6

MIX
Papier aus verantwortungsvollen Quellen
Paper from responsible sources
FSC® C105338

If you have any concerns about our products,
you can contact us on
ProductSafety@springernature.com

In case Publisher is established outside the EU,
the EU authorized representative is:
Springer Nature Customer Service Center GmbH
Europaplatz 3, 69115 Heidelberg, Germany

Printed by Libri Plureos GmbH
in Hamburg, Germany